Investment Treaty Law and Climate Change

Tomás Restrepo Rodríguez

Investment Treaty Law and Climate Change

Springer

Tomás Restrepo Rodríguez
Universidad Externado de Colombia
Bogota, Colombia

This work was supported by Universität Hamburg

Dissertation to obtain the doctoral degree in law at the Faculty of Law from the University of Hamburg, Albrecht Mendelssohn Bartholdy Graduate School of Law (AMBSL). Pursuant §13 OF AMBSL Doctoral Degree Regulations, this Dissertation was approved for publication by
First reviewer: Prof. Dr. iur. Stefan Oeter
Second reviewer: Prof. Dr. Dr. h.c. Marian Paschke

ISBN 978-3-031-18654-7 ISBN 978-3-031-18655-4 (eBook)
https://doi.org/10.1007/978-3-031-18655-4

This Springer imprint is published by the registered company Springer Nature Switzerland AG
The registered company address is: Gewerbestrasse 11, 6330 Cham, Switzerland

Acknowledgements

Special thanks to Prof. Dr. iur. Stefan Oeter, Prof. Dr.Dr. h.c. Juan Carlos Henao and Prof. Felipe Navia; to Hamburg Universität and Universidad Externado de Colombia.

Contents

Chapter 1
Introduction

Whether they are intended to promote or restrict climate interest, climate change regulatory measures implemented in host states may lead to investment treaty disputes. Ordinarily, these disputes are framed as a conflict between the investor's private interest against the host state's public interest. However, this portrays only one of the dimensions in which interests interplay in investment treaty law (private dimension). The system could also be envisaged as a conundrum of global public interests interacting with each other (global public dimension). Accordingly, investment treaty-climate change disputes could be defined as conflicts and synergies between the global public interest of climate, on the one hand, and the global public interest of transnational investment on the other. Yet tribunals seem to ignore the influence of the international climate change law on these disputes, giving rise to a fragmentary approach that has to be overcome if we are to harmonize the climate change international legal framework with the investment treaty law. The harmonization of these two systems could be achieved through a positivist approach, either by balancing the antagonism between the host state and the investor by interpreting investment treaties—bilateral investment treaties (BITs)—in the context of climate change (orthodox approach) or by relying on the direct application of the climate change international legal framework to investment treaty-climate change disputes (heterodox approach). The embracement of the climate change international legal framework in investment treaty disputes is a logical consequence of envisaging the system's operation as inseparable from its environment and necessary for its survival.

Even if Part I of this book constitutes—to a certain extent—an introduction, as it examines the material and formal *relevance* of climate interest in investment treaty disputes as a prelogical interpretative phase, it is worth making some preliminary remarks regarding the (1) context; (2) problem and research question; (3) methodological package; (4) premises, scope, terminology, and value; and (5) hypotheses and methodology of this study.

T. Restrepo Rodríguez, *Investment Treaty Law and Climate Change*,
https://doi.org/10.1007/978-3-031-18655-4_1

1.1 Is This "Deal" Really Working? Investment Treaty Disputes in a Warming World (the Context)

As I write this introduction, the world is failing to achieve the goals of the Paris Agreement on Climate Change.[1] Although 2020 was expected to be the warmest year on record since 1850,[2] the Decision that emerged from the Conference of the Parties held in Madrid in December 2019 (COP-25) was reduced to a laconic call for action that

> Re-emphasizes with serious concern the urgent need to address the significant gap between the aggregate effect of Parties' mitigation efforts in terms of global annual emissions of greenhouse gases by 2020 and aggregate emission pathways consistent with holding the increase in the global average temperature to well below 2 °C above pre-industrial levels and pursuing efforts to limit the temperature increase to 1.5 °C above pre-industrial levels.[3]

The lightness of this statement contrasts with the severity of the scientific community's findings. Indeed, these confirm that as of 2017, anthropogenic emissions have already increased the global temperature by around 1.0 °C above preindustrial levels and that this average is likely to rise to 1.5 °C between 2030 and 2050 if emissions continue to grow at the current rate.[4]

Failing to achieve the Paris Agreement's goal of leveling global warming leads to dystopian scenarios that significantly aggravate the consequences of climate change. According to the opening speech of the Chair of the Intergovernmental Panel on Climate Change (IPCC) in COP-25, these consequences imply that

> [t]he world will suffer from stranded assets [. . .] the financial sector will face greater uncertainty [. . .] [f]ood security will be threatened [. . .] [t]he world will face increased risks of losses of biodiversity and ecosystem services [. . .] the sustainable development goals such as no poverty, zero hunger and life on land will be compromised [. . .] [t]here will be little room for ecosystem-based adaptation, blue carbon ecosystems, sustainable fisheries and sustainable land management [. . .][5]

The Chair's opening speech also highlights that global warming is threatening the existence of humanity and that we must act to protect everyone's short- and long-term health, safety, and security.[6]

Yet the climate change international legal framework is insufficient; the system does not have international courts protecting the global climate interest, and the determination of the content of climate change obligations seems to be based on each country's preferences rather than on global consensus. Furthermore, a recent

---

[1] IPCC (2018).

[2] Graham (2019).

[3] United Nations (2019).

[4] IPCC (2018).

[5] Hoesung (2019).

[6] Ibid.

antiglobal approach, together with a climate-denial speech,[7] pushed the effects of anthropogenic emissions to the backseat of the United States' political agenda, to the point that the United States submitted a notice of withdrawal from the Paris Agreement.[8]

Unlike the lack of political commitment to execute the Paris Agreement, international policies designed to protect the global liberalization of markets, guarantee equal conditions for local and transnational corporations, and promote transitional investment are enhanced through a complex system of international treaties and international organizations. Commercial, free-trade, and investment agreements,[9] followed by the effectiveness of (soft law) codifications[10] of international commercial law and arbitration mechanisms,[11] in addition to the "Convention on the Recognition and Enforcement of Foreign Arbitral Awards,"[12] provide private economic interests with a structure of international law that ensures their efficacy at a global scale.

Within this constellation of commercial and economic interests in international law, the investment treaty law and investment arbitration have gained momentum, providing foreign investors with the instruments they need to protect their interests against regulatory instability. This protection has a fundamental role in boosting investments in climate-friendly technologies and in the growth of the green economy, which has also gained momentum.[13] Even if—in principle—multinational corporations are not burdened with obligations under the climate change international legal framework, they are at the center of the transition to cleaner sources of energy and, in general, to the reduction of greenhouse gases. The particular vulnerability of renewable energy investors to regulatory changes evinces the importance of protecting the former against the latter.[14] In brief, the synergetic relationship between the transition to a greener economy and the protection of foreign investment seems evident.

However, the relationship between the investment treaty law system and the transition to a greener economy may also be conflictive. The costs of protecting

[7] Collomb (2014).

[8] Pompeo (2019).

[9] Three hundred three regional trade agreements (RTAs) are in force as of January 17, 2020: WTO (2020); 2336 bilateral investment treaties (BITs) in force and 314 treaties in force with investment provisions (TIPs) as of January 17, 2020: UNCTAD (2020).

[10] For example, Principles on Choice of Law in International Commercial Contracts; Principles of European Contract Law (PECL); Draft Common Frame of Reference (DCFR); UNIDROIT Principles of International Commercial Contracts.

[11] For example, the International Chamber of Commerce (ICC), International Centre for Settlement of Investment Disputes (ICSID), International Centre for Dispute Resolution (ICDR), Organisation for the Harmonisation of Business Law in Africa (OHADA), and Arbitration Institute of the Stockholm Chamber of Commerce (SCC).

[12] UNCITRAL (1958).

[13] Infra. Chap. 4, Sect. 4.1.1.1.

[14] Infra. Chap. 4, Sect. 4.1.1.2.

investors whose activity is associated with global warming might spill over to the international community. In the case of investment treaty-climate change disputes, burdening the host state with all the costs of the measures to tackle global warming by imposing full compensation on the investor could equate to transferring the costs of climate change from the polluter to the host state. Furthermore, the contingency of being liable to pay damages for the adoption of climate change measures may drive host states to stagnate in the promotion of Paris Agreement goals. This phenomenon of abstaining from regulating to avoid costs is known in the literature as *regulatory chill.*[15]

In this regard, the investment treaty law has been thought to incur in excessive protection of investor interests and to undermine the fact that regulatory measures from host states address public interests,[16] such as climate change. This concern is based on decisions such as *Metalclad*, where the Tribunal established that the expropriation standard covers those regulatory measures that deprive the owner "in whole or in significant part, of the use or reasonably-to-be-expected economic benefit of property even if not necessarily to the obvious benefit of the host state." [17] A similar approach that illustrates the problem is *Asurix*, where the Tribunal affirmed that it is indifferent to its reasoning "whether the measure concerned is legitimate and serves to a public purpose," [18] or *Tecmed*, which, concerning legitimate expectations, posited that the host state must "act in a consistent manner, free from ambiguity and totally transparently in its relations with the foreign investor," [19] so that the investor "may know beforehand any and all rules and regulations that will govern its investments [. . .]."[20]

Other critics make a point of an exaggerated award of damages.[21] In *Oxy v. Ecuador*, for example, the host state was found liable to pay US $1.77 billion [22] as compensation for the unilateral termination of an oil contract whose operation was located in the Amazon region.[23] In 2012—the year when the award was rendered—Ecuador's total tax revenue was US $17.4 billion; thus, the amount of compensation

[15] Infra. Sect. 4.1.2.

[16] Infra. Sect. 4.2.2.

[17] *Metalclad Corporation v. The United Mexican States* (2000), para. 103.

[18] *Azurix Corp. v. The Argentine Republic* (2006), para. 310.

[19] *Técnicas Medioambientales Tecmed, S.A. v. The United Mexican States* (2003), para. 154.

[20] Ibid.

[21] Rosert (2014).

[22] *Occidental Petroleum Corporation and Occidental Exploration and Production Company v. The Republic of Ecuador* (2012). In 2015, an ICSID-appointed committee lowered the amount to US $1.1 billion (*Occidental Petroleum Corporation and Occidental Exploration and Production Company v. The Republic of Ecuador* (2015)).

[23] Ibid.

was higher than 10% of this revenue.[24] Similarly, in *Yukos v. Russia*, the US $50 billion [25] award was equivalent to 12% of Russia's tax revenues in 2014.[26]

Without intending to generalize these examples, they serve to trigger a discussion on whether in a warming world the investment treaty law system constitutes more of a barrier to promoting the global climate interest than it does a mechanism that boosts energy transition and the investments necessary to tackle climate change. In other words, the question is whether this "deal" of protecting investments through BITs is really working in terms of promoting, preventing, mitigating, and adapting to climate change.

The purpose of this book is not to answer the economic question of whether the investment treaty law system is a "good deal" in the big picture, a question that may be almost impossible to answer even for economists. However, one's rationale and intuition can point to the fact that a disbalance in awards that tilts in favor of investors over global public interests and the determination of compensations that are out of proportion by reference to the host state's economy—on the one hand—and decisions ignoring the role of foreign investment in enhancing energy transition—on the other—jeopardize the goal of leveling global warming. A deal on these terms would certainly not work for host states, nor would it work in terms of promoting global climate interest in a warming world.

1.2 Is It Possible to Achieve Harmonization? Fragmentation of Contemporary Young Systems in the Same Context of Global Warming (*the Problem and the Research Question*)

In *Copper v. Ecuador*, which concerns the termination of a mining concession—as a result of the investor's breach of the obligation to submit an Environmental Impact Assessment EIA and perform consultations with communities—the investor requested compensation under the expropriation and fair and equitable treatment (FET) standards. [27] Although the host state argued that it was not responsible because the measure pursued the protection of the environment—including the protection against deforestation—the Tribunal found that the host state was liable to compensate damages.[28] The Tribunal, in this case, did not take into consideration the fact that the protection against deforestation promotes the global climate interest.

---

[24] OECD (2019).

[25] *Yukos Universal Limited (Isle of Man) v. The Russian Federation* (2014).

[26] The World Bank (n.d.).

[27] *Copper Mesa Mining Corporation v. Republic of Ecuador* (2016), paras. 6.14, 6.66, 6.63.

[28] Ibid. paras. 4.25, 432.

In *Charanne v. Spain*, regarding the cancelation of renewable energy support schemes in the context of the 2007–2008 world financial crisis, [29] the Tribunal found the host state not liable, arguing that an investor cannot expect the local legislation to remain frozen.[30] Despite the fact that the cancelation of these schemes affects the goal of leveling global warming, the award did not make any reference to climate change when assessing whether the host state breached the fair and equitable treatment standard.

Both awards—*Copper v. Ecuador* and *Charanne v. Spain*—represent investment treaty disputes that require climate change considerations. In one case, the host state regulatory measure reduces deforestation; in the other, the measure reduces the production of renewable energies. Accordingly, the decisions that the tribunals adopted in these disputes may have—positively or negatively—affected the achievement of the goal of leveling global warming. However, the attitude of these tribunals was to remain silent as regards the role that the climate change context or the climate change international legal framework would have played in the reasoning of these awards. This is not to say that the tribunals should have adjudged these rulings differently but to underline that there is a trend of ignoring climate change issues in investment treaty disputes, a general trend to ignore the synergies and conflicts between the global climate interest and the global interest of transnational investment.

A problem of fragmentation appears to have emerged as a result of neglecting the global climate interest in investment treaty law rulings. This does not come as a surprise as the investment treaty law and the international climate change law are both young and contemporary systems structured within international law that involve different rationales.

On the one hand, the investment treaty law—at the outset, the result of developed countries' intention to preserve their interests in decolonized states[31]—began to expand in the 1990s with the massive adoption of BITs and free-trade agreements (FTAs),[32] which reflect neoliberal rational models that promote global economic growth by protecting foreign investment and free trade. Similarly, on the other hand,—and given the findings of the scientific community concerning the seriousness of the threat of global warming[33]—the international climate change law system dates back to the 1990s with the adoption of the UNFCCC in 1992[34] and expanded through different cooperation mechanisms promoting the prevention, mitigation, and adaptation to global warming,[35] whose ultimate expression is the Paris

[29] *Charanne (The Netherlands) And Construction Investments (Luxembourg)*, Stockholm Chamber of Commerce (2016).

[30] Ibid. para. 490.

[31] Infra. Chap. 4, Sect. 4.2.2.

[32] Supra. n. 9.

[33] IPCC Working Group (1992), p. 195.

[34] United Nations Framework Convention on Climate Change (1992)

[35] United Nations Framework Convention on Climate Change.

Agreement of 2015.[36] In short, while the rationale of the investment treaty law system is economic growth through the promotion of transnational investment, the climate change international legal framework's rationale is to protect humanity and the ecosystem against global warming.

Although the different rationales of the two systems might involve a clash of interests, the investment treaty law and climate change law are immersed in the same warming world. Thus, investment treaty-climate change disputes need to be resolved in the context of the worldwide goal of preventing, mitigating, and adapting to climate change. Accordingly, the research question that I intend to answer in this book is whether the investment treaty law system could be harmonized with the climate change international legal framework and the climate interest that lies beyond it. I attempt to achieve this harmonization within the limits of the operation of the investment treaty law system.

1.3 Shielding Harmonization from Arbitrariness: Reductionist Legal Positivism (Methodological Package[37])

In this research, I do not prioritize the application of climate interest over other interests by supposing its intrinsic superior value. One of the most critical risks that an international legal scholar faces when addressing a legal problem that involves intersections between isolated systems—such as this one—is how to break (or glue) the fragmentation without engaging in political activism. Despite the fact that recognizing that the particular place that one occupies in the world inexorably influences the normative choices and conclusions that one makes, this study is shielded from "suspicion" by proposing a harmonization of the international investment law system and the climate change international legal framework under the particular (rejuvenated) conception of positivism, called "reductionist legal positivism."[38]

Far from positing that positivism constitutes an absolute and exclusive cognition of the intricated concept of international law[39]—and far from rejecting the role of

[36] United Nations Framework Convention on Climate Change (2015).

[37] "it is vain to seek to distinguish legal methods and theories of law, for none is really independent from the other. Theories of (international) law come with certain methodological moves while various sets of legal methods found in literature and the practice imply and presuppose certain choices in terms of theory of law [...] Methodological packages [...] must be understood as a set of cognitive tools, techniques of argumentation, evaluative criteria and communication protocols meant to secure persuasiveness within a certain audience adhering to a certain concept of law". D'Aspremont (2016), pp. 263, 178.

[38] Ibid. pp. 106–107.

[39] "These scholarly enterprises cannot be lumped together with "uncritical orthodox" positive approaches, for they have included to move away from consensualism, the latter seen as nothing

dynamic-process-based approaches or the effects of deformalization in its formation[40]—reductionist legal positivism is based on the identification of formal sources of international law stemming from the operation of law-applying authorities in order to distinguish *law* from *non-law*.[41] Accordingly, this methodological package adopts the Hartian thinking that the pedigree of *law* is derived from its social recognition:

> [T]he international legal order is identified through formal criteria enshrined in the rules of law-making (the 'sources of law'), albeit in a different way than the current model offered by the *mainstream* doctrine of sources. They have maintained the doctrine of sources at the centre of the modes of cognition of law, thereby claiming that rules of the international law system ought to be ascertained via the formal pedigree defined by the doctrine of sources (. . .) these authors have embraced the social approach to law well known in the analytical jurisprudence according to which the meaning of formal pedigree indicators is found in the law-applying authorities (. . .).[42]

It is important to stress the idea that *reductionist legal positivism* in international law (as a development of Hart's analytical positivism) does not provide the positivist method intended to test the existence of sources of law in terms of deontic validity;

more than a form of natural law. These attempts have simultaneously recognized the arbitrary character of their scholarly approach and have come to terms with the idea that positivism was only one of the many ways to cognize international law. Some of their views are fundamentally value-relativistic with regard to the methodology and the possible content of positive regulation": Ibid. 106. "[. . .] this book has sought to shed some light on the illusions of formalism which shroud the mainstream theory of the sources of international law. Unearthing such mirages from formalism has been construed here as a prerequisite to initiating a rejuvenation of formal-law ascertainment in the age of pluralized normativity": D'Aspremont (2011), pp. 266, 222.

[40] "Another characteristic that they sought to address is the illusion of formalism which shrouds the mainstream theory of the sources of international law. Some of these scholars have also recognized the benefits of the insights of TWAIL and feminist critiques as well as studies on the dialogue between law-applying authorities, especially since these works can be used to contribute to the clarification of ascertainments mechanisms of international rules [. . .] to them static formalism in itself does not provide satisfactory descriptive framework to capture these new forms of exercise of public authority and comes with poor descriptive and evaluative frameworks. They accordingly accept that deformalization may be a necessary step to make sense of a reality unable to be fully capture with formal categories. In their opinion, law can also be considered a process, and law-making process can be diverse and include different actors": D'Aspremont (2016), pp. 108–109. "[. . .] it has been shown that the use of specific linguistic indicators for the sake of law-ascertainment is not a loggerheads with the use of non-formal nor-making procedures, for the practice of resorting to written linguistic indicators contemplated here does not require international act to be devised in accordance with a formally-defined procedure[. . .]": D'Aspremont (2011), p. 222.

[41] Ibid. pp. 195–220.

[42] D'Aspremont (2016), pp. 108–109.

thus, the sources' membership to the system is not derived from the prescriptive power of a superior rule. In contrast, the validity of the sources—under the reductionist outlook—derives from their social recognition and, more specifically, from the "social practice of the law-applying authorities."[43] Therefore, the examination of the structuring and modeling of the system of formal sources in investment treaty law that follows is not intended to preserve the purity of the system and, thereby, does not controvert the validity of the *sources of law* that characterize the system's operation against superior *norms*.

Particularly—*within the movement of reductionist legal positivism*—this theory is based on Jean Aspermont's approach in his book *Formalism and the Sources of International Law*.[44] Apart from embracing the source and social thesis as "linchpins" of his argument,[45] the author dynamizes the interpretation of international law by providing a main role to *law ascertainment*.[46] In this regard, he unravels that the traditional manner of addressing an international legal problem through *content interpretation* is only one of the layers that the interpreter should explore.[47] On top of content interpretation, the determination of what rules of international law apply to the case at hand may define the outcome of the legal problem.[48] Moreover, he affirms that while the process of content interpretation is generally constrained by Articles 31 and 32 of the Vienna Convention on the Law of Treaties (VCLT), the process of law ascertainment appears to be deployed with considerable discretion.[49]

Against this background, there are two caveats within my methodology. First, any finding or conclusion of this study embracing the direct or indirect application of the climate change international legal framework should be buttressed by the system of formal sources—as construed and modeled in practice by investment treaty arbitration tribunals—and, second, the investment treaty law and climate change law could be harmonized via a content interpretation of the BIT in the context of climate change and by a direct application of international climate change rules as a result of an exercise of law ascertainment.

[43] D'Aspremont (2011), p. 195.
[44] Ibid.
[45] Ibid. p. 7.
[46] Ibid. pp. 148–193.
[47] Ibid. pp. 157–160.
[48] Ibid.
[49] Ibid.

1.4 Harmonization Within the Limits of the Operation of the Investment Treaty Law: Enhancing the Efficacy of the Global Climate Interest Through the *Private* and *Public* Dimensions of the System (*Premises, Terminology, Scope*, and *Value of the Study*)

Taking the normative choice that fragmentation is undesirable is a premise that underlies the effort of harmonizing the investment treaty law with the climate change law. Likewise, adopting the methodological package of *reductionist legal positivism* is translated into the premise that the operation of the investment treaty law system defines the level of penetration of the global climate interest. These two simple premises—harmonization is desirable and harmonization is limited by the system—determine the outcome of the book from beginning to end.

Although this work could be framed within the studies that address the tension between the host state's public interest and the investor's private interest in investment treaty disputes—since in many cases host state interest overlaps with the promotion of the global climate interest—harmonizing the investment treaty law and international climate change law covers a broader scope. In this study, the harmonization of these two systems captures the fact that—in the context of investment treaty disputes—it is often the investor's interest that is attuned to the global public interest of leveling global warming, instead of the host state's interest. Disputes concerning the disarticulation of renewable energy support schemes (RESSs) confirm this statement.[50] In this regard, the concept of harmonization that I adopt is intended to cover both conflicts and synergies between the protection of the investor and the global climate interest.

This understanding of harmonization provides an innovative perspective when compared to most other studies that deal with climate change considerations in investment treaty law disputes. These studies seem to identify the problem of fragmentation only with the tension between the host state's public interest—represented in the climate interest—and the private interest of the investor (*the public-private dilemma*).[51] This book steers away from the *public-private dilemma* as the main problem and focuses on aligning the global climate interest with the protection of the transnational flow of investments and other global interests that may encroach on investment treaty-climate change disputes. The dilemma to be addressed here is how to provide *stability* to investors in the green industry and at the same time encourage host states to take measures for preventing, mitigating, and adapting to the risk of global warming.

Consequently, two combinations stemming from the interrelation between the global climate interest and the global interest in the flow of investment arise throughout the whole research: first, the global climate interest coming into tension

---

[50] Infra. Chap. 4, Sect. 4.1.1.2.

[51] Infra. Sect. 3.2.3.

with the global interest in transnational investment when the host state takes measures that promote the leveling of global warming by affecting the investor's interest and, second, the global climate interest and the global interest of transnational investments—on the one hand—coming into tension with other global interests—on the other—when the host state takes measures affecting investors in climate-friendly industries in order to promote other interests, such as social, security, economic, or human rights interests. Since the first premise of this study is that the fragmentation between the investment treaty law and the climate change international legal framework is undesirable, in my interpretation throughout the book, I attempt to favor the global climate interest when dealing with these two tensions (combinations) in investment treaty disputes, within the limits of the system's operation.

The second premise—harmonization is limited by the system—also sets forth a different perspective from most of the works on the harmonization of the investment treaty law and climate change law. While some critical studies claim that the system entails an innate imbalance in favor of protecting investments that renders any effort of harmonization useless,[52] other studies suggest that harmonization might be achieved through comparative public law interpretations, substantive validity theories that grant a superior value to (international) public interests, or topical rhetoric that identifies investment treaty law with international public law.[53] Unlike these approaches, the adoption of *reductionist legal positivism* ensures that the answer to the harmonization of the investment treaty law with climate interest is found in the system's operation, in turn shaped by the decisions made by investment treaty tribunals.

It is worth noting that not all studies that deploy comparative public law, substantive validity theories, or topical rhetoric to achieve harmonization completely disregard systemic considerations that may redirect to positivism, but their main line of argumentation appears to be outside the real operation of the investment treaty law system, or it is not solid enough to justify the application of the global climate interest, according to the system's operation.[54] By contrast, the positivist approach that I adopt by mirroring the system's operation may potentialize the community of international investment lawyers' and investment treaty tribunals' acceptance of the harmonization.

In the same vein, this study may be distinguished from others that suggest that the investment treaty law could be harmonized with the climate change international legal framework by modifying BITs. The scope of this work is restricted to the system's operation as it is today. Apart from the fact that reaching a global multilateral investment agreement integrating climate considerations or introducing climate change considerations in more than 3000 BITs appears unrealistic, it might also

[52] Infra. Sect. 4.2.2.

[53] Infra. Sect. 3.2.

[54] Ibid.

be unnecessary since the investment treaty law system already embraces mechanisms to integrate the global climate interest.

The line of argumentation that I propose is that, in connection with BIT drafting techniques, the Investment Tribunal's practice captures global public interests within its realm (*syncretism*). Thus, as long as it is determined that the global climate interest is relevant in the particular case at hand, the climate change international legal framework may influence the process of interpretation either through content interpretation or law ascertainment. While the former—content interpretation—is understood in this publication as the activity of finding a balance of interests between the investor and the host state, as entrenched in the BIT's text in the context of global climate change, mirroring the interpretation of contracts (*the private dimension*), the latter is understood as the application of rules from the climate change international legal framework as a result of an exercise of balance between the global climate change interest and other global interests that may enter into conflict with it (*the public dimension*). The division between *relevance*, *law ascertainment*, and *content interpretation* is reflected in the main structure of the book.

1.5 Relevance, Content Interpretation, and Law Ascertainment in Investment-Treaty-Climate-Change-Related Disputes: Pre(logical)interpretative Phase, Orthodox Approach, and Heterodox Approach (*the Content and Hypotheses*)

This book is divided into three parts. The first part examines the *relevance* of the climate change international legal framework in investment treaty disputes as a natural pre(logical)interpretative stage. The second part focuses on BIT's *content interpretation*, which is the orthodox approach to solving the fragmentation between the investment treaty law system and the international climate change legal framework. And the third part tackles this fragmentation via a heterodox approach that is grounded on the direct application of climate change principles through law ascertainment. Each of the chapters that unfolds these three parts is intended to provide the overarching hypothesis that the investment treaty law system could be harmonized with the global climate interest only if this stems exclusively from the system's operation.

Part I is entitled *Relevance of the Climate Change International Legal Framework in Investment Treaty-Climate Change Disputes*. The purpose of this part is to examine whether the practice of tribunals allows (or compels) conceiving the application of the climate change international legal framework, although the promotion of climate interests is certainly not within the rationale of the system. The starting point of the examination, as illustrated in Chap. 2, is that the Tribunal's power to choose the climate change international legal framework as applicable law is constrained by the BIT's choice of law and the verification of the *relevance* of the global climate interest to the case at hand. The verification of the formal (Chap. 3) and material relevance (Chap. 4) of the climate change international legal framework to the issue is presented as a necessary pre(logical)interpretative phase that precedes *content interpretation* and *law ascertainment.*

Chapter 3—formal relevance—posits that conceptualizing investment treaty law in operational terms as a process of *multi(syncretic)lateralization* embeds the protection against global climate change within its realm. While the first section of the chapter builds on Stephan Schill's theory of the *multilateralization of international investment law* as the basis for embracing the global climate change interest within its scope and unearths the need to understand *multilateralization* in tune with the *syncretism* of the investment treaty law (Sect. 3.1), the second section contrasts the benefits of adopting the notion of *multi(syncretic)lateralization* against other theories intended to harmonize (global) public interests with the investment treaty law, namely, comparative public law interpretations as well as (international) public law substantive validity theories and topical rhetoric (Sect. 3.2).

Chapter 4—material relevance—is intended to reveal, from an external perspective, that international investment tribunals address climate change conflicts. On the one hand, this chapter takes an economic approach that unearths the synergy between the rationale of the investment treaty law and the growth of the green economy and illustrates the chilling effects of the investment treaty law on the host state's climate action (Sect. 4.1). On the other hand, the chapter takes a sociopolitical perspective that profits from the Global Administrative Law (GAL) and the postmodern critical study approaches to reveal the influence of the investment treaty law in the global and local climate change policy, as well as the inherent disbalance of the system that favors investments over the climate interest (Sect. 4.2).

Part II is entitled *Gravitational Power of the Climate Change International Legal Framework over the BIT's Content Interpretation.* The objective of this part is to demonstrate that it is possible to promote the global climate by interpreting the BIT's content, particularly, that of expropriation (Chap. 6) and FET (Chap. 7) standards. This interpretation takes into consideration the premise expressed in Chap. 5: the BIT's content is shaped not only by (nonbinding) prior decisions but also by the application of the interpretation principles of the VCLT and the principle of deference to host state authorities.

Chapter 6—orthodox approach to expropriation—contends that the sacrificial-lamb doctrine represents a balanced interpretation of the expropriation standard that harmonizes with the climate change international legal framework. The first section of the chapter analyzes the divergent interpretations by investment tribunals of the expropriation standard—the sole-effects doctrine and the *police-power doctrine*—as well as their limitations in order to provide a satisfactory answer to harmonization (Sect. 6.1). Subsequently, in the second section, I argue that the *sacrificial-lamb theory*—as a renewed, tempered version of the *sole-effects doctrine*—conforms to the pliability of the expropriation standard as well as to the global climate interest (Sect. 6.2).

Chapter 7—orthodox approach to the FET—suggests that the notions of the *proportionality* and *reasonableness* of host state measures may lead to an interpretation of the FET standard harmonizing with the climate change international legal framework. While the first section of the chapter unearths the pattern in investment treaty disputes related to renewable energy support schemes (RESSs) that define the *(un)reasonableness* and *(dis)proportionality* of host state measures through the elements of *(non)capriciousness*, *(un)predictability*, and *(in)essentiality* (Sect. 7.1), the second section develops the argument that deploying the same pattern of RESSs to all climate-change-related disputes concerning the elements of *(un)reasonableness* and *(dis)proportionality* is consistent with the pliability of the FET standard and embraces the climate change context (Sect. 7.2).

Part III is entitled *Osmosis of the Climate Change International Legal Framework into Investment Treaty Law Through Law Ascertainment*. My intention in this part is to prove that global climate interest could be promoted through the direct application of the climate change international legal framework, specifically through the application of the *precautionary principle* (Chap. 9) and the *common but differentiated responsibilities and respective capabilities* (CBDRRC) principle (Chap. 10). This exercise of law ascertainment—as explained in Chap. 8—stems from the premise that, even if the operation of the investment treaty law is open to the application of rules pertaining to the multiverse of international law, this openness might be limited by the BIT's choice of law and the special provisions of the BIT—*lex specialis*.

Chapter 9—heterodox approach through the defense of necessity—explores the direct application of the precautionary principle to investment treaty-climate change disputes by deploying the defense of necessity. In the first section of the chapter, I describe the evolution of investment treaty decisions embracing global public interest measures within the scope of the international defense of necessity, and I argue that the flexibilization of the conditions of Article 25 of the Draft Articles on State Responsibility (DASR)—*essentiality of the interest, imminent peril, only way to safeguard the interest, non-contribution to the situation*—leads to the application of the precautionary climate change principle (Sect. 9.1). In contrast, the second section turns our attention to the relationship between *necessity* and the general exceptions as well as the adequacy of the precautionary climate change principle to different models of general exceptions (Sect. 9.2).

Chapter 10—heterodox approach through the proportionality of compensation—proposes that the direct application of the *common but differentiated responsibilities and respective capabilities* principle can determine the imposition, reduction, or exclusion of compensation in investment treaty-climate change disputes. In this regard, while in the first section of the chapter I analyze the suitability of CBDRRC to the practice of the investment treaty law by verifying the limitations of the full compensation principle (Sect. 10.1), in the second section I envisage the operationalization of the CBDRRC in investment treaty-climate change disputes stemming from the topical connections between the fragmented systems and the deployment of international liability devices that facilitate the adoption of a differential approach to compensation in cases concerning climate change (Sect. 10.2).

Apart from concluding that it is possible to harmonize the investment treaty law and international climate change law by applying the orthodox approach to the expropriation and FET standards and by the direct application of the climate change precautionary principle and the CBDRRC principle—heterodox approach—I suggest that tribunals are expected to soon openly address climate change disputes in their rulings. This optimism is based on the increasing recognition of global public interests in the investment treaty law, the likelihood of promoting the global climate interest within the operational limits and without jeopardizing the identity of the investment treaty law system, the rejuvenation of the epistemic community of international investment lawyers, the need to recognize the common concerns of humankind in investment treaty disputes for the survival of the investment treaty law itself, and the unacceptability of disconnecting the interpretation of the investment treaty law from the environment, which includes a warming world.

References

Collomb J (2014) The ideology of climate change denial in the United States. Eur J Am Stud. https://journals.openedition.org/ejas/10305. Accessed 26 July 2022

D'Aspremont J (2011) Formalism and the sources of international law. OUP

D'Aspremont J (2016) Epistemic forces of international law. Edward Elgar

Graham M (2019) 2020 set to extend series of earth's warmest years. Met Office. https://www.metoffice.gov.uk/about-us/press-office/news/weather-and-climate/2019/2020-global-temperature-forecast. Accessed 26 July 2022

Hoesung L (2019) Opening statement COP 25. https://unfccc.int/sites/default/files/resource/IPCC%20COP25%20opening%20Chair%20statement.pdf. Accessed 26 July 2022

IPCC (2018) Summary for policymakers. https://www.ipcc.ch/site/assets/uploads/sites/2/2019/05/SR15_SPM_version_report_LR.pdf. Accessed 26 July 2022

IPCC Working Group (1992) Climate Change 1992 Supplementary Report to the IPCC Scientific Assessment, p 195. https://www.ipcc.ch/site/assets/uploads/2018/05/ipcc_wg_I_1992_suppl_report_full_report.pdf. Accessed 26 July 2022

OECD (2019) Revenue statistics in Latin America and the Caribbean 1990–2017, p 176. https://www.oecd.org/tax/tax-policy/revenue-statistics-latin-america-and-caribbean-2019-launch-version.pdf. Accessed 26 July 2022

Pompeo M (2019) On the U.S. withdrawal from the Paris Agreement. Press Statement US Department of State. https://www.state.gov/on-the-u-s-withdrawal-from-the-paris-agreement/. Accessed 26 July 2022

Rosert D (2014) The stakes are high: a review of the financial costs of Investment Treaty Arbitration. IISD. https://www.iisd.org/sites/default/files/publications/stakes-are-high-review-financial-costs-investment-treaty-arbitration.pdf. Accessed 26 July 2022

The World Bank (n.d.) Tax revenue (% of GDP) - Russian Federation. https://data.worldbank.org/indicator/GC.TAX.TOTL.GD.ZS?locations=RU. Accessed 26 July 2022

UNCTAD (2020) Investment Policy Hub. https://investmentpolicy.unctad.org/international-investment-agreements. Accessed 26 July 2022

United Nations (2019) Decision 1/CP.25 Chile Madrid time for action. The conference of the parties. https://unfccc.int/resource/cop25/1cop25_auv.pdf. Accessed 26 July 2022

WTO (2020) Regional trade agreements. https://www.wto.org/english/tratop_e/region_e/region_e.htm#facts. Accessed 26 July 2022

International Legal Instruments

UNCITRAL (1958) Convention on the Recognition and Enforcement of Foreign Arbitral Awards. https://www.uncitral.org/pdf/english/texts/arbitration/NY-conv/New-York-Convention-E.pdf. Accessed 26 July 2022

United Nations Framework Convention on Climate Change (1992) FCCC/INFORMAL/84 GE.05-62220 (E) 200705. https://unfccc.int/resource/docs/convkp/conveng.pdf. Accessed 26 July 2022

United Nations Framework Convention on Climate Change (2015) Adoption of the Paris Agreement FCCC/CP/2015/L.9/Rev.1. https://unfccc.int/resource/docs/2015/cop21/eng/l09r01.pdf. Accessed 26 July 2022

Part I
Relevance of the Climate Change International Legal Framework in Investment Treaty-Climate Change Disputes: Pre(Logical)Interpretative Phase

Chapter 2
Iura Novit Arbiter and the Application of the Climate Change International Legal Framework to Investment Treaty Disputes

Arguing that the climate change international legal framework is part of the universe of the investment treaty law has to be supported by the reasoning of investment treaty tribunals. Although the investment treaty law is not isolated from the rest of international law, no international law instrument can make its way into the system without a proper filter, and the international climate change legal framework is not the exception.

Reductionist legal positivism adheres to the Hartian proposal that the law is to be found in the practice of law-applying authorities. Thus—according to the methodology that I have chosen to adopt—the first question to address in this research is whether the practice of tribunals allows (or compels) conceiving the application of the climate change international legal framework, even though the promotion of climate interests is certainly not the system's rationale.

In this regard, it appears that prior to ascertaining the law that applies to the particular controversy or interpreting its content, the tribunals' practice consists of identifying the sources that—within the multiverse of international law—may be applicable to the dispute. It also seems that the setting of this framework is driven by the tribunal's power to choose the applicable law, but it is limited to the BIT's choice of law and the *relevance* of the international instruments to the case at hand.

The principle of *iura novit arbiter*[1]—which entitles the tribunal to find *the law* that suits the facts (*da mihi factum, dabo tibi ius*[2])—supports the idea that the climate change international framework could be integrated into the investment treaty law

[1] The expression *iura novit arbiter*—which means *the arbitrator knows the law*—is deployed in the context of international investment arbitration as a synonym or adaptation of the principle *iura novit curia*: *Quiborax S.A., Non Metallic Minerals S.A. and Allan Fosk Kaplún v. Plurinational State of Bolivia* (2015), para 92. *Vestey Group Ltd v. Bolivarian Republic of Venezuela* (2016), para. 118.

[2] This Roman Law principle could be translated as *give me the facts, I give you the law*.

T. Restrepo Rodríguez, *Investment Treaty Law and Climate Change*,
https://doi.org/10.1007/978-3-031-18655-4_2

system. [3] The tribunals' operation seems to follow a general trend[4] that arbitrators do not limit themselves to the law invoked by the parties.[5] Moreover, requests of annulment under the International Centre for Settlement of Investment Disputes (ICSID) Convention have been generally rejected when applicants argued that the ruling was based on provisions that were not presented by the parties in their pleadings.[6]

However, if the parties to the dispute coincide in the relevance of the climate change international legal framework to said dispute, tribunals might have no inconvenience in endorsing this coincidence. It appears that, in practice, when parties to the dispute refer to BIT, national, and/or international law provisions—or any other instruments—and concur in the suitability of these sources, tribunals would support this correspondence. [7]

In contrast to the discretion of tribunals to endorse or ignore the sources invoked by the parties,[8] they cannot neglect the legal force of a BIT's choice of law.[9] Thus, if the BIT includes the climate change international legal framework within the scope of the applicable law, the tribunal would be compelled to conceive its application. In fact, failing to apply the *law* of the BIT constitutes a basis for annulment under Article 52(1)(b) of the Washington Convention, namely, a manifest excess of

[3] *Quiborax S.A., v. Bolivia*. (2015) para. 92; *Vestey Group Ltd v. Bolivarian Republic of Venezuela* (2016), para. 118; *Bogadov v. Moldova* (2005), para. 14; *Daimler Financial Services AG v. Republic of Argentina* (2015), para. 295; *Jan Oostergetel v. The Slovak Republic* (2013), para. 141; Metha-Tec v. Republic of Uzbekistan (2012), para 287; *Venezuela Holdings, B.V., et al* (case formerly known as *Mobil Corporation, Venezuela Holdings, B.V., et al.*) v. *Bolivarian Republic of Venezuela* (2014), para. 223.

[4] De Brabandere (2014), p. 106: "[...] the application of the principle *jura novit curia* has not uniformly been accepted in non-ICSID arbitrations [...] However, several tribunals in non-ICSID arbitration have also upheld the inherent power of investment tribunals to apply legal arguments that have not been advanced by the parties, which may prove that in treaty-based investment arbitration, even outside the ICSID framework the principle should be applied".

[5] See. note 3.

[6] Dolzer and Schreuer (2012), p. 306; *Compañiá de Aguas del Aconquija S.A. and Vivendi Universal S.A. v. Argentine Republic* (2002), para. 82-5; *Wena Hotels Ltd. v. Arab Republic of Egypt,* (2002), paras. 66–70; *Klöckner Industrie-Anlagen GmbH and others v. United Republic of Cameroon and Société Camerounaise des Engrais,* (1985), para 91. *RSM Production Corporation v. Grenada* (2009), para. 23.

[7] *RosInvestCo UK Ltd. v. The Russian Federation* (2013), para. 249; *Liman Caspian Oil BV and NCL Dutch Investment BV v. Republic of Kazakhstan* (2010), paras. 151, 29. *Romak S.A. (Switzerland) v. The Republic of Uzbekistan* (2009), paras. 166–167.

[8] Viñuales (2015), p. 110; *Enron Corporation and Ponderosa Assets, L.P. v. Argentine Republic* (2010), para. 375; *CME Czech Republic B.V. v. The Czech Republic* (2003), para. 411.

[9] "The nature of those [investment arbitration] claims varies from treaty to treaty—as demonstrated below, some treaties give broad authority to arbitral tribunals to resolve 'investment disputes', while others are more limited in the scope [...] Because treaties varies it is important to examine each agreement in its entirety to identify the authority that the tribunal enjoys to resolve certain claims and to apply different laws to those claims. In short, the applicable law provision must also be read in conjunction with any scope limitations included in the treaty": Bjorklund (2014), p. 264.

power.[10] This legal force seems to be grounded on the argument (fiction?[11]) that the investor implicitly consents to bestow competence to the tribunal through a request for arbitration under the BIT[12] and that the host state explicitly consented by adopting the BIT.[13]

The choice of law does not mathematically solve the intricacies of integrating climate change international law when setting the legal framework. If, on the one hand, a choice of law clause establishing the applicable rules of international law as the governing law demarcates limits by closing the door to instruments that are not international law, it leaves a door open for the tribunal to choose between a myriad of legal sources that belong to the fragmented multiverse of international law.[14] It is in this scenario that the examination of relevance emerges as necessary to define whether the climate change international legal framework is part of the applicable law.

The facts and allegations that support the parties' pleas (*the issue*)—rather than the law invoked by the parties—lead to finding the applicable law. As a result, the reading of the facts and allegations would indicate whether the use of the international climate change law is suitable in the controversy. In this vein, the Decision on the Request of Annulment in *Daimler v. Argentina* expressed that

> an arbitral tribunal is not limited to referring to or relying upon only the authorities cited by the parties. It can, *sua sponte*, rely on other publicly available authorities, even if they have not been cited by the parties, provided that the issue has been raised before the tribunal and the parties were provided an opportunity to address it (...) the fact that the Tribunal cited in the Award decisions or authorities not previously cited by the Parties, but which were publicly available, and **were relevant to the issues** on which the Parties had made extensive submissions, does not constitute a serious departure from a fundamental rule of procedure.[15]

Likewise, in *Jan Oostergetel v. Slovak Republic*, the Tribunal reached a conclusion that "[i]f and when the issue arises, it will determine whether the applicable international law should be limited to general principles of international law or whether it includes customary international law."[16] In another more recent case, where the claimant disagreed with the Tribunal's competence to address the EU *inter-se argument*—which is a discussion on the (lack of) jurisdiction for investment disputes between the EU member states—it was held that "the rules of EU law relevant in the matter are part of the applicable law" [17] not only for granting

[10] Gaillard (2004), pp. 223–241.

[11] Di Benedetto (2013), pp. 59–60; Schreuer and Malintoppi (2009), p. 571.

[12] In Van Harten's words: "As such, the investor's consent is more akin to a decision of an individual to go to court to resolve a complaint against the State under domestic administrative law or constitutional law". Van Harten (2007), p. 126.

[13] Pauwelyn affirms that this particular way of defining the consent to arbitrate finds its origins in "The silent revolution of *AAPL v. Sri Lanka*", which is a case of 1990: Pawelyn (2014), p. 31.

[14] Viñuales (2015), p. 108.

[15] Ibid. paras. 295–296.

[16] *Jan Oostergetel v. The Slovak Republic* (2013), para. 141.

[17] *Blusun S.A., Jean-Pierre Lecorcier and Michael Stein v. Italian Republic* (2016), para. 277.

jurisdiction, but they are also applicable in general to the dispute because the parties' consent for the dispute to be governed under international law "does not exclude any relevant rule of EU law."[18]

Confirming that relevance is crucial for setting the legal framework is not so surprising. Indeed, it seems to confirm a *prelogical* need for the interpretation of the investment treaty law. Certainly, when an investment tribunal addresses a case, the natural starting point would be not so different from the starting point of any other legal case: framing the dispute within the different topics that may affect the solution. These topics would, therefore, indicate which applicable law might be. If the submissions from the parties reveal that the issue involves climate change concerns, it would be reasonable for the tribunal to consider whether the climate change international legal framework could lead to a solution within the boundaries of the investment treaty law system.

Accordingly, the analysis of relevance for setting the framework involves a twofold operative connection between the issue and the investment treaty law. While the investment treaty law (as a practice of the tribunals) defines whether the issue can be resolved within its own system, the issue points to the particular instruments of international law that may govern the investment treaty law dispute. For this purpose, two interconnected challenges that appear evident are addressed in Chaps. 3 and 4. The first involves conceptualizing, from an internal perspective, the investment treaty law as a system that incorporates the global climate interest (ergo the climate change legal framework is relevant because it is formally part of the legal system). The second supposes a demonstration, from an external perspective, that climate change disputes can be formed within the investment treaty law system (ergo the climate change legal framework is relevant because there are climate change disputes in investment treaty arbitration).

References

Bjorklund A (2014) Applicable law in international investment disputes. In: Giorgetti C (ed) Litigating in international investment disputes. Brill Nijhoff, p 264

De Brabandere E (2014) Investment treaty arbitration as public international law. CUP

Di Benedetto S (2013) International investment law and the environment. Cheltenham, pp 59–60

Dolzer R, Schreuer C (2012) Principles of international investment law, 2nd edn. OUP, p 306

Gaillard E (2004) The extent of review of the applicable law in investment treaty arbitration. In: Gaillard E (ed) Annulment of ICSID Awards. Juris Publishing/IAI, New York, pp 223–241

Pawelyn J (2014) Rational design or accidental revolution? In: Douglas Z et al (eds) The foundations of international investment law. OUP, p 31

Schreuer C, Malintoppi L (2009) The ICSID convention: a commentary, 2nd edn. CUP, p 571

Van Harten G (2007) Investment treaty arbitration and public law. OUP, p 126

Viñuales J (2015) Foreign investment and the environment in international law. CUP

[18] Ibid. para. 278.

Chapter 3
Formal Relevance of the Climate Change International Legal Framework in Investment Treaty Disputes: Defining the System as a Process of *Multi(syncretic)lateralization*

Conceptualizing investment treaty law in functional terms as a system stemming from the process of *multi(syncretic)lateralization* embeds the promotion of the global climate interest within its realm. The point of departure of this assertion is Schill's refined observations on the multilateralization of the investment treaty law, which reveals how tribunals' decisions, along with bilateral investment treaty (BIT) drafting techniques, regulate foreign investment in a way that provides a "systemic solution to systemic problems."[1] As part of such systemic problems, the tension between the global public interest in tackling climate change and the interests of the parties in question in an investment treaty dispute urges a systemic solution that suits the *legal framework for the functioning of the global economic system.*[2]

It is crucial to underscore that defining investment treaty law as a legal framework for regulating transnational investment should not be understood in functional terms as restricted to the operation of promoting investments by providing a level playing field to investors. The tribunals' practice and the drafting techniques of BITs demonstrate that the operation of the investment treaty law system comprehends, apart from protecting the investor's interests, the application of international law instruments and international law norms that represent global public interests.

This process of interaction between systems, which (sometimes) embed colliding values framed within the same axis of validity, evokes the idea of *syncretism*. If, on the one hand, the concept of *multilateralization* discloses that investment treaty law is a system for the functioning of the global markets, on the other hand, the concept of *syncretism* that the system's operation is characterized by the communication of BIT provisions with other international law instruments, without losing its identity or autonomy. In this scenario, the climate change international legal framework emerges as a constitutive element of the investment treaty law system,

[1] Schill (2015), p. 1835.

[2] Schill (2009), p 281.

T. Restrepo Rodríguez, *Investment Treaty Law and Climate Change*,
https://doi.org/10.1007/978-3-031-18655-4_3

and accordingly, tribunals are compelled to envisage its application when it is relevant to the dispute. Accordingly, this chapter firstly describes how the concept of *multilateralization* provides a theoretical framework that makes it possible to integrate the global public interest of leveling global warming into investment treaty law (Sect. 3.1), and secondly, it reveals *syncretism* as an essential property of the investment treaty law system that compels tribunals to identify the climate change legal framework as applicable law (Sect. 3.2).

The conceptualization of *multi(syncretic)lateralization*—described in the two sections that follow—leads to detaching the climate change-investment law relationship from the public-private interest dilemma, and to unfolding the climate change-investment law relationship into two underlying tensions: the tension between the global climate interest against the flow of transnational investment, when the host state takes measures that promote the objective of leveling global warming against the investor's interest; and the tension between the global climate interest and the global interest in transnational investments, on the one hand, against other global interests, on the other, when the host state takes measures against investors in climate-friendly industries in order to promote other global interests (Sect. 3.3).

3.1 *Multilateralization* of the Investment Treaty Law: Envisaging the Global Climate Interest in Investment Treaty-Climate Change Disputes

Defining the investment treaty law as a system is central to claiming that the climate change international legal framework is formally relevant. In the absence of such a system, BITs would be perceived as private negotiations mean devoid of common interest considerations. In this hypothetical disconnected scenario of insulated treaties, the responsibility of any tribunal would be limited to deciding the issue exclusively on the basis of the parties' negotiations, and the only challenge that the interpreter would face is to give meaning to the text of the agreement.

Schill's insightful description of the *multilateralization* process links around 3000 BITs and 850 investment arbitration decisions within one framework—within one investment treaty law system that he defines in operational terms as "autopoietic, self-referential, and normatively closed"[3] and in functional terms as "providing a legal framework for the functioning of the global economic system."[4] Although the notion of a multilateralized system, characterized by the promotion of community interests,[5] captures the formal relevance of the climate change legal framework (Sect. 3.1.1), Schill's definition needs to be cleared from ambiguity in order to

[3] Ibid. p. 280.

[4] Ibid. 281.

[5] Schill (2015), pp. 1834–1835.

preserve the system's operative autonomy while ensuring its openness to other (sub)systems (Sect. 3.1.2). Thus—based on Di Benedetto's critics—I posit, on the one hand, that a social epistemological-constructivism reading of Schill's operational definition reveals that the self-referential closedness of the system is a condition to its openness toward the climate change international legal framework and, on the other hand, that the functional definition cannot be disconnected from the promotion of global public interests—including climate interest—because this would contradict the same concept of *multilateralization* and would ignore the real operation of the investment treaty law system.

3.1.1 The Process of Multilateralization: Describing Investment Treaty Law as a Legal System That Regulates Community Interests

Schill defines *multilateralization* as the "paradoxical phenomenon that international investment law is developing towards a multilateral system of investment protection on the basis of bilateral treaties."[6] In this regard, the concept of *multilateralization* synthesizes the conflicting paradigms of bilateralism and multilateralism by revealing that the development and implementation of BITs, coupled with the functioning of investor-state arbitration, have led to proper institutions and principles, which reflect the inception of *community interests* in international investment law[7]—rather than mirroring a simple negotiation between the parties to the treaty.[8] In this sense, international investment law is not perceived anymore as a group of uncoordinated bilateral treaties shaped as "*quid pro quo* bargains,"[9] which are the subject of inconsistent and independent decisions,[10] but instead, it is grasped as a set of standardized substantive protection, which is the subject of interreferential investment arbitration rulings[11] and which as a whole constitutes a system of law.[12]

The author supports his thesis by observing the operation of the investment treaty law from three different angles: bilateral investment treaty making,[13]

[6]Ibid. p. 1834.

[7]Ibid. pp. 1834, 1825.

[8]Ibid.

[9]Ibid. p. 1835.

[10]"Unlike genuinely bilateral treaties, BITs do not stand isolated in governing the relation between two States; instead they develop multiple overlaps and structural interconnections that create a relatively uniform and treaty-overarching legal framework for international investment relations based on uniform substantive and procedural principles with little room for deviation": Ibid.; 1821.

[11]Ibid. pp. 1830–1834.

[12]Ibid. p. 1835.

[13]Ibid. pp. 1822–1826.

multilateralization devices in investment treaties,[14] and investment treaty arbitration.[15]

Concerning bilateral investment treaty making, he contends that the convergence of the standardized model that is reproduced in the generality of BITs is the product of states' "intentional planning"[16] toward "establishing uniform investment rules"[17] that echo "multilateral aspirations."[18] The multilateral character of BIT standardized provisions—according to Schill—reflects previous failed attempts at adopting multilateral investment treaties, whose substance was later replicated in national model treaties of western countries, which, in the end, were transformed into BITs.[19] As a result, he concludes that "this BIT-making in multilateral processes illustrates that the bilateral form of BITs was not a reaction to the need for flexibility but a vehicle for implementing multilateral aspirations."[20]

Regarding specific devices of multilateralization, Schill bases his findings on the functioning of most favored nation (MFN) clauses and on the dynamic of nationality planning (treaty shopping).[21] Apropos the former, the author expresses that since MFN clauses compel the host state to grant the investor protected under the BIT the same treatment as that provided to investors from other nationalities, in this case, BIT conditions propagate from one to the other, resulting in harmonization and multilateralization.[22] With respect to the latter—the multilateralization effects of treaty shopping—Schill noted that the openness of the concepts of investment and investor in BITs, coupled with the unrestricted ability of corporations to legally settle in any state, means that "virtually any investor from virtually any country is capable of opting into virtually any BIT regime."[23] If any investor can be benefited from any BIT, bilateralism becomes an impossible aspiration and ends up giving way to multilateralism.[24]

Schill's third observation discloses the multilateralization of the investment treaty law as a result of the tribunals' interpretation, *in pari materia*, and the use of a precedent in investment-treaty-based arbitration.[25] The first phenomenon consists of the tribunals' practice of using BITs adopted by third parties as a reference for

[14] Ibid. pp. 1826–1829.

[15] Ibid. pp. 1829–1834.

[16] Ibid. pp. 1822–1826.

[17] Ibid. p. 1823.

[18] Ibid. p. 1824.

[19] Ibid. pp. 1822–1824.

[20] Ibid. p. 1824.

[21] Ibid. pp. 1826–1829.

[22] Ibid. pp. 1826–1827.

[23] Ibid. p. 1828.

[24] Ibid.

[25] "Tribunals assume a norm creating function in adjudicating investor state disputes (...) employ rationales and argumentative structures that suggest the existence of an overarching body of international investment law that has merely found its expression in bilateral treaty relationships": Schill (2009), p. 278.

interpreting the BIT that governs the controversy at hand.[26] This practice of "cross-treaty interpretation"[27] gives rise to a uniform framework of principles that overarches the operation of investment arbitration; that is to say, it *multilateralizes* the investment treaty law.[28] The second phenomenon—the use of a precedent—relates to the tribunals' practice of grounding decisions on prior rulings of investment arbitration tribunals or, otherwise, providing reasons for deviating from them.[29] Analogous to the effects of interpretation *in pari materia*, the use of a precedent entails a form of interpretation that falls beyond the analysis of the parties to the BIT, forming a "standard setting device or an instrument of system-wide law-making."[30] Despite the fact that Schill labels this manifestation of multilateralization as the *use of precedent*[31] and the fact that he labels the tribunals' repeated elaborations as *jurisprudence constante*,[32] he makes clear that the latter "becomes more authoritative than a formal source of international law."[33] Even if the use of these expressions might be subject to disagreement, the point to stress here is that backing decisions with prior rulings "presuppose[s] the existence of a treaty-overarching framework of international investment law."[34]

Schill's three observations support, in general terms, my claim that the global climate change interest (and the climate change international legal framework) is formally relevant in investment treaty arbitration. Taking Schill's results to their ultimate consequences, it might not be an error to posit that in a multilateralized context, investment treaty disputes are resolved not only by reference to the parties' interests—those of the investor and host state—but also by reference to *community interests* at large—including the global climate interest.

Despite evidence that the scope of the investment treaty law reaches interests beyond the parties' interests, making it possible to affirm—in general terms—that the climate change international legal framework is *relevant* to the system, there could be still concerns as to what extent the *multilateralization* process involves an *openness* to legal (sub)systems that—in principle—constitute different entities and, if so, how this *openness* would affect the *autonomy* of the investment treaty law as a particular system. In other words, the fact that the operation of the investment treaty law shows that there is a system equipped with principles and standards of its own that reflect community interests does not inexorably entail that the system captures,

---

[26] Schill (2015), p. 1829.
[27] Ibid. p. 1833.
[28] Ibid. pp. 1829–1830.
[29] Ibid. pp. 1830–1834.
[30] Ibid. p. 1831.
[31] Ibid. p. 1830.
[32] Ibid. p. 1831.
[33] Ibid. p. 1829.
[34] Ibid. p. 1833.

within this notion of community interests, any interest embedded in other international law (sub)systems—such as human rights, international environmental, or climate change law. Criticism against Schill's definition makes it possible to dissipate the ambiguities it may have concerning the system's capacity to integrate the principles and rules of international law while preserving its autonomy.

3.1.2 Criticism to the Definition Multilateralization: Understanding Community Interest as a Notion That Embraces the Global Climate Interest

Multilateralization—as conceptualized by Schill—has been thought to have the potential to result in a self-contained approach to the system that is biased in favor of investors' interests.[35] In fact, the author's hypothesis that the investment treaty law "operates as part of an *autopoietic, self-referential, and normatively closed* system of law (...)"[36] that "contributes to providing a legal framework for the functioning of the global economic system"[37]—if not read through the lens of *constructivist social epistemology*—may put at odds my statement about the process of multilateralization showing the relevance of the global climate interest in investment treaty disputes.

According to Di Benedetto, the language deployed, when defining the operation of the investment treaty law, "attests"[38] that Schill's unitary vision of the system embodies the significant risk of detaching it from international law.[39] Di Benedetto adds that if one analyzes the language (*autopoietic, self-referential, and normatively closed*[40]) in connection with the "theory's ideological assumptions"[41] that the legal framework is oriented toward the concretion of the "process of economic globalization"[42] and "a particular form of western liberalism,"[43] one might conclude that the system is "sharply oriented towards protecting the investor"[44] and excluding other layers of interests, such as environmental protection.[45]

[35] Di Benedetto (2013), pp. 41–44.
[36] Schill (2009), p. 280.
[37] Ibid. p. 281.
[38] Di Benedetto (2013), p. 41.
[39] Ibid.
[40] Schill (2009), pp. 280–281.
[41] Di Benedetto (2013), p. 42.
[42] Ibid. p. 42.
[43] Ibid. p. 43.
[44] Ibid. p. 42.
[45] Ibid. p. 43.

The criticism concerning *multilateralization*, then, seems to be directed against Schill's description of the system's operation (as such) and the functional definition that he proposes. Both disagreements deserve to be analyzed in order to support my claim that *multilateralization* proves the relevance of the climate change legal framework in investment treaty law. Thus, before examining the correctness of defining the investment treaty law as a "legal framework for the functioning of the global economic system"[46] and the place that climate change interests may occupy in this conceptualization (Sect. 3.1.2.2), it is convenient for methodological reasons to examine the system's operational description in terms of *autopoiesis* (Sect. 3.1.2.1). Understanding the system's operation in this way leads us to claim that, far from insulating investment treaty law, *multilateralization* is in fact an essential condition for conceiving the climate change legal framework as part of it.

3.1.2.1 Does the Operational Definition of *Multilateralization* Isolate the Investment Treaty Law? Understanding the Self-Referential Closedness of the System as a Condition to Its Openness to the Climate Change International Legal Framework

The criticism over the operational definition of *multilateralization* is not insurmountable and is, to a certain extent, unfair. The expressions *chosen* by Schill to define the operation of investment treaty arbitration as "autopoietic, self-referential, and normatively closed system of law"[47] or as containing "autonomous and overarching principles of investment"[48] protection that do not involve *per se* detachment from the rest of international law and international relations, as Di Benedetto contends.[49] Conversely, neither in a mere rhetorical word-game nor articulated in a vernacular language are the concepts behind Schill's words deployed to express the idea of a *system of law* without discarding any cross-fertilization between systems. The discovery of the investment treaty law as a *system* is central to Schill's entire discourse. As he puts it, *multilateralization* "provides theoretic foundation to a claim that is often made, but rarely explicit conceptualized, namely that it is possible to speak of a system of international investment protection."[50]

Understanding the *systems theory* under Luhmann's constructivist social epistemological approach[51] is what allows us to read Schill's *multilateralization* in the appropriate key. In this context, the term *autopoiesis*—originated in disciplines such as biology, mathematics and neurophysiology—has permeated sociology and legal

[46] Schill (2009), p. 281.
[47] Ibid. pp. 280–281.
[48] Schill (2015), p. 1836.
[49] Di Benedetto (2013), p. 41.
[50] Schill (2015), p. 1835.
[51] Luhmann (2009), p. 180.

theory in order to describe the capacity of a system to subsist and reproduce itself through self-reference as an essential feature for its own identity.[52]

On this subject, Luhmann explains that the "appropriateness of the transposition"[53] of *autopoiesis* to the legal theory is buttressed by two assumptions from the systems general theory: conceiving "differentiation as the establishment of system-environment relationships in systems"[54] (theory of system differentiation) and "the assumption that such differentiation is possible only through the establishment of a self-referential closedness in the systems becoming differentiated."[55] In other words, the self-referential closedness of the system—characterized by proper operations, which are the result of referring to prior operations and have the effect of modeling future operations[56]—is the foundation for distinguishing the system from the environment, but, paradoxically, self-referential closedness cannot be conceived without the environment, which makes it possible to repack the system's operation in (artificial) functional terms. In sum, "the concept of the autonomy of the legal system (...) refers only to the system's operative closedness, as a condition for its openness."[57]

Against this background, the deployment of the expressions "autopoietic, self-referential, and normatively closed system of law"[58] to describe the operation of the investment treaty law—based on the constructivist social epistemology approach—means that any legal problem concerning the investment treaty law is addressed by referring to the outcome of its operations and to the potential significance of the answer to the operation of the investment arbitration treaty law system ("[i]n each of its own operations it has to reproduce its own operational capacity"[59]).

Schill's description of the operational features of the investment treaty law system—wrapped up within the concept of *multilateralization*—is not only necessary to stress the autonomy of the system; it also enables the openness of the system. This is so because it is crucial to identify the investment treaty law as a universe—as opposed to the multiverse of international law and international relations—in order to allow them to interact; i.e., the self-referential closedness of international investment arbitration opens the interaction with the system of international law and international relations. Otherwise—without self-referential closedness—investment arbitration law would not exist and would not be distinguishable from international law, and in the end, it would not be possible to talk about communication between the system of investment arbitration and the multiverse of international law.

---

[52] Ibid. pp. 181–182.

[53] Ibid. p. 182.

[54] Ibid. p. 183.

[55] Ibid.

[56] "In each of its own operations it has to reproduce its own operational capacity": Ibid. p. 184.

[57] Ibid. p. 185.

[58] Schill (2009), p. 280.

[59] Luhmann (2009), p. 184.

Considering the foregoing, describing *multilateralization* in terms of *autopoiesis*, self-referentiality, and closedness is essential to asserting that the international climate change legal framework is *relevant* in investment treaty law. The concept of *multilateralization*, by providing autonomy to the investment treaty law, renders conceivable its interaction with global public interests that are embodied in systems pertaining to the multiverse of international law—such as the global public interest of leveling global warming. In brief, self-referential closedness entrenched in the concept of multilateralization entails the international investment law's openness to the climate change international legal regime, in the sense that it can be distinguished by reference to the latter, and, moreover, it reveals to what extent the latter models the former.

By reference to the environment of the system (international law and international relations), the *autopoiesis* of the investment treaty law system could be rewrapped in functional terms. Reframing the autopoietic operation of the process of *multilateralization* in the language of the system of international law and international relations identifies international investment treaty law and, at the same time, provides a theoretical framework for unearthing whether the climate change international legal framework is part of it. Consequently, having shown that Schill's description of the system's operation, instead of isolating investment arbitration law, enables communication with international law and international relations, I will examine whether defining investment treaty law in functional terms as "a legal framework for the functioning of the global economic system"[60] excludes the climate change international legal framework.

3.1.2.2 Does the Functional Definition of *Multilateralization* Exclude International Global Interests from Investment Treaty Law? Understanding the Functioning of the Global Economic System as Comprehensive of Climate Change Interests

Schill's definition of the investment treaty law as "a legal framework for investor-state relations in the global economy"[61] or a "legal framework for the functioning of the global economic system"[62] is broad enough to comprehend the inclusion of the promotion and protection of global public interests within its scope. In fact, theoretically, there is nothing in this definition that implies that the legal framework for governing global economic investor-state relations cannot integrate elements promoting investments that encourage the global climate interest or counter rational elements that limit the protection provided to investors in order to avoid undermining the global climate interest.

---

[60] Schill (2009), p. 280.
[61] Schill (2015), p. 1835.
[62] Schill (2009), p. 280.

Yet the definition is also broad enough to reject the formal relevance of the climate change international legal framework if the functioning of the global markets is considered to be exclusively driven by market forces. In this regard, the author's elaboration of the definition seems to take a normative choice toward favoring an understanding of the functioning of the global economic system from an unrestricted global neoliberal economic approach. It may therefore be that the criticism against the elaboration of this functional definition—rather than against the definition itself—might have a point or, at least, shows that there is some ambiguity that needs to be clarified.

The author's elaboration of what the purpose of the legal framework for the functioning of the global economic system is (or should be?) commences with a functional approach but ends up taking a normative choice in favor of an economic model that might contradict the operation of a multilateralized system. The author is right to unfold the operation of the framework in functional terms, as founded on the concepts of "equality and not discrimination,"[63] as well as in pointing out that the investment treaty law provides a "level playing field"[64] to investors. However, the author's assertion that the investment treaty law is an instrument for the efficient allocation of capital "in an increasingly liberal fashion,"[65] disregarding any limitations to the free functioning of the global markets, may destroy the essence of *multilateralization*: if the community interest that *multilateralization* protects focuses exclusively on protecting unrestricted direct foreign investment capital flows, then would it not be better to achieve market efficiency without having any system of international investment law at all?

It might be contradictory to talk about a multilateralized system when the only community interest that is relevant is the promotion of foreign direct investment. The achievement of this objective does not require a multilateralized system because this goal would be better served by interpreting BITs as private bargains, where the offer and demand for capital find an optimal balance. Since the global market itself promotes an efficient allocation of resources within the capital-exporting states and the capital-importing states, we do not need to envisage a systemic interpretation to protect the movement of investments.

In sum, if the market forces lead to the best possible deal, it may be useless for tribunals to open up the interpretation to systemic considerations as a means to promote the efficient allocation of investments beyond the parties' agreement. Limiting "the functioning of the global economic system"[66] to "a particular form

[63] Schill (2015), p. 1825.

[64] Ibid.

[65] "(...)enables market to unfold" (...) "but have a broader function in creating institutions that back up an international market economy in which capital flow can in an increasingly liberal fashion between different national economies to wherever it is allocated most efficiently" (...) "stabilizes and structures the economic activities of foreign investors and requires host states to their behavior to the rule of law standards, thereby enabling market forces to unfold": Ibid. p. 1835.

[66] Schill (2009), p. 280.

of western [neo]liberalism"[67] is not compatible with the statement that "Investment treaties are (...) not designed to function such as private law contracts that order the relationship between limited number of parties and contain the exchange of specific transactions"[68] nor with the claim that treaties "have a constitutional function in providing legal framework for investor-state relations in the global economy."[69]

Moreover, the real operation of the investment treaty-multilateralized law system suggests that there are considerations of global public interest that coexist with and limit the free flow of investments in the investment treaty law; carved-out clauses and general exceptions in BITs, as well as decisions that directly or indirectly apply international law norms, confirm this statement.[70] This is not to undermine the role the efficient allocation of resources plays as a key function of the system, nor that the global public interests of equality and non-discrimination are not protected under the legal framework. The idea to stress here is that the spillovers and externalities inherent in the functioning of the global liberal markets also fall within the radius of the system, which also promotes and protects other global public interests.

The rereading of Schill's functional definition of the investment treaty law, which I am proposing, highlights the fact that *multilateralization* entails a legal framework for the functioning of the global markets that encapsulates global public interests. This approach, thus, emphasizes that investment treaty arbitration is not manifested as an image of the "invisible hand" but rather comprehends the restrictions entrenched in international law. It is within this operational framework that climate change international legal instruments find a place in investment treaty law as applicable law.

Thus, in descriptive terms, my perspective represents an innovation of Schill's conception of *multilateralization*. Although his definition of the system as a "legal framework for the functioning of the global economic system"[71] is still in line with my hypothesis, devoid of the integration of the values entrenched in the system of international law to the investment treaty law legal framework, the *multilateralization* process would not only contradict its own objective of serving *community interests*, but would also fail in acknowledging that the operation of the tribunals, as well as the drafting techniques for BITs, shape the multilateralized investment treaty law system by reference to global public interests.

---

[67] Di Benedetto (2013), p. 43.

[68] Schill (2015), p. 1835.

[69] Ibid.

[70] See. Infra. Part III.

[71] Schill (2009), p. 281.

3.2 Syncretism of the Investment Treaty Law: Envisaging the Global Climate Interest in Investment Treaty-Climate Change Disputes

Encasing international public interests—including climate change interests—and the fragmented multiverse of international law—including the climate change international legal regime—within the operation of the investment treaty law system would be better labeled as a process of *multi(syncretic)lateralization*.

Multi(syncretic)lateralization is a paradoxical phenomenon for two reasons. While the first paradox is obedient to Schill's finding that bilateral treaties are modeling international investment law as a multilateral system (*multilateralization*), the second paradox claims that the multiverse of international law is modeling the investment treaty law system by constituting part of the latter (*syncretism*). Accordingly, while the first paradox enables the operation of the international treaty law to be defined in operational and functional terms as an autopoietic legal system for the functioning of global markets, the second paradox complements this conceptualization by stressing that the investment treaty law fosters the allocation of foreign direct investment in connection to and limited by the promotion and protection of global public interests.

Yet *syncretism* is not the first doctrinal attempt to comprise (global) public interests within the investment treaty law system. In general terms, the most prominent theories in this regard are grounded on the classification of investment treaty law/arbitration as (international) public law. The current state of the art is represented in particular by Schill's comparative public law methodology (Sect. 3.2.1), Van Harten's public law substantive validity theory (Sect. 3.2.2), Kulick's global public interest theory (Sect. 3.2.3), and De Brabandere's international public law approach to investment treaty arbitration (Sect. 3.2.4).

Against this backdrop of theories, it is worth considering whether it is not sufficient to simply state that investment treaty law is public law, comparative public law, or international public law in order to prove the formal relevance of global public interests to the system rather than labeling the process as *multi(syncretic)-lateralization*. The answer to this question allows me, on the one hand, to unravel how the notion of syncretism differs from the current state of the art when it comes to the formal relevance of global public interests, including climate change interests, and, on the other hand, leads me to propose that *syncretism* gives rise to the possibility for a direct application of climate change provisions (Sect. 3.2.1); preserves the identity of the investment treaty law while recognizing the formal relevance of the international climate change legal framework (Sect. 3.2.2); gives ground to the reformulation of the (global) public interest challenge by framing the challenge within the dynamic three-cornered tension between the investor, the state, and the global public interest, including climate interest (Sect. 3.2.3); and integrates the climate change legal framework into investment treaty law beyond the topical rhetoric of public law (Sect. 3.2.4).

3.2.1 *Distinguishing Multi(syncretic)lateralization from the Comparative Public Law Methodology: Allowing the Possibility for a Direct Application of the Climate Change International Legal Framework*

Considering the idea of global public interest in investment treaty disputes as integral to the operation of a *syncretized* system might not be in line with the consequences that Schill attributes to *multilateralization*. He affirms that "[...] the claim that investment law is continuously multilateralizing allows reconstructing international investment law as *forming a subsystem of international law*";[72] however, this assertion appears to refer only to the reaffirmation that investment treaty arbitration is a system that forms part of the fragmented multiverse of international law, rather than aiming to recognize the authority of other international law instruments entrenching global public interests in international investment treaty disputes.[73]

Nevertheless, it would be incorrect (and unfair) to accuse Schill of rejecting any influence of public interests in investment treaty law. Indeed, it is beyond any doubt that he is one of the leading authors promoting the publicization of international investment treaty law through the adoption of a comparative public law methodology.[74] His attempt to provide a public law approach to investment treaty law by making an analogy to domestic and international public law systems is not directly linked to his theory of *multilateralization* but is buttressed by the notion that the system could be classified to a certain extent as (international) public law, as well as grounded on the understanding that investment treaty law is an instrument of global governance.[75]

Schill's adoption of the comparative public law methodology is a consequence of labeling investment treaty arbitration as (international) public law. In his work, while distinguishing commercial arbitration from investment treaty arbitration, the author posits the characteristics that enable him to claim that the latter is to a certain extent (international) public law, namely, that the system "involves questions about the scope and limits of the Host States regulatory powers"[76] (subject matter of the disputes[77]), that "the rights invoked [originate] from obligations the Host-state has assumed under an international treaty"[78] (nature of the obligations[79]), that "foreign investors and Host states stand in hierarchical relationship of super- and

---

[72] Schill (2015), p. 1836.

[73] Ibid.

[74] Schill (2010a, b).

[75] Schill (2010a, b).

[76] Ibid. p. 14.

[77] Ibid.

[78] Ibid. p. 15.

[79] Ibid. p. 14.

subordination"[80] (relationship of the parties[81]), and that "arbitral jurisdiction in investment treaty arbitration [. . .] involves a unilateral offer by the host state, given in an investment treaty, in generalized and prospective form [. . .]"[82] (consent to arbitration[83]).

Certainly, the characteristics of the investment treaty law described by Schill conform to some of those commonly present in the academic category of public law. Nonetheless, one cannot derive any authority of (global) public interests in international treaty arbitration from this conclusion since the authority of any formal source is grounded on the practice of the applying authorities rather than on any analogical reasoning, at least from the social practice approach. This might be the reason why Schill does not go as far as to claim that the similarities between investment arbitration and other public law systems provide a pedigree of formal sources for other instruments that could be labeled as international public law;[84] i.e., his theory remains a nonexclusive method of content interpretation rather than a general theory of framework setting, or it could be simply understood as a means to propose that the investment treaty law should be modified.[85]

Relying on comparative public law may not be sufficient to bridge the gap between investment treaty law and (international) public law because it reduces the operation of the latter to the content interpretation of the BIT. Thus, comparative public law would not be enough to legitimize the direct application of international climate change provisions to investment treaty law disputes. Conversely, since a syncretic investment treaty law system presupposes that global public interests—as well as the legal instruments in which they are embedded—are a constitutive part of the system, the direct application of climate change provisions is conceivable. Furthermore, while choosing a comparative public law method of interpretation would not be compulsory for tribunals (and could be even considered outside the limits of Article 31 of the Vienna Convention on the Law of Treaties (VCLT)), *syncretism* emphasizes the authority of international law provisions by integrating

[80] Ibid. p. 15.

[81] Ibid. p. 14.

[82] Ibid. pp. 15–16.

[83] Ibid. p. 14.

[84] "Comparative Public Law analysis serves several purposes: It helps (1) to concretize and clarify the interpretation of the often vague standards of investment protection and determine the extent of state liability in specific context; (2) to balance investment protection and non-investment concerns; (3) to ensure consistency in the interpretation and application of investment treaties (. . .) (4) to ensure cross-regime consistency and mitigate the negative effects of fragmentation (. . .) (5) to legitimize existing arbitral jurisprudence; and (6) to suggest legal reform of investment treaty making or changing to arbitral practice (. . .): Ibid. p. 25.

[85] "Comparative public law and the national and international regimes one may draw on do not control the interpretation and the application of international investment treaties (. . .) To the extent that investment treaty obligations leave no room for doubt, the ambit of comparative public law will be limited to *a lege ferenda* perspective": Ibid. p. 27.

global public interests in investment treaty law and compels tribunals to conceive the climate change international legal framework provisions if material to the dispute.

I want to stress that I am not contending that it is *per se* incorrect to determine the content of BITs through a comparison with (international) public law. Yet neither is it incorrect to use other interpretative methods or instruments or even ignore comparative public law as a method for interpretation. Concerns about the social validity of content interpretation are misplaced as long as the interpretation does not enter into conflict with Article 31 VCLT, which is clearly consolidated into the social practice of investment treaty tribunals. On the contrary, as *syncretism* underlines that the system's operation evinces that global public interests—and the corresponding international legal instruments—are regarded in social terms by tribunals as applicable law, denying these instruments their binding authority would be invalid.

It may be useful to provide an example in order to illustrate the practical difference between adopting a comparative public law approach and *multi (syncretic)lateralization*. As proposed below,[86] this study considers the possibility of assessing compensation in investment treaty-climate change disputes based not only on the impact of the damage that the investor suffers but also on the host country's development and economic situation. For this purpose, I find it apposite to directly apply the "principle of common but differentiated responsibility." *Multi (syncretic)lateralization* allows the direct application of this principle since the provisions of the climate change legal framework are impregnated with authority, thanks to the essentiality of global public interests for the operation of investment treaty law. In contrast, if a *comparative public law* methodology is deployed, the application of the "principle of common but differentiated responsibility" would not be possible unless the reading of the BIT provisions indicate so.[87] As BIT provisions do not normally give any weight to the development and economic situation of host states, a comparative public law methodology appears insufficient to drive the inclusion of the "principle of common but differentiated responsibility" into the investment treaty law system.

It could be argued that instead of relying on *syncretism* as an essential feature of the system, the insufficiency of the comparative law approach as a means to incorporate the application of the climate change legal framework into investment treaty disputes might be alleviated by simply stating that investment treaty law is (international) public law. Accordingly, in the following section, I demonstrate that *syncretism* is better equipped at validating the application of climate change provisions to investment treaty law than public law substantive validity theories.

[86] Infra. Chap. 10.

[87] Infra.

3.2.2 *Distinguishing* Multi(syncretic)lateralization *from (International) Public Law Substantive Validity Theories: Preserving the Identity of International Investment Law*

Van Harten's 2007 study on "Investment Treaty Arbitration and Public Law"[88] is, without a doubt, one of the most frequently cited studies on the integration of public law into investment treaty arbitration. Despite the fact that his thesis could be framed within the global administrative law movement[89] and that the focus of his criticism is the lack of openness, independence, and accountability of the investment treaty arbitration procedure,[90] his considerations of the interpretation and application of *law* certainly impact the assessment of the validity of global public interests in investment treaty law.[91]

Occasionally, Van Harten's work appears to be positing that public law is only an instrument for the interpretation of BIT standards content, placing the BIT as the system's center of validity.[92] This perspective would be equivalent to Schill's comparative public law and deserves the same criticism. Yet, in many cases, the author seems to push the authority of public law beyond content interpretation by suggesting that it constitutes the applicable law framework. Since his theory appears to locate the axis of validity within the concept of public law, I argue that his approach could be labeled as a substantive validity theory (Sect. 3.2.2.1) that—unlike *syncretism*—threatens the identity of the investment treaty law system (Sect. 3.2.2.2).

3.2.2.1 Identifying Van Harten's Public Law Approach to Investment Treaty Arbitration as a Substantive Validity Theory That Is Incompatible with Formalism

Van Harten's theory confers substantive validity to the characterization of investment treaty law as public law. Throughout the book, he attempts to unearth the

[88] Van Harten (2007), p. 126.

[89] The author "show[s] that private arbitrators have a new-found power to review and discipline states, and to convey an impression of how this power interacts with the lives of ordinary people and the way they are governed": Ibid. p. 3.

[90] ". . .systems unique use of private arbitration in the regulatory sphere conflicts with cherished principles of judicial accountability and independence in democratic societies. . ." (. . .) "disputes. . .presenting the major challenge to public principles of judicial accountability, openness, and independence": Ibid. pp. 4–6.

[91] Ibid. pp. 143–151.

[92] "the view that the appropriate approach is to accept the regulatory context for investment disputes, and thus the relevance of public law, within the boundaries set by inter-state bargain of an investment treaty": Ibid. p. 6.

public law essence of investment treaty arbitration as a sort of supra-reality that lies beyond the "complexity of the system,"[93] which in his words "(...) may tend to mask its character in this respect."[94] As a result—concerning the applicable law and interpretation—the author asserts that "[...] the appropriate way to implement the adjudicative function under an investment treaty is to adapt a public law framework to the novel context of regulatory adjudication under international law."[95] This process of adaptation implies that the domestic public law framework trumps international law principles,[96] that liability rules and the assessment of damages should mirror domestic public law,[97] and that the public law framework compels tribunals to interpret BITs while taking into account public interests (countervailing interests[98]—in his words).[99]

Regarding the role of public interests in investment treaty law, Van Harten's logic could be simplified into the following syllogism: since all public law adjudication processes take public interest considerations into account (major premise) and international investment treaty law is a public law adjudication process (minor premise), investment treaty law takes public interest considerations into account (conclusion).[100]

Van Harten's major premise is intended to dissipate the confusion of law and morals by dressing what I understand as a moral or ethical principle as a natural fact: any public law adjudication process compels the operator to promote and protect public interests. However, he cannot do so without some logical contradiction. On the one hand, he states that it is natural for investment treaty arbitration—as with any public law adjudication process—to take into account public interests, but, on the other hand, he claims, throughout his work, that investment treaty arbitration, in practice, often does not take into account public interests.[101] If it is natural for any

[93] Ibid.

[94] Ibid.

[95] Ibid. p. 143.

[96] "(...) the individualization of claims, especially in as far-reaching a form as under investment treaties, transforms international adjudication by expanding the degree into which engages the regulatory sphere, thus superposing the analytical framework of public law. This means that the international principles which originate in adjudication of disputes must be modified or jettisoned": Ibid. p. 135.

[97] Ibid. p. 150.

[98] "regulatory concerns (..) including the countervailing interests that are represented in public opposition to particular business activity; the implications of a particular investment for the host economy; the impact on the host's states ability to prioritize social stability, public health and morals, and environmental concerns over objectives of economic efficiency and investor's confidence": Ibid. p. 144.

[99] Ibid.

[100] In this regard, the author posits "there is no excess of jurisdiction" (Ibid.) in taking into account public interests in investment treaty arbitration, given that these concerns are an "integral part" (Ibid.) of any system of public law adjudication.

[101] "Unfortunately some arbitrators and many commentators seem to act aggressively when protecting investors while downplaying the implications for the respondent state. This convert

public adjudication process to recognize public interests, how could investment treaty arbitration—in many cases—not recognize public interests (how could something "be" and "not be" simultaneously?). Either (1) public interests are not natural to public adjudication processes, or (2) investment treaty arbitration is not a public adjudication process at all. While answer number "1" contradicts his major premise (*any public law adjudication process compels the operator to promote and protect public interests*), the second answer contradicts his minor premise (*International investment treaty law is a public law adjudication process*). This is why Van Harten's major premise can only be understood as a moral imperative, where his syllogism would be better reformulated as follows: since all public law adjudication processes *should* take public interest considerations into account (major premise) and international investment treaty law is a public law adjudication process (minor premise), investment treaty law *should* take public interest considerations into account (conclusion).

Since the major premise—public law adjudication processes *should* take public interest considerations into account—is a moral or ethical criterion, Van Harten's work struggles to separate *law* from *non-law* or normative claims from descriptive claims. The public interest substantive value thesis clashes with the methodology of this book, just as positivism clashes with naturalism.[102] As a result, I cannot accept—based on the sole idea that investment treaty law is public law and without making any reference to the system's operation—that the climate change international legal framework is formally relevant in investment treaty disputes. Moreover, if we accept the substantive validity of public law, we risk erasing the distinction between the investment treaty law system and any other (sub)system that could be labeled as public law.

3.2.2.2 *Syncretism* as a Means to Preserve the Identity of the Investment Treaty Law in Contrast to the Public Law Substantive Validity Theory

Reformulated or not, Van Harten's major premise that all public law adjudication processes take (should take) public interest considerations into account might well

investment treaty arbitration into a regime that governs the public sphere by private law rules or rights-based norms and, as such (...)": Ibid. p. 150 "(...) [T]his analogy of commercial arbitration is open to the criticism that over-emphasizes principles of private law...at the expense of principles that typically apply in public law adjudication (...) The problem arises when this analysis is transferred into the realm of treaty arbitration, as in numerous awards (...)": Ibid. p. 125.

[102] Several fragments of Van Harten's book denote the intrinsic value that public law infuses in arbitration: "part of the public character of investment treaty arbitration is the recognition that the aims of investment treaties also depend on how values other than property rights and the sanctity of contracts are advanced within the adjudicative process": Ibid. p. 150. "(...) [T]he appropriate way to implement the adjudicative function under an investment treaty is to adapt a public law framework to the novel context of regulatory adjudication under international law": Ibid. p. 144.

not support the conclusion that *investment treaty law takes (should take) public interest considerations into account*. This is so because taking public interests into account is an essential feature of *law* rather than an exclusive characteristic of public law; i.e., the author falls into the logical fallacy of turning a correlation based on a common *causa* into a causal correlation, *cum hoc ergo propter hoc* (with this, then because of this). The statement proposed as a conclusion—investment treaty law takes (should take) public interest considerations into account—is the outcome of a coincidence between public law and investment treaty law—law in general takes public interests into account—rather than the consequence of the latter being public adjudication.

All *laws* involve public interests. Private law—the natural antagonist of public law—limits the private interest of its agents and their autonomy, for instance, by referring to good faith, the concept of *ordre public*, the doctrine of *l'abus de droit*, or the regulation of absolute nullities, not to mention the constitutionalization or publicization of private law experienced in the past three decades, apparent in classic private law disciplines such as family law or property law. Even the *pacta sunt servanda* doctrine or the protection of property rights—foundational principles of private law—embeds public interests, such as certainty, predictability, autonomy, or the functioning of a market economy. Therefore, if one cannot distinguish public law from private law in view of the protection of public interests, one could not categorically affirm that the entry of public interests into investment treaty law results from its public law character.

Against this backdrop of public interests in private law, if one wishes to continue on the path of substantive validity, one would need to re-reformulate Van Harten's syllogism to its minimal expression: since all law adjudication processes take (should take) public interest considerations into account and international investment law is a law adjudication process, then investment treaty law takes (should take) public interest considerations into account.

The re-reformulated syllogism, as well as Van Harten's original syllogism, might appear simplistic, and it would also erase any boundaries between systems. It is simplistic because it does not provide any explanation on how to determine what public interests are part of the investment treaty law. Formulating that investment treaty law is public law (or law) is so ethereal that the operator would not be able to discover when to deploy or where to look for these public interests: are national constitutions or national jurisprudence applicable? Which cases should be resolved by reference to international law sources, and which should be resolved by mirroring national law? Are international public adjudication bodies relevant? Is it possible to rely on the law of one group of western countries to find out the principles of public law? Or, perhaps, are public law manuals embedded with authority? In these terms, the substantive public law theory, instead of resolving the problem of "taking into account" public interests, ends up deferring the problem: if investment arbitration is law or public law that contains public interests, then where is the law or public law that contains those public interests? In Van Harten's defense, he asserts that "such a detailed review is beyond the scope of the present [his] work."[103] Again, some level of formalism is required here.

More acute than its simplicity is the practical problem which the substantive validity of public law (or law) may pose to the identity of the investment treaty law system. Notably, perceiving public law (or law) as the source of validity jeopardizes the operation of the investment treaty law system and vacates the content of the latter. Moreover, given that under this theory any public law (or law) is relevant as part of the system without limitations, arbitrators might have absolute freedom to choose between a myriad of fragmented and undetermined rules, principles, and systems, which in the end is the same as not having a system. Devoid of any formal means to identify investment treaty law, investment arbitration tribunals would merely operate as public law tribunals or law tribunals. One may even ask—given the incapacity to make any judgment on the accuracy of applicable law under this theory—whether a tribunal in such conditions is still a law tribunal.

The vacating-system effect of Van Harten's substantive validity thesis is, somehow, softened when he limits the application of public law "principles developed domestically and, to a lesser extent, in the international sphere"[104] to cases that are in need of "resolving ambiguity in the scope and content of the standards of review or the appropriateness of an award of damages as a remedy against the State."[105] Although he recognizes BITs as having some level of authority, he does not seem to model investment treaty law as an autonomous system, or, in any case, he cannot do so without contradiction. If—as he argues—national public law principles supersede international law principles,[106] if public law is the exclusive instrument of interpretation,[107] and if the rules for liability as well as those for compensation are ascertained by mirroring national public law,[108] then there is not much room left for a system of international treaty law.

Unlike Van Harten's public law substantive validity theory, *multi(syncretic)-lateralization* reaches a conclusion that global public interests are entrenched in investment treaty law because the system's operation recognizes international law as governing law—without denying that the center of validity is still the investment treaty law system. Indeed, *syncretism* presupposes that the principles and values of the system of international law are an integral part of the investment treaty law system but, at the same time, are limited by the latter's operation and by its own principles and methods for dealing with clashing interests. By doing so, my proposal

[103] Ibid. p. 144.

[104] Ibid. p. 143.

[105] Ibid. p. 144.

[106] "(...) the individualization of claims, especially in as far-reaching a form as under investment treaties, transforms international adjudication by expanding the de degree into which engages the regulatory sphere, thus superposing the analytical framework of public law. This means that the international principles which originate in adjudication of disputes must be modified or jettisoned": Ibid. p. 135.

[107] Ibid. p. 144.

[108] Ibid. p. 150.

avoids risking the identity of the system and simultaneously preserves global public interests within its realm.

In this vein, when I claim that the climate change international legal framework is formally relevant as a constitutive part of the investment treaty law framework, I am not stating that the former is hierarchically superior to the latter or that its application is boundless and should always trump rules that protect other equally important interests. For example, as regards compensation for expropriation—as is shown below[109]—I argue that BITs that contain provisions specifying standards for their assessment—such as "market value"—supersede climate change provisions since the use of the principle of *lex specialis* is a regular practice of investment arbitration tribunals. Similarly, when a measure that favors climate is discriminatory or arbitrary, we claim (below)[110] that the host state is liable to pay compensation even if the measure promotes the leveling of global warming for the sole reason that the practice of tribunals emphasizes that public interest does not justify discriminatory or arbitrary measures.

Importantly, and despite having been written over 10 years ago, Van Harten's work is, to date, one of the most influential and critical interdisciplinary studies on investment treaty arbitration. By emphasizing that investment treaty arbitration involves regulatory relationships that require deviation from the traditional methods of adjudging in classic international law, it has the virtue of warning about the miss-application of rules and principles from private tort law or state to state responsibility controversies to investment treaty disputes, without the proper filter or adaptation. His work is not only a precursor to Schill's comparative public law method, but it also inspired[111] Kulick to assume "the public interest challenge"[112] and formulate his general "global public interest theory,"[113] which I examine in the next section.

3.2.3 Distinguishing Multi(syncretic)lateralization from Kulick's Global Public Interest Theory: Reformulating the (Global) Public Interest Challenge

The public interest challenge—proposed by Kulick—stems from the need to find tools to safeguard "common interests"[114] or "third party interests"[115] in investment treaty disputes,[116] an endeavor that—with some reservations on how the author

[109] Infra. Chap. 10.

[110] Infra. Chap. 7, Sect. 7.2.1.

[111] Kulick (2012), p. 6.

[112] Ibid.

[113] Ibid. p. 5.

[114] Ibid. p. 2.

[115] Ibid.

[116] Ibid.

determines global public interest—certainly matches the purpose of this research. Additionally, the influence of *multilateralization* in his work is as palpable as it is in this work. The author acknowledges that he profits from Schill's assessment of the investment treaty law as a "harmonious regime"[117] and/or a "de facto multilateral system,"[118] and, as I do in this study, he attempts to take *multilateralization* to a "different level"[119] that incorporates global public interests (*syncretism*).

In line with Van Harten's observations, Kulick describes the system's operation as leaning in favor of the investor and against the interests of the host state.[120] Accordingly, the author envisages the public interest challenge within the bilateral tension between the host state's public interest and the protection of the investor's interests.[121] From this point onward, the author analyzes, from multiple angles, the tension that pendulates from global administrative and constitutional law[122] to substantive validity theories,[123] from formalism[124] to the economic analysis of law,[125] and back to comparative law.[126] He ends up providing an answer to the challenge through a general theory of global public interest[127] and the adoption of proportionality tests.[128]

I argue that even if Kulick's global public interest theory is initially grounded on the substantive validity of public law, he ends up supporting his conclusions—to a considerable extent—through the operation of the investment treaty law system. Instead of contradicting *syncretism*, this eclecticism reinforces the hypothesis that, in formal terms, the promotion of global public interests is an essential feature of the system (Sect. 3.2.3.1). However, I propose a reformulation of Kulick's public challenge by reframing the challenge as a three-way tension between the host state, the investor, and the global community's interests, as well as a redefinition

[117] Ibid. p. 6.

[118] Ibid.

[119] Ibid. p. 7.

[120] "Once we accept this notion of supremacy of international law within the integrated system of the law applicable in an investment dispute, and once we acknowledge that such applicable international law mostly consists of the law enshrined in BITs, we face a simple fact: BITs deal mainly and often exclusively with investor rights": Ibid. p. 51.

[121] "Thus, any public interest measure undertaken by the host State that conflicts with the – often very broad – BIT provisions guaranteeing investor rights, will be trumped in an investment dispute. In other words, if the host State applies the public interest measure in conflict with a BIT provision, the investor has a claim under the realm of international investment law – irrespective of the urgency and the importance of applying the said measure. I shall call this problematic the public interest challenge": Ibid. p. 51.

[122] Ibid. pp. 77–85.

[123] Ibid. pp. 94–98, 127–154.

[124] Ibid. pp. 154–167.

[125] Ibid. pp. 57–65.

[126] Ibid. pp. 99–117.

[127] Ibid. pp. 127–153.

[128] Ibid. pp. 168–224.

of the challenge through the question of how to align the investor's and host state's interests with global public interests (Sect. 3.2.3.2).

3.2.3.1 The Eclecticism of the Global Public Interest Theory: Reinforcing the Thesis That the System's Operation Is Characterized by Global Public Interests

The global public interest theory posits that investment treaty law has an "inherently public law character"[129] that calls for "balancing the individual with the public interest."[130] These interests, according to the author, "find legal translation"[131] in the system through general principles and international customary law, which might be shaped in the form of the host state's defenses against the investor.[132] Kulick concludes from this observation that "customary international law and principles of law encapsulate global public interest in the realm of investment law, which is equivalent to public interest in a domestic public law situation."[133]

Kulick's thesis is normally labeled as another expression of comparative public law interpretations or substantive-value public law thesis,[134] and to a certain extent, it is. Yet his work seems to be far more sophisticated than this. Comments on his work perhaps dismiss the fact that—unlike Van Harten—he tries to move within the boundaries of the system. A different perspective reveals that the global public interest theory is an attempt to introduce a balancing of public interests stemming from the system's operation.

Kulick's reasoning is still positivist reasoning. First, instead of jumping directly to the conclusion that global public interests are within the realm of the investment treaty law because of the their public law nature, he admits that public interest needs to be accommodated to the primacy of international law over national law—which is an operational premise of the system.[135] Additionally, he does not only claim that the deployment of international public law principles and customary international law in investment treaty law implies the application of global public interests,[136] but he also confirms this statement with a tribunal decision that advocates for the recognition of

[129] Ibid. p. 5.

[130] Ibid.

[131] Ibid.

[132] Ibid.

[133] Ibid. p. 163.

[134] De Brabandere (2014), p. 5.

[135] "International law's prominence has now arrived at a stage where Tribunals not only acknowledge its supremacy over domestic law within the realm of international investment law, but also grant the former the role as the primary source in an investment dispute. I call this development the "internationalization" of international investment law": Kulick (2012), p. 4.

[136] Ibid. p. 163.

"public policy"[137] and "the values of the international legal system."[138] He also justifies the adoption of the proportionality principle in investment arbitration through the text of Article 42 of the International Centre for Settlement of Investment Disputes (ICSID), Articles 31(1) and Article 31(3)(c) of the Vienna Convention, and Article 38(1) of the Statute of the International Court of Justice (ICJ)[139]—sources that are generally accepted in the tribunals' practice.

Against this background, I dare to affirm that Kulick's positivist reasoning and observations not only match my reductionist legal positivism methodology, but to a significant extent, they also confirm the global public interests' permeation of the investment treaty law, and they reiterate the coherence of the affirmation that the climate change legal framework is formally relevant to the system. His methodology recognizes tribunals' practice as *law* and accommodates his public law (moral) orientation to the limits of this *law* and proves that this *law* captures global public interests.

The adoption of a positivist methodology as the basis of the global public interest theory reveals that Kulick's call for balancing interests in investment treaty arbitration grounded on its "inherently public law character" appears unnecessary and is subject to the same criticism as Van Harten's public law substantive validity thesis and Schill's comparative public law theory.[140] The solution that he provides to the public interest challenge would be the same by merely referring to the system's operation.

3.2.3.2 Reformulation of the (Global) Public Challenge Interest Stemming from *Syncretism*

My proposal of *syncretism* differs from Kulick's reasoning. While the latter intends to provide *international-based* defenses to the host state in order to level the disbalance in favor of the investor through the *global public interest theory*,[141] we simply suggest that global public interests are embedded within and are limited by the investment treaty law system. This contrast presupposes different solutions. Comparing *syncretism* and the global public interest theory with the use of a couple of examples might illustrate the similarities and differences in their solutions.

Example 1 According to Kulick's logic, if the host state closes a coal plant that is operated by a foreign investor as part of a plan to comply with the objectives of the Paris Climate Change Agreement, the host state may successfully oppose the defense that the measure was taken in the global public interest. This deduction

137 Ibid. p. 165.

138 Ibid.

139 Ibid. pp. 169–171.

140 Supra.

141 Kulick (2012), pp. 4–5.

emerges from the global public interest theory's premise that when the public interest of the host state matches the public interest of the international community, it is opposable against the investor's claim.[142] In this regard, *syncretism* would arrive at the same conclusion by simply arguing that since one of the system's properties is the recognition of global public interests specified in international law, then a measure pursuing a global public interest, such as protection against climate change, may be opposable.

Example 2 However, the answer differs when the interest that matches the international community interest is the investor's interest rather than the host state's interest. According to Kulick's theory, if the host state closes a wind park against the objectives of the Paris Climate Change Agreement, the investment would not be regarded as pursuing a global public interest, nor would the investor's interest have the possibility of overcoming the host state's defense. Under Kulick's approach, whether the host state's measure goes against the interests of the international community would be irrelevant in this case. This is so because in Kulick's system, the concept of global public interest is restricted to the hypothesis where the public interest behind the host state's measure matches the international community's public interest[143] and excludes the hypothesis where the investor's interest matches the interests of the international community. Conversely, under *syncretism*, since a general attribute of the system is recognizing international law and the interests of the international community as part of it, climate would be recognized as a global public interest, which might supersede the host state's interest. This is so because *syncretism* is intended to be neutral and recognizes the formal relevance of global public interests independently of the party whose interest runs in parallel with the global public interest.

As shown in the examples above, while under *syncretism* an international community interest might always have an impact on the outcome of the controversy, disregarding which of the parties benefits from aligning their own interest with the international community interest, under Kulick's theory, only the host state might benefit from aligning its interest with the community's international interest.[144] His theory conveniently accommodates the system in favor of the host state's interests and disregards the fact that the real operation of the international treaty law system may entail the investor also pursuing international global public interests.

Undoubtedly, Kulick's answer is *very smart* from a problem-based perspective and brings forth the fact that national interests are relevant in investment treaty law, as long as they are captured as global interests by international law. In his perspective, the disbalance between the protection of investor and host state interests originates from the primacy of international law over national law.[145] Thus, the

[142] Ibid.

[143] Ibid.

[144] Ibid.

[145] "International law's prominence has now arrived at a stage where Tribunals not only acknowledge its supremacy over domestic law within the realm of international investment law, but also

author proposes a leveling of the host state's position to that of the investor by conferring an international character to public national interests.[146] The way in which he internationalizes national public interests is by premising that when national public interest matches community international interest, they become a global public interest that could be deployed as a host state's defense against the investor's claim.[147] As a result, "the way Global Public Interest concerns will play out doctrinally is through defenses from the host State (...) it is simply inevitable that such matter will very often be introduced as a defense against an infringement of investor rights."[148]

Kulick's elaboration, thus, solves the public challenge of safeguarding "common interests"[149] or "third party interests,"[150] only if it is understood that promoting these interests is beneficial to the host state. But this might not be correct from a systemic point of view. International law and BITs incorporate several "common interests"[151] or "third party interests"[152] that match investors' interests.[153] Environment, climate change, labor rights, and human rights, among other global public interests, could be promoted through investments, as could other global public interests more closely linked to the functioning of global markets, such as the *pacta sunt servanda* principle, the protection of property rights, or the certainty and predictability of the system. Kulick's public challenge, then, should be reformulated in the understanding that "common interests"[154] or "third party interests"[155] cover a broader universe that encompasses the interests of both host states and investors.

Syncretism captures the wide universe of global public interests and provides Kulick's public challenge with a new shape and meaning—a new shape because it would no longer be modeled as a static two-cornered tension between the host state's public interest and the investor's interest, but rather as a dynamic three-cornered tension between the investor's interest, the host state's interest, and the global public interest. Such as in any triangular relationship, where the subjects interrelate in more than one way, the interests in an investment treaty dispute may conflict and overlap in various ways. For instance, some BIT provisions may appear to oppose the global public interest, yet some other clauses might appear to harmonize with the latter; very often—as Kulick affirms—the host state's interest promotes the global public

grant the former the role as the primary source in an investment dispute. I call this development the 'internationalization' of international investment law." Ibid. p. 4. See also Ibid. pp. 46–52.

[146] Ibid. pp. 153–154.

[147] Ibid.

[148] Ibid. p. 4.

[149] Ibid. p. 2.

[150] Ibid.

[151] Ibid.

[152] Ibid.

[153] See. Infra.

[154] Kulick (2012), p. 2.

[155] Ibid.

interest, but seldom these interests play against each other; similarly, investors' interests could either be in line with or oppose global public interests, not to mention that while the parties to the treaty intend to establish a collaboration (cooperation) relationship through BITs, the resultant relationship between the investor and the host state degenerates into competing interests.

Syncretism gives new meaning to Kulick's public challenge because the key question would no longer be how to internationalize the host state's public interest in order to level the investor and the host state's position in investment treaty law, but rather how to align the protection of investor and host state interests with global public interests.

3.2.4 Investment Treaty Arbitration as International Public Law: Revealing the Climate Change Legal Framework as Part of the Investment Treaty Law System Beyond Topical Rhetoric

A similar effort to that of Van Harten and Kulick is the attempt made by De Brabandere to link the (global) public interest with the procedural aspects of investment treaty arbitration in his book *Investment Treaty Arbitration as Public International law*. As happens with Van Harten's study, even if the intention of his book is to emphasize the influence of international public law in the investment treaty arbitration procedure, Chap. 4[156] of his book—dedicated to the applicable law and noninvestment considerations in investment treaty arbitration—inexorably carries consequences for the interpretation of substantive law.[157] Furthermore, the last and most prominent section of this chapter is dedicated to the applicability of human right's obligations to investment treaty law.[158]

However, Van Harten and De Branbandere's approaches differ considerably. While the former's theory is grounded on the substantive value of the concept of *public law* in investment treaty arbitration, the latter adopts a functionalist method based on the system's operation, which leads him to conclude that

> the fact that contemporary investment arbitration is a preponderantly public international law dispute settlement mechanism does, indeed, result from the procedural aspects of the dispute settlement method, and at the same time has influenced the procedure.[159]

---

[156] De Brabandere (2014), pp. 122–147.

[157] De Branbandere admits that "Jurisdiction, however, is very different from the question of the law applicable to the dispute, but they are without doubt connected, especially when one deals with specific regime-related dispute settlement mechanisms." Ibid. p. 134.

[158] Ibid. pp. 129–147.

[159] Ibid. p. 11.

De Brabandere's posture parallels the *multi(syncretic)lateralization* approach in terms of method. The starting point of both theses is the real functioning of the system as deployed by tribunals. However, while I avoid encapsulating the system's operation into a single topic, the author epitomizes the system as an "integral part of public international law"[160] or as "essentially public international law."[161]

The drawback of epitomizing the system as international public law is that by focusing the interpretation of the system on the topic, the interpreter might move away from formal law as defined by investment treaty arbitration tribunals. If we take De Brabandere's functional definition to an extreme, it restricts the discussion on the law and content interpretation to the premises of international public law, disregarding that the system's axis of validity consists of the BIT, and that the tribunals' practice is what defines the limits for framework setting, law ascertainment, and content interpretation. For example, *rhetorically*, one could argue that Article 38(1) ICJ, on the sources of international law, is an international public law instrument applicable to investment treaty arbitration, despite the fact that the practice of the tribunals contests its binding character.[162] This topic-based approach involves the risk of positioning the role of rhetoric in the front seat of the investment treaty law and tends to erode the identity of the investment treaty law, as do substantive validity theories.[163]

The erosion of the system's identity as a result of taking De Brabandere's topical approach to the extreme, however, is clearly not one of the objectives of his book. He makes sure to limit the effects of using international public law as a topic, firstly, by consciously excluding any substantial validity approach from the scope of his study;[164] secondly, by not pretending to defend a general theory of public interests in investment treaty law;[165] and, thirdly, by not denying that international investment arbitration has a "private law dimension."[166]

Moreover, when dealing with the application of "non-investment norms in investment treaty arbitration"[167]—where his posture necessarily has consequences beyond procedural aspects—De Brabandere remains cautious in the extension of his theory by adopting a positivist approach. The author grounds his conclusion that "there is no reason for the tribunal to exclude *ipso facto* human rights considerations as a *matter of applicable law*"[168] on the fact that treaties, in general, contain

---

[160] Ibid. p. 4.

[161] Ibid. p. 12.

[162] For example, Liman Caspian Oil BV and NCL Dutch Investment BV v. Republic of Kazakhstan (2010), para. 172.

[163] De Brabandere (2014), p. 13.

[164] Ibid.

[165] Ibid.

[166] Ibid. p. 9.

[167] Ibid. p. 129.

[168] Ibid. p. 134.

provisions that "open the door"[169] to their applicability; furthermore, the author supports this statement with a backdrop of treaties (ICSID, Energy Charter Treaty (ECT), North American Free Trade Agreement (NAFTA), USA's Model BIT).[170]

Naturally, *topic* is useful for describing the *law*, and its value within the dimension of law argumentation is enormous. However, *topic* exists in a different dimension than positive *law* and should not be confused with it. One must be careful in differentiating when one uses international public law as a *topic* from when one refers to international public law as positive *law*.

Without ignoring that De Brabandere expressly manifests that his work is not intended to develop a general theory of investment treaty law, the functional analysis of international investment arbitration that he embarks on corroborates my vision that investment treaty law is characterized in practice by elements of international public law. No less evident is the fact that De Brabandere and Kulick—despite approaching international investment law from different angles—coincide in finding global public interests in the practice of investment treaty tribunals. This common observation, which allows us to affirm that the international climate change legal framework is formally relevant to investment arbitration beyond topical rhetoric, is shared by other authors who explore the influence of international environmental law and/or international climate change law, as shown in the concluding remarks.

3.3 Concluding Remarks on *Multi(syncretic)lateralization* and the Formal Relevance of the Climate Change International Legal Framework in the Investment Treaty Law

The operational characteristics of the investment treaty law show that the climate change legal framework is a constitutive part of the system; thus, it is formally relevant (1). This conclusion stems from the conceptualization of the system's operation as a process of *multi(syncretic)lateralization.*

It is important to stress that *multi(syncretic)lateralization* is not a proposal as to how investment treaty law *should* operate. On the contrary, it describes investment treaty law as it *is*—an autopoietic system for the functioning of global markets permeated by global public interests, including climate interest (2). Envisaging investment treaty law as a process of *multi(syncretic)lateralization* captures two essential features of its functioning: *multilateralization* and *syncretism.*

Schill's finding that investment treaty law is a multilateralized system that comprises international community interests is crucial for claiming that the climate change legal framework is formally relevant. First, unearthing the self-referentiality

[169] Ibid. p. 135.
[170] Ibid. pp. 135–136.

and closedness of the investment treaty law makes it possible to encapsulate its operation in an autonomous autopoietic system that not only distinguishes itself by reference to the climate change international legal framework (3) but, even more importantly, has the potentiality to establish communication with the climate change international legal framework (4).

Second, the international community interest pursued by *multilateralization* cannot be restricted to the enhancement of the flow of foreign direct investment, but rather, it should be extended to the promotion of other global public interests, such as climate (5). Otherwise, this restriction would hinder the real operation of the investment treaty law system (6)—which demonstrates the relevance of global public interests through the application of principles, customs, and provisions of international law that embody global public interests—and would vacate the content of the multilateralized system (7) since enhancing the flow of foreign direct investment would be better achieved by allowing market forces to reach the most efficient bargain by themselves, instead of relying on a system. For these reasons, I use Schill's description of the investment treaty law as a "legal framework for the functioning of the global economic system"[171] in my work to reflect the *gravitational power* and *osmosis* of the global public interests into the legal framework (8). This permeation of global public interests is made explicit through the concept of *syncretism*.

Syncretism—as an essential feature of the investment treaty law—evokes the idea of a system containing values that are apparently irreconcilable. This concept emphasizes the paradox that the global public interests embedded in international law—such as climate—characterize investment treaty law, while the system preserves its autonomy (9).

It is enough to scan the social practice of investment treaty arbitration tribunals and BIT drafting techniques to verify that the balancing of global public interests is a foundational feature of the system (10). Global public interests—such as the flow of transnational capital, the *pacta sunt servanda* principle, the protection of legitimate expectations, good faith, certainty, predictability, nondiscrimination, equality, equity, sustainable development, the protection of health, the protection of the environment, the protection of human rights, the protection of world heritage, and the preservation of the international public order, among others—conflict and align themselves within the same investment treaty law system. In this context, the formal relevance of the climate change international legal framework emerges as a logical consequence of the interplay between global public interests that characterize the operation of the investment treaty law (11).

Uncovering the formal relevance of global public interests and the climate change legal framework in investment treaty law is to a certain extent, then, a reinstatement of the obvious (12). The observations of De Brabandere and Kulick point in the same direction; even if the theory of the latter is impregnated with some public law substantive validity considerations and the former relies on the concept of

[171] Schill (2009), p. 281.

international public law as a *topic* that characterizes the functioning of the system, both acknowledge that the operation of the investment treaty law promotes global public interests.

Other authors who have specifically embarked on the study of the relationship between climate change law and investment treaty law, without the intention of providing a systemic theory that explains the relevance of global public interests in investment treaty law, take a positivist approach by accommodating the climate change international framework to the system's operation through formal benchmarks. That is the case of Baetens and Vadi, who defend the international climate change law's application to the system based on Article 31 VCLT[172]—which is a provision that is almost always invoked in the practice of investment treaty arbitration tribunals.[173] Other tactics—inspired by the doctrines or theories tribunals have (not very consistently) adopted—also figure among the tools used by academics to envisage the application of the climate change international framework, e.g., the contextual interpretation of the investor's legitimate expectations, the defense of necessity, and the doctrine of police power.

A deeper study of these specific attempts to achieve communication between the climate change legal framework and investment treaty law is developed in Parts II and III. The message I would like to deliver in these concluding remarks by highlighting other authors' approaches concerning the application and the interpretation of the climate change international legal framework in investment treaty law is a double message. On the one hand, I intend to express that there is significant interest in finding an answer to the global public interest challenge—in general—and the application of the climate change legal framework—in particular—through a formal approach (14). On the other hand, I attempt to reaffirm that the observation that global public interests and the climate change legal framework are formally *relevant* to the system is not the result of wishful thinking but conforms to the system's operation (15).

Among the attempts to capture the relevance of global public interests—including the climate—the conceptualization of *multi(syncretic)lateralization* is innovative for several concatenated reasons.

Firstly, by adopting a formal-positivist approach, the operation of global public interests in investment treaty law does not need to be explained through substantive validity law theories positing that the system is public law or public international law or by analogy to one or the other, neither through topical rhetoric (16). In this vein—secondly—even if *syncretism* presupposes that the balancing of global public interests characterizes the operation of the investment treaty law, the system preserves its identity (17). This is because *multi(syncretic)lateralization* acknowledges that the center of validity remains in the system, which has its own axiomatic principles as well as its own means of reproducing and renewing itself.

---

[172] Infra. Chap. 6 text to notes 22; Chap. 7, text to notes 14, 194, 200; Chap. 9 text to notes 246–247.

[173] Infra. Chap. 6 text to notes 16–18, Chap. 7 text to notes 172, 198.

Thirdly, *multi(syncretic)lateralization*, being a general theory that unearths the formal relevance of global public interests in investment treaty law, renders conceivable the direct application of international law principles or provisions belonging to other (sub)systems; i.e., the relevance of global public interests influences not only the operation of *content interpretation* but also the operation of law ascertainment (18).

Fourthly, while the dilemma regarding (global) public interests in investment treaty law has been ordinarily framed as a two-cornered tension between the state's public interest and the investor's private interest, *syncretism* gives a more appropriate shape to the tension by finding that the global public interest is theoretically different from the investor's interests as well as to the state's interests. This supposes, then, a dynamic three-cornered tension that involves overlaps and collisions between the parties' interests and the global public interest (19). Accordingly—and fifthly—this new framing between global interests and the parties' interests makes it possible to reformulate Kulick's public interest challenge: the dilemma would no longer be about how the host states' public interest reaches the pedigree of global public interest in order to be relevant in investment treaty law but about how to align both the parties' interests with global public interests (20)—which are (*per se*) formally relevant to the system.

In brief, *multi(syncretic)lateralization* explains the formal relevance of global public interests—including climate—through a general theory on the functioning of the investment treaty law grounded on a positivist-formal approach that, while preserving the autonomy and identity of the system, acknowledges that global public interests characterize the system through the stages of law ascertainment and content interpretation. It also leads to the reformulation of the (global) public interest challenge stemming from a three-cornered tension between the investor's interest, the host state's interest, and the global public interest.

This conceptualization of the investment treaty law as a *multi(syncretic)* system reveals that the climate change international legal framework is formally relevant in investment treaty law (21). Since the interplay between global public interests is an essential feature of the system, the interaction between climate interests and other global public interests is feasible in investment treaty law. Moreover, the relevance of the international climate change law is not restricted to the operation of content interpretation but rather extends to the operation of law ascertainment. Therefore, if material to the dispute, the direct application of provisions belonging to the climate change legal framework is perfectly valid in investment treaty law (22). However, as the center of validity is kept in the investment treaty law system, the applicability of climate change provisions is limited by the operative principles of the latter (23).

The reshaping of the (global) public interest challenge as a three-cornered tension suggests that the global climate interest could collide and/or overlap with either the investor's interest or the host state's interest (24). Even if the global climate interest often conflicts with the investor's interests, these interests may become aligned. Likewise, even if the global climate interest is often in line with the interest of the host state, these interests may come into conflict. The complex relations that derive from reshaping the tensions lead to the question on how to align the interests of each of the parties with the global climate interest in two hypotheses (25): (1) the global climate interest comes into tension with the global interest of transnational investment, when the host state takes measures that promote the objective of leveling global warming affecting the investor's interest; and, (2) the global climate interest and the global interest of transnational investments—in one corner—come into tension with other global interests—in the other corner—when the host state takes measures affecting investors in climate-friendly industries in order to promote other interests.

References

De Brabandere E (2014) Investment treaty arbitration as public international law. CUP
Di Benedetto S (2013) International investment law and the environment. Cheltenham
Kulick A (2012) Global public interest in international investment law. CUP
Luhmann N (2009) Law as a social system. Revista - Journal- Derecho & Poder Diritto & Potere, Right & Power, Recht und Macht, Direito & Poder, Droit & Pouvoir
Schill S (2009) The multilateralization of international investment law. OUP
Schill S (ed) (2010a) International investment law and comparative public law. OUP
Schill S (2010b) International investment law and comparative public law — an introduction. In: Schill S (ed) International investment law and comparative public law. OUP
Schill S (2015) Multilateralization: an ordering paradigm for international investment law. In: Bungenberg M et al (eds) International investment law: a handbook. Nomos/CH Beck/Hart, Baden-Baden/München/Oxford
Van Harten G (2007) Investment treaty arbitration and public law. OUP

Chapter 4
Material Relevance of the Climate Change International Legal Framework in Investment Treaty Law: Unveiling Climate Change Disputes in Investment Treaty Law

According to the practice of investment treaty tribunals, the formal relevance of the climate change legal framework needs to be accompanied by an examination of its material relevance in order to determine whether it is applicable to a specific dispute.

Certainly, some of the issues that international investment tribunals address implicate climate change vicissitudes. It is even possible to assert that investment arbitration tribunals take climate change decisions.[1] This assertion has to be proven from a perspective external to the system that, naturally, relies on interdisciplinary approaches to investment treaty law. As Oeter puts it, "[f]or applying the law, you must understand the political, social and economic context in which the law applies—and understanding such context is not a task which might be mastered purely with legal instruments."[2] Therefore, interdisciplinary studies are not only a tool from which the scholarship and law adjudicators "could profit"[3] but also a necessary tool for assessing the materiality of climate change into investment treaty law.

The purpose of this chapter is, then, to demonstrate, from an external perspective, how the functioning of the investment treaty law system affects the global climate interest. Admittedly—in practice—only the facts, allegations, and context of each

[1] Viñuales (2015), p. 23.

[2] Oeter (2008), p. 2. A similar approach but concerning the study of political economy is undertaken by Manger: "At the center of this interdisciplinary effort is the argument that as political economists move into the analysis of the *content* of international investment agreements, rather than merely explaining their existence, they must take international law more seriously than they have done so far. On the other hand, for the legal mind, much can be learned from applying the empirical and theoretical apparatus of social science to the study of public and customary international law. However, this requires compromising a positivistic view of legal scholarship and at least partially accept what economists (unfortunately) call 'positive political economy'—*i.e.* to explain the development of law as the outcome of political struggles, in which actors strategically create and dismantle, bolster and strengthen international law as they see it in their interests": Manger (2010), pp. 41–42.

[3] Oeter (2008), p. 2.

T. Restrepo Rodríguez, *Investment Treaty Law and Climate Change*,
https://doi.org/10.1007/978-3-031-18655-4_4

concrete case may define whether it is proper to ascertain the climate change legal framework as applicable law or whether it is correct to interpret the bargain of the parties to the treaty within the context of global warming. However, the need for an assessment *in concreto* neither obstructs the plausibility of tracking general patterns nor invalidates empirical data that may indicate that climate change interests are material in investment arbitration.

Importantly, the content of this chapter is not intended to address each of the disciplines and models that may prove the materiality of climate change law to investment treaty law. The multiplicity of disciplines and methodologies for studying the relationship between climate change and foreign investment would require a titanic effort that falls outside of my capacity and ambitions. On the contrary, the evidence of materiality that follows is, in general terms, based on international relations, which—as a result of methodological and esthetical choices—are further dissected into economic (Sect. 4.1) and sociopolitical (Sect. 4.2) analyses. The study undertaken in these two subsections affirms, by demonstrating material relevance, the necessity and importance of the research question: how to align the climate change international legal framework with(in) investment treaty law.

The following snapshot of international relations concerning the relevance of climate change in investment treaty law embraces both rational and constructivist models. Following again Oeter's thoughts on the role of international relations in international law, these models should be perceived as complementary rather than opposing each other.[4] Indeed, if on the one hand the "rationality of real life actors is to a large degree dominated by context specific 'social constructions of reality'"[5] and "cannot be analyzed without paying attention to these specific factors,"[6] on the other hand, these specific "factors (...) might be phrased also as expressions of 'bounded rationality.'"[7] Therefore, if, on the one hand, the economic analysis on the collisions and synergies between the climate change international legal framework and investment treaty law (Sect. 4.1) profits more from rational models than constructivist models and if, on other hand, the sociopolitical analysis on the relationship between the two regimes (Sect. 4.2) profits more from constructivist models than from rational models, both subsections reflect the "quite productive division of labour"[8] between these international relations' models to the point of, at times, blurring the distinction as to whether the assumptions on which I rely belong to one model or the other.

[4]Ibid. pp. 6–7.

[5]Ibid. p. 6.

[6]Ibid. p. 8.

[7]Ibid.

[8]Ibid. p. 7.

4.1 Economic Approach to the Relationship Between the Investment Treaty Law and the Climate Change Law: Framing the Conflicts and Synergies

It is common in the (law and) economic literature to come across the statement that the impact of the investment treaty law system on the global climate interest is "double-edged"[9] or "two-sided."[10] While the investment treaty system has positive effects on leveling global warming because it fosters foreign direct investment in the "emerging green economy,"[11] it has negative effects by "chilling" host governments' potential to promote regulation addressing global warming. In this regard, one can risk asserting that there is a great consensus—if not an absolute consensus—on the "double-edged"[12] effect.

Conversely, there seems to be dissensus in the scholarship on the outcome of weighing the positive effects and negative effects of adopting investment treaties.[13] Although this disagreement may cloud the possibility of taking sides in favor of or against the convenience of the general investment treaty law system, rational behavioral models and empirical research suggest that the arbitrator's choices in investment treaty arbitration cases may encourage or discourage investors and/or host states to promote the goals of the climate change legal framework. In other words, the lack of clarity on the general global effects of the system on climate change does not contradict the fact that the rationale of the investment treaty law (Sect. 4.1.1) and the *regulatory chill* (Sect. 4.1.2) suggest that international investment treaty law disputes could take the form of climate change disputes.

4.1.1 The Rationale of the Investment Treaty Law and Climate Change: Synergetic Relationship

Why do countries adopt investment treaties? A simple and direct answer to this question would be that investment treaties promote foreign investments. However, this answer falls short in terms of explaining the popularity of bilateral investment treaties (BITs) if it is not supported by an explanation as to why host states are willing to constrain themselves in exchange for investments and why the investment treaty law system is a tool that can promote investments.

[9] Alschner and Tuerk (2013), p. 217.

[10] Gehring, Kent (2013), p. 189.

[11] Ibid. pp. 187–190.

[12] Alschner and Tuerk (2013), p. 217.

[13] Johnson et al. (2018a).

Economists, in general, agree that foreign investment is *per se* positive for the (global and local) economy.[14] In an era of globalization, it is almost perceived as a dogma that the channeling of capital from capital-exporting countries to countries where investments are needed increases foreign investors' prospects and boosts local economies—a win-win situation in which foreign investments are said to, simultaneously, promote employment, technology, and economic growth in the capital-importing countries and to benefit capital-exporting countries with the returns (which are meant to be repatriated).[15]

Turning from a fossil-based economy to a green-based economy requires large investments and technologies that developing countries (and even some developed countries) are unable (or at odds) to provide without the support of foreign capital and foreign experts.[16] However, foreign investors are not willing to sign blank checks for this purpose without protecting themselves against the risk of host states negatively affecting the profitability of the investment through expropriations, unequal or unfair mistreatments, or disruptions on the investor's capabilities to freely transfer funds to other countries, among other possibilities.[17] The depolitization of investor-state relationships seems to be an essential feature for the efficient flow of capitals into the green economy.

Investment treaty law is an instrument with which to mitigate political and regulatory risks interfering with the investor's plans,[18] gravitating naturally toward making a profit.[19] The burden on host states to honor the substantive protections established in BITs, in conjunction with the investor's faculty to bring their claims to (theoretically) politically neutral international tribunals, incentivizes foreign investment.[20] Consequently, the state's choice to limit its own autonomy by bounding itself to protect foreign investors' interests above the regular domestic legal framework is the result of considering that the benefits from attracting foreign investment are higher than the risk of facing international investment disputes. In the end, it seems that, in general, states consider it to be a good (or necessary) deal to adopt investment treaties.

---

[14] Ibid.

[15] Price (2017), pp. 126–127.

[16] "[T]he gradual transition to a low-carbon economy requires, first and foremost, technology and financing": Alschner and Tuerk (2013), p. 221.

[17] For more examples, see Mohajeri (2013), p. 4; Erkan (2011), p. 24.

[18] Thurnell (1977), p. 4; Lax (1983), p. 9.

[19] "What is the objective of making an investment. That is the key issue and is simple. From the point of view of international investors, they invest capital to make a profit. It is obviously not a good thing if, while you expect to make a profit, you lose your own capital": Erkan (2011), p. 24.

[20] "Why do countries sign BITs at all? One often unspoken assumption is that since IIAs provide legal protection for investors beyond that extended by local courts, investments are made that would otherwise be "too risky". This implies that BITs should have a measurable effect: *Ceteris paribus,* two countries that have a BIT in force between them should have more bilateral FDI flows (or rather more flows from the predominantly capital-exporting to the capital-importing partner) than the same country pair without such a treaty". Manger (2010), p. 52.

Projects designed to lower greenhouse-gas emissions are particularly prone to being disrupted by governmental decisions. In consideration of this political risk, the relationship between the investment treaty system and the climate change international legal regime seems synergetic[21]—the protection of green investors and climate change objectives flow in the same direction. Furthermore, viewed from this angle, tribunals' decisions concerning climate change aspects are not only relevant for shaping the content of the investment treaty law, but they also appear as necessary to encourage green-industry investors and regulators to contribute to leveling global warming.

The growth of the green economy anticipates conflicts between host states and investors. Indeed, the race for investments in renewable energies between governments has already generated a considerable number of disputes regarding the modification of renewable energy support schemes. Before examining the vulnerability of these schemes to political risks and presenting some examples in this respect (Sect. 4.1.1.2), it is worth briefly depicting why it is economically rational for host states and investors to develop the objectives of the climate change legal framework (Sect. 4.1.1.1).

4.1.1.1 The Momentum of the Climate Change Economy

Combatting climate change is attractive to investors. Naturally, this appeal resides not in philanthropic feelings but in the prospects of profiting from emerging climate change markets.[22] However, the appeal for host states might not be so different. Even if one supposes that climate change is a public interest that deserves priority treatment within the national policy for the very basic reason of safeguarding the future of humanity and preserving ecosystems, governments might (need to) think about the climate change challenge in terms of socioeconomic development and consider other plethora of public needs.[23] The assessment of macroeconomics does play an important role in the host state's decision to promote climate change industries, as well as in modeling the intensity and specificity in which national development plans on climate change are deployed.

The 2018 Global Climate Action Summit press release—based on *The 2018 Report of the Global Commission on the Economy and Climate*—[24] conveys fairly good prospects for the climate change economy.

---

[21] Viñuales (2015), pp. 27–28.

[22] Gehring and Kent (2013), p. 189.

[23] For example, in the USA, any change in the energy mix toward renewable energies should weigh the fact that "151,000 people are employed in fossil fuel power generation, with an additional 887,000 people in extraction (74,000 in coal, 310,000 in gas, and 503,000 in oil)" or, for instance, in India, the climate change policy must take into account the fact that "Coal India, a state-owned enterprise (SOE) that produces 80% of Indian coal, employs more than 300,000 people": The new climate economy (2018), p. 39.

[24] Ibid.

It finds that a stepped-up transition to a low-carbon economy can:

- Result in $26 trillion in economic benefits worldwide through 2030
- Generate over 65 million new low-carbon jobs in 2030, equivalent to today's entire workforces of the UK and Egypt combined (...)
- Generate, through just subsidy reform and carbon pricing, an estimated US $2.8 trillion in government revenues per year in 2030—equivalent to the total GDP of India today (...)
- By a shift to more sustainable forms of agriculture combined with strong forest protection, deliver potentially more than US $2 trillion per year of economic benefits, generating millions of jobs, improving food security—including by reducing food loss and waste—and delivering over a third of the climate change solution (...)[25]

These promising figures, in principle, should incentivize investors and states to develop climate change industries.[26] It has been claimed that the green economy has gained momentum.[27] US$ 90 trillion is expected to be invested for the next 10–15 years in infrastructure to make cities more efficient, implementing renewable energies, developing sustainable food production and consumption, promoting sustainable freshwater use and protecting freshwater ecosystems, avoiding deforestation and land grabbing, as well as decarbonizing transportation and industrial production.[28] No less significant is the fact that 366 companies listed in "The Carbon Clean 200" that are valued over US$1 billion reported returns of more than 10% from clean energies[29] or that, in 2017, 10.3 million people globally were employed in the renewable energy industry.[30]

Yet today's market incentives for investors and states to engage in the project of leveling global warming are not devoid of changes tomorrow. Different political, economic, social, and even natural conditions may lead to the host government's decision to subvert positive climate policies. The recent explosion of claims concerning the modification of renewable energy support schemes (RESSs) illustrates the green economy's vulnerability before changes in host states' regulation, which is not a mere intuition but a real threat.[31]

4.1.1.2 The Particular Exposure of Renewable Energies Against Political and Regulatory Risks

Given the long-term nature of energy projects in general, renewable energy investments face risks that are also common to investments in conventional energies. However, the uncertainty in terms of the outcome of the adoption of new

[25] Globate Climate Action Summit (2020).

[26] The new climate economy (2018), p. 8.

[27] Ibid.

[28] Ibid. p. 8.

[29] Ibid. p. 55.

[30] Ibid. p. 56.

[31] See. Infra. 4.1.1.2.

technologies and the higher costs that renewable energies may suppose in comparison to conventional energy increases the political and regulatory risks for them.

Regarding energy projects, Cameron identifies two cycles—which also apply to renewable energies—that fairly define the issue of political and regulatory risks: the *obsolescing bargain* cycle and the *price cycle*.[32]

According to the *obsolescing bargain* cycle, there is a point during the relatively long life of a contract or license where the initial (attractive) conditions offered by the host state that incentivized the foreign investor to develop an energy project may no longer be perceived as a good deal by the government as the initial conditions may later appear obsolete.[33] There are several reasons that might lead the host government to reach this conclusion and encourage it to modify the conditions. For instance, there could be pressure from public opinion and the communities to get a better bargain, the development of more profitable and efficient forms of energy, etc. In fact, once the large amount of capital required for the development of the energy project has been channeled, the government may find no reason not to attempt to reach (impose) a new agreement with better conditions for it.[34]

Concerning the price cycle, a rise in commodity or energy service prices during the life of a contract or license may result in extraordinary returns for the foreign investor. As a consequence, the host state might attempt to profit from these unexpected returns, for example, by taxing or cutting subsidies.[35]

Renewable energy is more expensive than conventional sources of energy. This fact increases the risk of the former falling into the *obsolescing bargain* cycle. It could be that once the capital needed for the development of a project is invested, a country's changing economic situation alone or in conjunction with other factors—such as pressure from communities and industries (voters) to lower the prices,[36] the need to be competitive in the global market, the need for balancing the energy mix, etc.[37]—could stimulate the government to regulate or amend the agreement or license, resulting in damaging the prospects of renewable energy investors. As the host state (almost) always has the option of turning to cheaper sources of energy, there is always a risk of the initial deal appearing obsolete.

Additionally, renewable energy is more dependent on new technologies and innovation than conventional energy. This results in a riskier business, whereby

[32] Cameron (2010), pp. 4–7.

[33] Ibid. pp. 4–5.

[34] Ibid.

[35] Ibid. p. 5.

[36] Ibid. pp. 27–28.

[37] "(...) government may face challenging economic times and simply conclude that they cannot afford to support green initiatives. There may be more partisan reasons for changes in the level of commitments. For example, a new administration may be less committed to technological solutions to greenhouse gas mitigation and thus seek to withdraw a subsidy extended to CCS projects": Bankes (2012), p. 498.

the efficiency and payout of the operation are uncertain.[38] It is plausible for foreign investors to fail to obtain the expected return, but the opposite may also occur in the sense that the foreign investor may obtain extraordinary revenues.[39] The joy of hitting the jackpot in the latter case may be obscured by the government's will to capture a portion of the investor's earnings. In brief, the uncertainty in the outcome of the operation of renewable energy might end up falling into the *price cycle.*

Investment treaty disputes on the modification of solar energy schemes in Europe could be described as the consequence of the *obsolescing bargain* cycle and *the price cycle*. Following Directives 2001/77/EC and 2009/28/EC, the Member States adopted plans toward incentivizing the production of renewable energies, which included regulated tariffs, tax exemptions, priority access, and green certificates, among others.[40] However, the early achievement of targets,[41] in addition to the 2008 credit crunch[42] and the plummeting of the costs of solar energy production, induced governments to remodel and cut some of the initial (attractive) conditions for investment. The measures taken in this regard in Spain, the Czech Republic, Italy, and Bulgaria are clear examples of how political and regulatory risks can affect renewable energy projects.

Spain[43]

Real Decreto 661/2007 established a renewable energy support scheme (RESS) based on an appealing regulated tariff or a regulated premium offered to investors for a 25-year period (Articles 35–37 and 44, Annexes 7–8 and 12), which could be

[38] "While there are success stories for renewable energy technology, the timing of improvements needed for renewables to displace fossil fuels is uncertain. For example, better ways to produce solar photovoltaic panels are lowering panel production cost, but we still have costs of daily and seasonal intermittency. Cellulosic ethanol technology allows ethanol production from switchgrass rather than corn, but land availability still places constraints on switchgrass production. Better technology can lower renewable energy costs to some extent, but cannot change fundamental characteristics of energy sources": Timmons et al. (2014), p. 32.

[39] "(...) increased demand in a sector may stimulate technological innovations and reduce the unit costs of production making any subsidy, whether framed as a FIT or in some other manner, unnecessarily generous". Bankes (2012), p. 498.

[40] See. Restrepo (2017), pp. 101–137.

[41] While Spain's target for 2010's renewable electricity consumption was 12%, in that year the consumption was 13.8%: Behn and Fauchald (2015), p. 120. The Czech Republic's objective of producing 2.167 GWh of renewable energy was reached in 2011; 2 years early than expected: Tirado (2015), *p.* 8. Italy achieved 17% consumption of renewable energy in 2014, which was 2020's target: Bellantuono (2016), p. 4. In the case of Bulgaria, while the target of installed capacity was set at 300 MWp PV for 2020, the country's capacity rose to 1000 MWp PV in 2012: Vasani et al. (2015).

[42] Geels (2013).

[43] On the Spanish case: Tirado (2015), pp. 5–7; Behn (2015), pp. 120–123; Vasani (2015); Restrepo (2017).

extended to a second period (Article 36). However, a first set of modifications initiated in 2010 commenced to erode the RESS. *Real Decreto 1565/2010* (Article 29.2) canceled the application of the regulated tariff for the second period and compelled investors to adopt some technical requirements to control tension gaps that raised the costs, *Ley 14 de 2010* levied a 0.5 euro per megawatt access fee to the grid, and *Real Decreto 1614/2010* restricted the number of hours that would be subject to the regulated tariff (Article 2).

Two years later, the disarticulation of the RESS went further by taxing the access to the grid at 7% (*Ley 15/2012*, Article 8). In 2013, the possibility of adopting a regulated prime was abolished, and the factor for adjusting the tariff was lowered below the inflation index (Real Decreto 2/2013). However, the most damaging regulation for the renewable energy industry was *Real Decreto Ley 9/2013*—complemented by *Ley 24/2013* (Act 24/2013), Real Decreto 413/2014, and Orden Ministerial IET/1045/2014—which established a system for assessing the tariff based on model costs rather than real costs. Detaching the tariff calculation from real costs affected the investors' certainty on the affordability of the fixed costs. Consequently, investors in the RESS filed 33 claims against Spain. [44]

Czech Republic[45]

The Czech Republic's RESS was adopted through Act No 180/2005. On the one hand, this Act compelled distribution system operators to purchase the entire production of renewable energy producers for a period of 15 years (Articles 5 and 6.2). The Act also allowed producers to opt for feed-in tariffs and feed-in premiums (Article 6), which the regulator was entitled to adjust on a yearly basis but by no

[44] *Sevilla Beheer B.V.*; *Infracapital F1* S.À R.L. and *Infracapital Solar B.V.*; *Sun-Flower Olmeda Gmbh & Co Kg and Others*; *Alten Renewable Energy Developments Bv,*; *Eurus Energy Holdings Corporation And Eurus Energy Europe B.V.*; *Landesbank Baden-Württemberg and Others*; *Watkins Holdings S.À R.L. and others*; *Hydro Energy 1 S.À R.L. and Hydroxana Sweden Ab*; *Soles Badajoz Gmbh*; *Operafund Eco-Invest Sicav Plc And Schwab Holding Ag.*; *E.On Se, E.On Finanzanlagen Gmbh And E.On Iberia Holding Gmbh*; *Cavalum Sgps, S.A.*; *Jgc Corporation*; *Ks Invest Gmbh and Tls Invest Gmbh*; *Matthias Kruck*; *Cube Infrastructure Fund Sicav And Others*; *Baywa R.E. Renewable Energy Gmbh and Baywa R.E. Asset Holding Gmbh,*; *9ren Holding S.A.R. L*; *Steag Gmbh*; *Stadtwerke München Gmbh, Rwe Innogy Gmbh et Al.*; *Rwe Innogy Gmbh And Rwe Innogy Aersa S.A.U.*; *Masdar Solar & Wind Cooperatief Ua*; *Renergy S.À.R.L. v. Spain*; *Infrared Environmental Infrastructure Gp Ltd. et al*; *Nextera Energy Global Holdings B.V. and Nextera Energy Spain Holdings B.V.*; *Eiser Infrastructure Limited And Energia Solar Luxembourg S.À.R.L.*; *Antin Infrastructure Services Luxembourg S.À.R.L. and Antin Energia Termosolar B.V.*; *Rreef Infrastructure (G.P.) Limited and Rreef Pan-European Infrastructure Two Lux S.À.R.L.*; *Csp Equity Investment S.À.R.L.*; *Isolux Infrastructure Netherlands B.V.*; *Charanne (The Netherlands) And Construction Investments (Luxembourg)*; *The Pv Investors v. Spain*. All '*v Republic of Spain*' cases, source: International Energy Charter.

[45] On the Czech case: Tirado (2015), pp. 5–7; Behn, Fauchald (2015), pp. 123–124; Vasani (2015); Reuter (2015), p. 13; Perry and Karadelis (2014).

more than 5% (Article 6.4.). Besides this, a system of "green bonuses" was structured through the Act (Articles 1(d), 4.3, 4.7, 4.15, and 6).

Yet in 2010, some constraints were placed on the RESS. The government imposed taxes of 26% and 28% for green bonuses and feed-in tariffs, respectively. Furthermore, the annual threshold of 5%, which limited the government's power to alter the feed-in tariff or the feed-in prime, was derogated (Act 137/2010; Act 165/2012, Articles 11 (6)). Lastly, in 2014, the RESS was completely disarticulated (Act 165/2012, Act 310/2013).[46] These actions implemented by the host state motivated seven claims against the Czech Republic under the Energy Charter Treaty (ECT). [47]

Italy[48]

The key elements of Italy's RESS were the emission of green certificates (*Legge* 296/2006, Article 1) and the setting of a feed-in tariff (*Decreto legislative* 387/2003; *Decreto 19 Febbraio* 2007, Articles 3–6). Energy importers and producers were required to comply with a specific percentage of renewable energies within their portfolio. If it was not possible to reach the percentage, importers and producers were able to obtain green certificates in the market in order to fulfill the percentage (*Decreto 19 Febbraio* 2007).

A set of measures between 2010 and 2014 gradually disarticulated the RESS. First, *Decreto Ministeriale 6 Agosto 2010, n. 197*, slashed the price of the green certificates by 16% (Article 9); subsequently, *Decreto 5 Maggio 2011, n. 109*, canceled the green certificates scheme and reduced the benefits of the feed-in tariff and the feed-in prime, which were even more reduced later in 2014 by the *Decreto taglia-bollette* **(Decreto Legge 24 Giugno 2014, n. 91, Article 26**) and *Legge 11 Agosto 2014, n. 116*. These modifications generated six claims against Italy under the ECT.[49]

[46] Ibid.

[47] *Solar and Göde* (2013); *Natland Investment Group NV, Natland Group Limited, G.I.H.G. Limited, and Radiance Energy Holding S.A.R.L.* (2013); *Voltaic Network GmbH* (2013); *ICW Europe Investments Limited* (2013); *Photovoltaik Knopf Betriebs-GmbH v. Czech Republic* (2013). All "v Czech Republic" cases, source: International Energy Charter.

[48] Massari (2015), p. 13.; De Luca (2015); Bellantuono (2016); Tirado (2015), pp. 5–7; Behn and Fauchald (2015), pp. 123–124; Vasani (2015); Reuter (2015).

[49] Espf Beteiligungs Gmbh, Espf Nr. 2 Austria Beteiligungs Gmbh, And Infraclass Energie 5 Gmbh & Co. Kg (2016); Eskosol S.P.A. In Liquidazione (2015); Belenergia S.A. (2015); Silver Ridge Power Bv (2015); Greentech Energy Systems And Novenergia (2015); Blusun Sa, Jean-Pierre Lecorcier And Michael Stein (2014). All cases 'v Italy', Source: International Energy Charter.

Bulgaria[50]

The 2017 *Law for Renewable and Alternative Energy Sources and Biofuels* encouraged investors in renewable energies by providing a feed-in tariff scheme, long-term purchase contracts for renewable energy production, as well as by guaranteeing priority access.[51] The regulator was entitled to adjust the feed-in tariff within a range of 5%.[52]

In 2011, the *Law on the Energy for Renewable Resources* modified the RESS by limiting the term of the feed-in tariff's operation and reducing the term of the power purchase contracts. This law also abolished any threshold for the feed-in tariff's adjustment (Article 31). Later, in 2013, the government adopted the *Electricity Trading Rules*, which compelled renewable energy producers to sell energy in the electricity balancing market (Article 56.2). Lastly, in 2014 and 2015, the RESS was amended to restrict the preferential purchase price to a portion of the renewable energy production[53] and taxed wind and solar power production at 5% (Articles 35a and 35c). This set of measures motivated two claims against Bulgaria under the ECT.[54]

4.1.1.3 Incentivizing Climate Investments Through Investment Treaty Arbitration Decisions: Law and Economic Reasoning

The examples presented above on the modification of RESSs demonstrate that investment treaty law cannot avoid dealing with climate change issues. Moreover, tribunals' decisions in this respect play an important role in the future development of these schemes. Awards rejecting compensation may prevent future foreign investors from risking their capital in the green market. In contrast, if these awards are used to protect the investors' interests, new financial and operational resources can be expected to be channeled into renewable energies. In other words, concerning investments in the green economy, investor protection and climate protection go hand in hand; decisions protecting investments against political and regulatory risks might be factors that incentivize investors to promote the switch from a fossil-based economy to a green-based economy.

Yet the choice of incentivizing the green economy through tribunals' decisions on renewable energy supports the need for schemes—or any other dispute related to climate change—to be subject to the limits of the investment treaty law system. Tribunals' decisions are, of course, subject to the rule of law. In this sense, decisions that are only inspired by economic considerations are not valid decisions.

[50] Jetchev (2013); Vasani (2015); Draguiev (2015).

[51] Republic of Bulgaria, Ministry of Economy (2011).

[52] Jetchev (2013).

[53] International Energy Agency.

[54] *EVN AG v. Bulgaria* (2013); *ENERGO-PRO v. Bulgaria* (2015).

Nevertheless, the inherent economic impact of tribunals' decisions on climate change cannot be ignored by these tribunals. Since the *syncretic* system of investment treaty law comprises global climate interests [55] within its scope, economic considerations that may affect the achievement of the objectives of the international climate change legal framework can enter the system through the mechanisms of law ascertainment or content interpretation.[56] The link between economics and the law is unavoidable in investment treaty-climate change disputes.

It should be clear at this point that we are not intending to claim that tribunals should always rule in favor of investors when the investment is in the development of climate-friendly projects. If some weight must be given to the economic impact of tribunals' decisions, the law system limits the scope of the economic reasoning. For example, if a tribunal concludes that it is clear that the investor had no legitimate expectations in terms of the stability of renewable energy schemes, then it might be wrong to conclude that the host state's conduct went against fair and equitable treatment.[57] In brief, I am not stating that economic considerations should outweigh legal reasoning; nevertheless, I acknowledge that economic considerations form part of this legal reasoning.

Besides recognizing that tribunals incentivize the leveling of global warming when they protect the interest of the investor in green industries, it must be highlighted that a decision adverse to the host state to take a measure to tackle climate change may induce it to avoid taking similar actions in the future. In such circumstances, the (preventive) role of the investment treaty law reveals itself as counterproductive to the achievement of the climate change international legal framework's purpose. From an economic perspective, the "chilling" of the host state's climate change regulation is another factor that demonstrates that investment treaty law disputes can be perceived as climate change disputes, as will be presented in the next subsection.

4.1.2 Regulatory Chill and Climate Change: Conflictual Relationship

Regulatory chill is not really a phenomenon outside the rationale of the investment treaty law. In addition to the compensatory role of civil (state) liability, the preventive role occupies an important position within this discipline. With investment treaty law being a civil (state) liability system designed to foster foreign investments, the freezing of the host state's regulation could be understood as the system's success in mitigating political risks. Besides, the investment treaty law system is not—from a strictly formalistic-legal point of view—a restriction to the host state's

[55] Supra. 3.3.

[56] Ibid.

[57] Infra. 7.2.

freedom to model the regulation against foreign investors' interests. In the end, the host state can always choose to substitute the noncompliance with a BIT's obligation with the payment of damages. Consequently, the "chilling" of the host state climate regulation from a classic-civil liability standpoint does not seem to be a legal problem.

However, a broader analysis discloses that although investment treaty law instruments may not be envisaged to directly affect climate change regulation, they might indirectly hinder climate change policies.[58] The factual difficulties (or impossibility) for host states to, simultaneously, enact a climate-friendly regulation and to honor BITs or to pay compensation in case they decide to take climate change measures that affect investors may lead to the conclusion that the protection of the investor under BITs is undesirable.[59] After all, it might be that the protection of the foreign investor against policies aimed at leveling global warming might be perceived as a negative spillover from the system that calls for the adoption of counterrationale mechanisms.

Authors convey that the investment treaty law system may degenerate into a regulatory chill under two circumstances. While the first refers to the existence of BIT obligations and the risk of facing international treaty arbitration in the event of noncompliance, the second relates to a particular situation where the foreign investor actually persuades the host state by expressing its intention to file or by filing an investment arbitration claim if a given government measure is adopted or if a government measure is not withdrawn. In addition to these two forms of regulatory chill, it appears that multinational corporations may take advantage of their position by deliberately presenting a considerable number of claims against different host states as part of a global strategy of the industry for disincentivizing public interest measures, such as those related to climate change. While the first two variants of regulatory chill have been labeled by Tienhaara as the "internalization chill"[60] (Sect. 4.1.2.1) and "threat chill"[61] (Sect. 4.1.2.2), respectively, she calls the last one the "transboundary chill"[62] (Sect. 4.1.2.3). Although these categories are interrelated and may represent different expressions of the same phenomenon, we have chosen to

---

[58] "There is a common misconception that governments can be forced by ISDS tribunals to roll back regulations that have been put in place; this is not the case. As such, ISDS has no direct impact on health and environmental regulation. However, there is significant potential for indirect impacts to occur as a result of the effect ISDS has 'on the way in which host states exercise their regulatory powers'". Tienhaara (2018), p. 233.

[59] "The challenging of harnessing foreign investment to sustainable development is compounded by the fact that investment protection obligations in IIAs may constrain host countries' policy space necessary to regulate foreign investment in the public interest. From a sustainable development angle, therefore, IIA raise complex issues in terms of their impact, touch upon sensitive public policy concerns and affect wide range of stakeholders": Alschner and Tuerk (2013), p. 218.

[60] Tienhaara (2018).

[61] Ibid.

[62] Ibid.

keep Tienhaara's language because of its descriptive virtues and completeness, instead of trying to innovate or adopt other typologies that are equally acceptable.

4.1.2.1 Internalization Chill[63] and Climate Change Measures

The possibility of being subject to international investment arbitration for dishonoring BIT obligations may prevent host states from adopting climate change mitigation measures. Changes in the use of land or regulations compelling certain industries to reduce carbon emissions as well as measures imposing carbon taxes or implementing RESSs that place investors in renewable energy in a better position than investors in conventional forms of energy, among others, could be perceived as expropriatory or constituting an unfair, unjust, or differential treatment, which entitles the foreign investor to claim compensation.[64] As a result, the host state would find itself between the devil and the deep blue sea, faced with the difficult challenge of deciding whether to adopt a measure to promote the public interest of leveling global warming with the negative consequence of exposing itself to the risk of investment treaty arbitration or to avoid this risk with the negative consequence of not complying with its obligation to fight climate change. If the costs of investment arbitration risk affecting the host state's economy and development, its decision would likely be to refrain from adopting climate change measures and, thereby, not complying with the objective of reducing global warming.

Appealing to some examples may provide a more tangible idea of how big the risk of arbitration is for host states when dealing with measures related to climate:

- On February 19, 2016, Cosigo Resources Ltd. and two other mining companies filed a "Notice of Demand for Arbitration and Statement of Claim" for USD 16.5 billion against the Republic of Colombia under the United Nations Commission on International Trade Law's (UNCITRAL's) rules in accordance with the US-Colombia Trade Promotion Agreement. [65] The country's government designated a portion of the Amazon Forest as a natural park and prohibited mining activities in this area, disregarding the existence of a valid mining license.[66]
- In April 2009, Pacific Rim Cayman LLC sued El Salvador at the International Centre for Settlement of Investment Disputes (ICSID) under the Central America Free Trade Agreement (CAFTA) for USD 304 million for the delays in providing

[63] "Internalization chill is a process whereby regulatory progress is dampened across all areas that affect foreign investors because government officials are concerned about the risk of an investor-state dispute arising": Ibid. p. 234.

[64] Schill (2007), pp. 469–470; Alschner and Tuerk (2013), p. 222; Bankes (2012), pp. 501–503.

[65] *Cosigo Resources Ltd. et al. v. Republic of Colombia* (2016).

[66] Ibid.

a mining extraction license. El Salvador contested this, arguing that the delay was necessary for evaluating whether the activity was polluting water.[67]

- In June 2012, Ecuador was found liable to pay USD 1.77 billion in compensation for the damages caused to Occidental Petroleum Corporation. [68] The Ecuadorian government unilaterally terminated an oil and gas contract whose contract area was located in the Amazon region.[69]
- On March 30, 2009, Vattenfall registered arbitration at ICSID against the Federal Republic of Germany for EUR 1.4 billion founded on the delay in the provision of environmental permits to operate a coal plant.[70]

The extremely high compensation claimed or obtained by foreign investors in the examples above represents a real threat to the defendants' economies: USD 16.5 billion is equivalent to 20% of the Colombian national budget for 2016,[71] USD 300 million is equivalent to 2% of El Salvador's gross domestic product (GDP),[72] USD 1.77 million is equivalent to 10% of Ecuadorian tax revenues in 2012,[73] and EUR 1.4 billion is almost 0.01% of the total revenue of the German state in 2009.

In conjunction with the persuasiveness that the potential costs of compensation entail, the host state may be significantly induced to avoid, freeze, or withdraw climate regulation also to avoid the risk of assuming the legal costs of arbitration. Even if the host state is excused for paying damages, the sole costs of arbitration might be overly expensive. In a recent policy paper published by the *Columbia Center on Sustainable Investment*, it is estimated that the average costs of litigation for host states in investment arbitration disputes are around US$ 5 million, but in some cases—such as *Yukos* [74] or *Libananco* [75]—these costs could reach approximately US $30 million.[76]

Regulatory chill is also rooted in the inconsistency of investment arbitration decisions. The open nature of BIT obligations and the important role that international law plays in their determination, the assessment of liability, and the lack of the

[67] In October 2016, the tribunal found in favor of El Salvador. *Pac Rim Cayman LLC v. Republic of El Salvador* (2016).

[68] *Occidental Petroleum Corporation and Occidental Exploration and Production Company v. The Republic of Ecuador* (2012). In 2015 an ICSID-appointed Committee lowered the amount to USD$1.1. Billion *Occidental Petroleum Corporation and Occidental Exploration and Production Company v. The Republic of Ecuador* (2015).

[69] Ibid.

[70] In March 2011, the parties reached an agreement to suspend proceedings. *Vattenfall AB et al. v. Federal Republic of Germany* (2011).

[71] Asprilla (2016).

[72] http://www.argentinaindependent.com/currentaffairs/analysis/el-salvador-vs-pacific-rim-the-price-of-saying-no-to-a-gold-mine/.

[73] OCDE (2019), p. 176.

[74] *Yukos Universal Limited (Isle of Man) v. The Russian Federation* (2014).

[75] *Libananco Holdings Co. Limited v. Republic of Turkey* (2011).

[76] Johnson et al. (2018b), p. 11.

doctrine of precedent leave considerable space for interpretation. This fact leads to the adoption of opposing theories—sometimes pro-investor, and on other occasions pro-state—that increase stakeholders' and host state's uncertainty.[77]

Moreover, the adoption of approaches that clearly favor the position of the investor limits the host state's will to take measures to promote global interest in leveling global warming. The sole-effects doctrine, as deployed in *Metalclad*, [78] for example, is a serious reason for host states to have doubts about adopting climate change measures. In this case, *Metalclad* purchased a company that developed hazardous waste landfill operations and obtained all the environmental permits that were needed from Mexico's federal government. However, after a couple of years, the local government prohibited *Metalclad* from operating the landfill because it was allegedly polluting a water well next to its premises and causing health problems for the population. The tribunal ruled that Mexico was liable to compensate *Metalclad* because the North American Free Trade Agreement (NAFTA) protects the investor from any "interference in the use of property [...] even if not necessarily to the obvious benefit of the Host State."[79] This ruling was followed by the tribunals in the *Azurix* [80] and *Santa Helena* [81] cases. In *Azurix*, the statement in favor of the "sole-effects ruling" was even sharper: "The issue is not so much whether the measure concerned is legitimate and serves to a public purpose, but whether it is a measure that, being legitimate and serving a public purpose, should give rise to a compensation claim." [82]

Tecmed [83] appears as another example of a decision that might discourage host states from promoting the leveling of global warming. The tribunal establishes a clearly pro-investor interpretation of the investor's legitimate expectations. In this case—also related to hazardous landfill operations—*Tecmed*'s license was rejected for political reasons, and the local government did not provide any other alternative or guideline for the investor to obtain the approval. In this regard, the tribunal held that the conditions at the time of investment should remain untouched, with any

[77] 'Governments need reasonable certainty as to the interpretation of the obligations they have taken on in order to feel reasonably confident that bona fide public policy measures will not be found to have violated a broadly read provision in a treaty. And the system needs certainty that legal correctness in an arbitration decision will be an important factor in its review. Today, the opposite is the case on both counts. There is no certainty for governments due to the competing streams of jurisprudence. The impact of the uncertainty adds to the risk of regulatory chill, as the uncertainty surrounding a large damages award is a state's primary concern if an investor brings an arbitration claim. Mann (2013), pp. 521–544, 529, 532.

[78] *Metalclad Corporation v. The United Mexican States* (2000).

[79] Ibid. Para. 103.

[80] *Azurix Corp. v. The Argentine Republic* (2006).

[81] *Compañia del Desarrollo de Santa Elena S.A. v. Republic of Costa Rica* (2000).

[82] *Azurix Corp.* (2006) para. 310.

[83] *Técnicas Medioambientales Tecmed, S.A. v. The United Mexican States* (2003).

change in the investment framework being contrary to legitimate expectations;[84] this approach has been followed in a few other decisions.[85]

Although the phenomenon of internalization chill seems obvious, empirical research supporting its plausibility is limited. Moreover, given the plethora of factors that may influence governments' behavior in this regard and the intricacies of weighing these factors—not to mention the problems for gathering data—it seems very difficult to prove that there is an internalization chill at play concerning climate change measures. However, since the only difference between "internalization chill" and "threat chill" is that in the latter the risk of liability is higher because the foreign investor has manifested its intention to claim compensation—a difference of intensity—the limitation for empirical research on regulatory chill in general is lessened by the evidence of cases concerning climate change, where host states have not taken or maintained public interest measures because of the fear of assuming the costs that investment arbitration involves. This is explained in the following subsection.

4.1.2.2 Threat Chill[86] and Climate Change

The risk of investment treaty arbitration becomes tangible when a formal notice of arbitration has been served or when the host state understands that the foreign investor is serious about presenting a claim by any other means. The commencement or imminence of investment treaty arbitration as a result of the enactment or imminence of climate change regulation might undermine the host state's initial intention to reduce greenhouse gas emissions. It may be that the only way to stop the investor from taking further action toward compensation is the nonadoption or derogation of climate change measures.

The host state knows that the investor does not have much to lose by requesting compensation. The costs of arbitration would not be a barrier for the foreign investor to commence disputes since contingency fee agreements and third-party funding may shield the investor from this burden.[87] Furthermore, "contingency fees can be

[84]Ibid. para. 154: '[...] this provision of the Agreement [the fair and equitable treatment provision], in light of the good faith principle established by international law, requires the Contracting Parties to provide to international investments treatment that does not affect the basic expectations that were taken into account by the foreign investor to make the investment. The foreign investor expects the Host State to act in a consistent manner, free from ambiguity and totally transparently in its relations with the foreign investor, so that it may know beforehand any and all rules and regulations that will govern its investments, as well as the goals of the relevant policies and administrative practices or directives, to be able to plan its investment and comply with such regulations'.

[85]*CMS Gas Transmission Company v. The Republic of Argentina* (2005); *Oxy v. Ecuador* (2012).

[86]'Threat chill is the most familiar form of regulatory chill. It concerns the chilling of specific regulatory measures that have been proposed by governments following an investor's threat to arbitrate': Tienhaara (2018), p. 40.

[87]Mann (2013), p. 534.

reimbursed—including the slice of award owed to lawyers."[88] Under this reduced risk, in conjunction with firms' interest to make a profit,[89] there is a high probability that the host state would end up being sued unless it withdraws the climate change regulation in question.

The "threat chill" is increased by the fact that investors tend to inflate claims in order to strengthen their position in the negotiations with host states. Since there are no penalties in investment treaty arbitration for presenting inflated claims, there is no reason for investors not to take advantage of this situation in order to persuade host states not to take or withdraw climate measures.[90] A paper published in 2014 confirms that while the average compensation claimed is around US $622.6 million, the average compensation awarded is approximately $16.6 million,[91] which is a 3750% average inflation.

The examples of *Transcanada* and *Vattenfall* concerning the delay or rejection of environmental permits in the context of the decarbonization of the energy industry provide a fair idea of how the threat chill works.

Transcanada v. United States [92]

In 2016, the investor filed a notice of intent for US $15 billion compensation at ICSID under NAFTA.[93] The US government declined to grant *Transcanada* a permit, which was needed for building a pipeline from Alberta to Nebraska, grounded on the negative results of the project's environmental impact.[94] Moreover, John Kerry—the Secretary of State at that time—apart from expressing that the project raised a "range of concerns about the impact on local communities, water supplies, and cultural heritage sites,"[95] added that it "(...) would facilitate transportation into our country [US] of a particularly dirty source of fuel(...)"[96] and "... would significantly undermine our ability to continue leading the world in combatting climate change."[97] This happened in the context of the Paris Climate Change Agreement negotiations. Against this background, it has been said that the reason for

[88] Peterson (2017a).

[89] Ibid.

[90] Johnson (2009), pp. 11147–11150.

[91] Franck (2014).

[92] *TransCanada Corporation and TransCanada PipeLines Limited v. The United States of America* (2019).

[93] Peterson (2017c).

[94] Ibid.

[95] Ibid.

[96] Ibid.

[97] Ibid.

denying the permit was the intention of the US administration to show commitment to the common objectives established in the Paris Agreement.[98]

Two years later, in March 2017, the arbitration process was discontinued after the host state "signed and issued a Presidential Permit to construct the Keystone XL Pipeline."[99] Although there might be additional political reasons for issuing the permit, the threat of being liable to compensate in the amount of US$15 billion would certainly constitute one of the main motives for changing the decision to reject the project and for approving the construction of the pipeline.

Vattenfall [I] v. Germany[100]

As mentioned above,[101] in 2009, *Vattenfall* filed a request for arbitration for EU $1.4 billion at ICSID under ECT rules. The investor experienced delays in obtaining the environmental permits that were necessary to operate a coal plant in Hamburg by the Elba river.[102] Following "reports of the Intergovernmental Panel on Climate Change,"[103] the ministry of the state imposed additional requirements for the plant's operation.[104] From the investor's standpoint, this additional burden was motivated by the environmentalists' and the green party's pressure on the government to prevent the operation of the plant, given their concerns about CO_2 emissions and water pollution.[105] From the government's side, the deputy environment minister expressed that "[i]t's unprecedented how we are being pilloried just for implementing German and European Union (EU) laws."[106]

The following year, in August 2010, the parties announced that they had reached a settlement,[107] and in March 2011, the agreement between the parties was endorsed by the tribunal.[108] Accordingly, the permits were conceded to the investor, and in 2014, the coal plant commenced to operate.[109] Although the settlement agreement remains undisclosed, the change of decision on the permits was surely intended to avoid compensation.

[98] Grewal and Adkins (2016).

[99] Transcanada's parent company public announcement cited by Luke Peterson, 'Transcanada Says it has Permit for Keystone Pipeline, and ICSID Case is Swiftly Discontinued'(March 27, 2017b) IA Reporter.

[100] *Vattenfall AB v. Germany* (2011).

[101] Supra. 4.1.2.1.

[102] Peterson (2009).

[103] ISDS Platform, Case study: *Vattenfall v. Germany I.*

[104] Ibid.

[105] Ibid.

[106] Ibid.

[107] IA Reporter (2010).

[108] *Vattenfall AB v. Germany* (2011), para. 12.

[109] ISDS Platform, Case study: *Vattenfall v. Germany I.*

Even if the threat chill refers to the direct impact on the host state that has taken the measure, the threat of arbitration against climate measures can go beyond the borders of the defendant and intimidate other host states that deal with the same objective of leveling global warming. That is the case of the transboundary chill, which is described in the next subsection.

4.1.2.3 Transboundary Chill[110] and Climate Change Measures

Tienhaara's proposal to embrace the transboundary chill as a particular type of regulatory chill is grounded on the observation of the recent investment arbitration disputes promoted by the tobacco industries.[111] The author finds that the tobacco companies strategically filed investment arbitration claims in different jurisdictions to create a sort of globalized awareness on the negative consequences for host states of taking measures against the industry.[112] She also compares the tobacco industry to the fossil-energy industry, in an attempt to predict the potential of the transboundary chill concerning the adoption of climate change measures.[113]

Tobacco disputes refer to a group of claims that surged against measures restricting individualization signs on cigarette packaging (plain packaging) and/or imposing explicit graphic warnings on the packaging. These regulations—intended to disincentivize tobacco consumption in Uruguay and Australia—were challenged by Philipp Morris in an investor-state dispute settlement (ISDS). Although none of these claims were successful, it is argued that they bestow chilling effects. Tienhaara demonstrates this hypothesis using New Zealand's experience, where the government decided to suspend the adoption of the plain packaging regulation until the dispute between Philip Morris and Australia was resolved.[114]

If the tobacco industry is capable of globally disincentivizing public health measures through the instrumentalization of investment arbitration to these effects, the fossil-energy industry may adopt similar chilling maneuvers regarding climate change measures. This analogy, according to Tienhaara, is based on four reasons:

---

[110] "Investors pursue this type of chilling effect when a government adopts a policy that affects a form of investment common to many jurisdictions, is easily transferable, and is highly likely to be emulated by other governments. The most effective way to induce this type of chill is to actually launch an ISDS case in one or more jurisdictions (preferably the first jurisdiction that adopts a policy if an investment treaty is accessible). The aim from the investor's perspective in launching a case is not necessarily to prevent or delay the take-up of a policy in the jurisdiction in which the claim is brought—this jurisdiction might not even represent a particularly important market or investment location—but in all other jurisdictions in which the investor is active": Tienhaara (2018), p. 238.

[111] Ibid.

[112] Ibid.

[113] Ibid.

[114] Ibid. p. 238.

"both industries face 'existential threats' and are 'fighting for survival,'"[115] "neither industry has a particularly good reputation with the public,"[116] "both industries are being targeted for regulation at the international level,"[117] and "both industries are dominated by a small group of extremely wealthy and truly multinational corporation."[118]

The majority of the characteristics Tienhaara attributes to the fossil-energy industry are shared by agro-industries that are interested in land grabbing and deforestation. If we take palm oil production—the main agricultural product causing deforestation—as an example, we find that *Cargill*—the biggest palm oil producer and trader worldwide—is so strong that it could potentially deploy a number of global strategies to impede regulation against deforestation, including the use of investment arbitration. Therefore, it makes sense to envisage the emergence of a transboundary chill of climate change measures as a result of the agro-industries' strategic use of investment arbitration litigation as much as it is expected to arise from the fossil-energy industry's action.

In this scenario of transnational corporations investing all around the world and sophisticated corporate structures that facilitate "forum shopping," the transboundary chill is likely to become a normal threat to the objective of leveling global warming. However, even if it could be said that there is consensus in academia that investment treaty arbitration entails the risk of regulatory chill, there are still voices that doubt the real impact of the phenomenon,[119] and there are still questions on how arbitrators should make decisions in an economic context where investment treaty law boosts investments in the green economy but, at the same time, chills regulation by promoting the fight against climate change.

4.1.3 Considering the Economic Rationale of the Investment Treaty Law and the Risk of Regulatory Chill in International Investment Arbitration Decisions: Concluding Remarks on Law and Economics

Despite there being a group of economists positing that investment treaty law does little or nothing to increase foreign investment[120] and others who consider that the

[115] Ibid. p. 239.

[116] Ibid. p. 240.

[117] Ibid.

[118] Ibid.

[119] Schill (2007), p. 474.

[120] "Hallward-Driemeier, famously titled Do Bilateral Investment treaties attract FDI? *Only a Bit…and they could bite*. This study suggested that there was little to no impact in terms of attracting investment for developing countries and spawned an enormous number of studies over the next five years". Mann (2013), p. 528; "Allee and Peinhardt find that it makes no different for

regulatory chill does not exist or, at least, not to a relevant extent,[121] this skepticism occurs in only a minority of academics. On the contrary, there is general acknowledgment that BITs do increase foreign investment and that regulatory chill is a real threat against the enactment of (global) public interest regulation.

Although it is not my duty to enter this debate—as my aim is to profit from conclusions from economists rather than demonstrating the accuracy of the findings from economists—I justify the choice to conform with the mainstream approach, not only with the convincing reasonability of the rational models but also with the empirical evidence that we have presented above. On the one hand, I relied on the European experience with RESSs to show that BITs mitigate the risks of the obsolescing bargain and the price cycle for foreign investors in renewable energy. This experience reinforced my belief in the synergy between investment treaty law and the objective of the international climate change legal framework; i.e., one can expect that the adoption of BITs and international investment arbitration encourages investments in the green economy. On the other hand, I based myself on Transcanada and Vattenfall (I) to prove the threat chill that investment treaty arbitration embodies regarding the delay or rejection of permits because of climate change concerns, as well as on the case of New Zealand's suspension of the plain packaging for tobacco regulation to illustrate the transboundary chill that may affect climate change regulation, given the analogy between the tobacco industry, the fossil-energy industry, and the agro-industry. These examples confirm the hypothesis that investment treaty law conflicts with the objective of leveling global warming by chilling the development of climate change regulation.

Although we cannot deny that this evidence does not constitute full proof of the accuracy of the mainstream models concerning the rationale of investment law and regulatory chill, we find it better to accept the mainstream approach rather than falling into the nihilism behind the skeptical approach. In fact, the skeptical approach (in general) is grounded on the idea that it is impossible to demonstrate that investment treaty arbitration increases foreign investments in the host state and/or to demonstrate that investment treaty arbitration prevents countries from adopting public policy measures. The consequence of adopting this skeptical approach would be denying the benefits of undertaking economic studies. In contrast, I consider that rational economic models are crucial for interpreting the investment treaty law in context and for avoiding arbitrariness in legal reasoning.

Criticism against the mainstream theories is positive in terms of shedding light on the extent to which reality conforms to these models and provides opportunities for refinement. These studies suggest that the investment treaty law may need to be adjusted in order to promote the development of host states and enhance global public interests, such as climate. The existence of regulatory chill and the questions

investment flows whether an investment treaty contains a dispute settlement mechanism, although without it, the treaty clauses are practically non-enforceable for an investing firm" Manger (2010), p. 42.

[121] Schill (2007), p. 474.

on the real impact of BITs on the growth of investments may conflict with the idea that investment treaty law is always good for (national or global) development.

Schill challenges the doubts about the investment treaty law system by denying the existence of regulatory chill with respect to climate change under the discipline's current state of development.[122] In his view,

> (...) investment treaties do not restrict the host state's sovereignty in passing general measures for environmental protection as long as these measures are non-discriminatory, proportionate in reasonably balancing the interests of investors and the protection of the environment, and conform to due process. Overall, investment treaties neither obstruct nor chill state regulation that aims at reducing greenhouse gas emissions[123]

Although it might be true that under the circumstances described by Schill a tribunal may refrain from ordering compensation, it is also true—as shown above[124]—that the adoption of effect-based theories in several decisions[125] and the general inconsistency of the investment treaty law[126] may tilt the balance against the public interest of leveling global warming. Additionally, the evaluation of proportionality or reasonability still includes a considerable ingredient of subjectivity. Thus, the rejection of regulatory chill, as proposed by Schill, could be regarded

[122] "Although stricter emission standards entail additional costs for compliance, such measures generally do not impose disproportionate burdens, as restricting emissions will usually not deprive an investor of every economically viable use of its property or completely destroy its investment. This may, however, be the case if the regulatory framework changes without giving investors sufficient time to adjust their production and operation techniques to the new framework. Similarly, bans on producing specific products will usually qualify as non-compensable bona fide regulation. Such bans will usually constitute proportionate restrictions on the investor's use of its industrial facilities because a plant or factory can normally also be used to produce other products (...) Only rarely will product bans therefore deprive commercial property of any reasonable economic use. In sum, the concept of indirect expropriation therefore leaves broad leeway to host states to regulate production processes, to raise emission standards, and even to impose production bans on products that contribute specifically to greenhouse gas emissions. Thus, the concept of indirect expropriation cannot be regarded as chilling unilateral general regulation that aims at mitigating climate changes (...) The police power exception under the concept of indirect expropriation and the flexibility of the standard of fair and equitable treatment will usually enable states to regulate in the public interest without being liable for damages incurred by foreign investors (...) investment treaties should not lead to a chill on environmental regulation nor obstruct measures that are introduced in an attempt to mitigate climate change": Schill (2007), p. 474.

[123] Ibid. p. 470.

[124] See. Infra. Chap. 6, 6.1.2.

[125] Miles (2008), pp. 13–17.

[126] "Governments need reasonable certainty as to the interpretation of the obligations they have taken on in order to feel reasonably confident that bona fide public policy measures will not be found to have violated a broadly read provision in a treaty. And the system needs certainty that legal correctness in an arbitration decision will be an important factor in its review. Today, the opposite is the case on both counts. There is no certainty for governments due to the competing streams of jurisprudence. The impact of the uncertainty adds to the risk of regulatory chill, as the uncertainty surrounding a large damages award is a state's primary concern if an investor brings an arbitration claim": Mann (2013), p. 532.

as wishful thinking, or in the words of Tienhaara, it could be perceived as "overly simplistic."[127]

Criticism against the investment treaty law grounded on the regulatory chill can be seen as constructive. Awareness of the possibility that the development of a climate-friendly society could be jeopardized by the threat of investment arbitration should be taken as an opportunity to rethink about what investments should be protected through the system and what criteria or theories should be deployed to determine liability and compensation when dealing with climate change measures.

Again,[128] I must reiterate that I do not intend to superpose economics over the law. The law restricts the plausibility of founding a decision purely on the regulatory chill. For example, rejecting compensation to the investor for a climate change measure that appears by all means arbitrary would be against the practice of tribunals and, thus, invalid. In contrast, if—in a given case—a climate regulatory measure could be justified under a law theory or source that is socially accepted, the regulatory chill may be a valid argument.

As the climate change legal framework is part of the investment treaty law system through *syncretism*, the factual synergies and conflicts between the protection of the investor and the global climate interest in a given case are relevant to the investment treaty law. Investment arbitration tribunals should be aware that on some occasions, investor protection runs in parallel with the global investor's interests in lowering greenhouse-gas emissions and that on other occasions, investor protection runs against this interest. This means that tribunals should be aware that their decisions could—positively or negatively—affect whether the objectives of the climate change legal framework are met, and they should therefore promote these objectives within the limits traced by the treaty law system.

Beyond the economic perspective, the relevance of the climate change legal framework in investment treaty law could also be proven from a sociopolitical perspective. However, it is worth mentioning that as the conflictual or synergetic (three-cornered) relationship between the host state, the investor, and the global public interest in leveling global warming is still the center of the problem, the findings of the sociopolitical studies that follow share some common ground with the economic analysis, which is hardly surprising considering that we are dealing with social science/behavioral interaction.

---

[127] Tienhaara (2018), p. 235.

[128] Supra. 4.1.1.3.

4.2 A Sociopolitical Approach to the Relationship Between the Investment Treaty Law and the Climate Change Law: The Contribution of Global Administrative Law and International Critical Studies

Some jurists have undertaken interdisciplinary research linking investment treaty law with sociopolitical studies in order to provide external perspectives to the system, which often serve to support innovative theories to describe the legal framework and/or take the form of normative claims to reshape the law. This is the case of global administrative law (GAL) and the critical studies that are related to investment treaty law. In my work, I attempt to profit from the studies that these two disciplines have already completed in order to prove the relevance of the climate change international legal framework in investment treaty law. While GAL unearths the fact that investment treaty arbitration is an instrument of climate change governance (Sect. 4.2.1), critical studies unravel the unneutrality of the system against the global climate interest (Sect. 4.2.2). The section concludes that tribunals are required (allowed) to rely on the climate change international legal framework, not only to counterbalance the excess of investor protection and exercise their authority in shaping the climate change policy in accordance with the context of global warming but also to ensure the survival of the investment treaty law system (Sect. 4.2.3).

4.2.1 The Investment Treaty Law System as Climate Change Governance: The Contribution of Global Administrative Law

Global administrative law (GAL) understands investment treaty arbitration not only as a dispute settlement mechanism but also as an instrument of governance.[129] The influence of investment treaty law decisions on the management of the global commons—such as the climate—and public interest policies reveals that international arbitration tribunals have the power to determine the behavior of foreign investors, host states, other tribunals, adjudging authorities, and/or decision-makers.

[129] "The concept of global administrative law assumes that much of global governance can usefully be analysed as administration. Instead of neatly separated levels of regulation (private, local, national, inter-State), a congeries of different actors and different layers together form a variegated "global administrative space" that includes international institutions and transnational networks, as well as domestic administrative bodies that operate within international regimes or cause transboundary regulatory effects. The idea of a "global administrative space" differs from those orthodox understandings of international law in which the international is largely inter-governmental, and there is a reasonably sharp separation of the domestic and the international. In the practice of global governance, transnational networks of rule-generators, interpreters and appliers cause such strict barriers to break down.": Kingsbury and Schill (2009).

The tribunals' interpretation of BITs in conjunction with the application of international law, as well as the deployment of theories and choices to define the outcome of a given case, affects the way in which domestic law is modeled. It may, therefore, be that, in the end, the climate change regulation is dictated from the outside of the host state rather than from the inside. Furthermore, the replication of a tribunal's elaboration might become viral among international and local tribunals as a sort of exercise of global governance.

Conceiving investment arbitration tribunals as climate change policy makers is certainly liaised with the fact of host states not taking the adequate measures to promote the climate interest in order to avoid the costs of arbitration and compensation to foreign investors, as explained in the previous section dedicated to regulatory chill. The convenience of adopting BITs depends on increasing the income from foreign investment above the losses incurred through compensation to foreign investors. Thus, if a host state wants to obtain a "good deal" from the investment treaty law system, its administrative conduct must follow the standards of behavior set by the tribunals in their rulings. The adaptation of the climate change regulation to these tribunals' "good administration" standards indicates the factual governance of investment arbitration tribunals over host states. Subjecting, for example, local administrative proceedings concerning gas emission permits to the international arbitration tribunals' conception of stability, predictability, consistency, transparency, legitimate expectations, due process, proportionality, reasonability, etc. is a typical administrative regulation that, despite its adoption formally depending on the action of the national competent authority, finds its origins in the international investment tribunals *de facto* governance.

There is no international institution, such as a global administrative court, for reviewing the legality of climate change measures. However, the climate is a global public interest and a common concern that requires a sort of global management since the market itself is unable to provide satisfactory answers in terms of the leveling of global warming.[130] The international treaty arbitration system is one of the few—if not the only—international systems capable of governing host states' administrative action intended to meet the objectives of the international climate change legal framework. Given this power, rather than constituting "an unintended but effective tool for enforcing climate change law in mixed disputes,"[131] investment treaty arbitration should be conceived as an intended tool for these means.

[130] "As a public good, climate cannot easily be provided **by** the "invisible hand" of the market. Rather, a global economic system built on resource extraction and consumption has spurred climate change. Human induced climate change has been conceived as "the biggest market failure the world has ever seen". Because of the key features of public goods, "the market alone is often unable to ensure their efficient provision", requiring some form of governmental intervention. International regulation allows states to address the shortcomings of national regulation because nations and regions may not fully, or sufficiently, appreciate the value of climate as a global public good": Vadi (2015) pp. 1286, 1292–1293.

[131] Ibid. p. 1318.

Suspicion of parties' intention in investment-climate treaty disputes should be a concern for tribunals. Foreign investors in nonclean sources of energy and investors interested in deforestation—among others—may attempt to manipulate the investment treaty law system to suit their needs.[132] It is crucial, therefore, for arbitrators to be aware that their decisions and the process supporting their decisions are susceptible to becoming (degenerating into) an instrument for corporative policy making against the protection of global public interests, such as the climate. Less likely but fairly conceivable all the same, host states (such as the OPEC countries) may also attempt to influence global governance through investment arbitration tribunals against renewable energy support scheme projects.

While the law and economic studies on the synergies and conflicts between investment treaty law and the international climate change legal framework—above— [133] focus on the rational choices for maximizing the host state's and investor's wealth, the conception of the investment treaty law system as climate change governance turns the focus to the power of shaping the global and local policy regarding climate change. Whichever the approach is, either governance or economics, the facts are still the same: the global climate change objectives could be promoted or demoted by a tribunal's decision. Thus, the climate change legal framework should be relevant in investment arbitration.

In contrast to the economic and governance perspectives, critical studies—albeit it is unfair to box them into one category—share the characteristic of unearthing the fact that the investment arbitration system may not be neutral or compatible with the climate change international legal framework. This is discussed below.

4.2.2 *The Incompatibility of Investment Arbitration Law with the Climate Change Legal Framework's Objectives: The Contribution of Postmodern Critical Studies*

BITs presuppose—in principle—obligations only for the host state. Accordingly—even if we have argued that the protection of global public interests is within the scope of the investment treaty law—it seems that in its origins, the foreign investor

[132] "Using BITs skillfully and drawing on some expansive interpretations tribunals have given of what is covered as an investment by particular BITs, it would be possible to structure many assets in Western economies through offshore companies in ways that would bring them under BIT protection and thus enable investors to challenge measures taken by traditionally capital-exporting countries. BIT protection and investor-State arbitration could thus become increasingly attractive for private economic actors as a valuable safeguard against possible policy choices by Western governments". Kingsbury and Schill (2009), p. 4.

[133] Supra 4.1.

and the foreign investment were exclusively the subject and the object of protection under these international instruments.[134]

In this respect, critical studies have unveiled an asymmetry between the negotiation power of parties to a BIT[135] and a disbalance[136] in the investment treaty law system in favor of the investor, which reproduces imperialist patterns of domination of developed countries over developing countries and of commoditization of the environment.[137] In this postcolonial-neoliberal rationale, climate concerns struggle to be acknowledged as a global public interest, which is relevant to the investment treaty law system.[138]

When addressing critical studies on investment treaty law, it is impossible not to mention Kate Miles, whose work has contributed enormously to the understanding of the discipline.[139] She locates the roots of investment law in the seventeenth century and traces its evolution until today.[140] The study of this evolution concludes that investment law can be comprehended as an appendix to the European countries' process of colonization[141] and as a heritage from the "oppressive protection of commercial interests"[142] over non-European countries. Masked under the shadow of a neutral-apolitical regime[143]—she claims—the investment treaty law system

[134]Miles (2015), p. 10.

[135]Price (2017), p. 128; Manger (2010), p. 54.

[136]"BITs and other bilateral trade agreements (...) are also seen as insidious because they are frequently conducted in unequal contest in closed session between unequal partners—mainly a developed country and a developing or least developed country. With the stronger partner having an established objective to protect its own economic and trade interests, and having little or no interest in making allowance for the divergent or conflicting national and developmental interests of the weaker partner, the resultant agreement is both inevitable and predictable. However, such argument ignores the fact that BITs and other bilateral and regional investment agreements also exist with similar frequency between comparatively equal partners amongst developed countries". Cfr. Price (2017), p. 127.

[137]Miles (2015).

[138]"Sands makes the comment that in *Santa Elena* international investment rules took precedence over national and international norms of environmental protection. The award did not make any concession to competing value systems within international law, one of which is that it is in the interests of all states to reduce global rates of biodiversity loss. This dismissive approach also has relevance for attempts to address other global environmental issues such as climate change. It is indicative of perceptions held within the investment community of the primacy of investment rules even when faced with environmental issues of global concern and host state obligations under multilateral environmental agreements. The *Santa Elena* approach suggests that in any conflict between the interests of investors and climate change mitigation measures, arbitral tribunals will give very little weight, if any, to global environmental objectives such as the reduction of carbon emissions": Miles (2008), p. 18.

[139]Miles (2008, 2015).

[140]Miles (2015), p. 1.

[141]Ibid.

[142]Ibid. p. 2.

[143]Ibid.

excludes the host state's interest from its sphere of protection,[144] which is manifested by

> (...)its lack of responsiveness to the impact of investor's activity on the local communities and environment of the host state, the alignment of home state interest with those of the investor, the categorization of public welfare regulation as treaty violation, and the commoditization of the environment in host states for the use of foreign entities.[145]

Transnational corporations appear to be the winners in the myopic disbalance of the investment treaty law system.[146] Even if on the surface it sounds fair that in exchange for foreign investments the host state commits to mitigating political risk by adopting a BIT, once we look under the surface of the system, we may find that the citizens end up internalizing investors' cost-savings. Certainly, investors look to operate in the most cost-efficient manner possible, which entails implementing production in the country that offers the best conditions, such as weaker environmental and climate change regulations. Developing countries racing for investment might sacrifice the environment to attract investment and reinforce their commitment to maintaining weaker environmental regulations through BITs. This means that the burden of lowering costs, which initially lies in transnational companies, is transferred to the citizens, who would bear the undesirable consequences of low environmental and climate costs.

Given that there are doubts about the potential of BITs to increase foreign investments in developing countries, the adoption of these instruments has been promoted by arguing that in our contemporary globalized world, these countries are not only capital importers but also capital exporters that require protection.[147] However, checking the statistics of international investment arbitration reveals that a clear majority of claimants belong to developed countries and that most of the defendants are developing countries.[148] The United Nations Conference on Trade and Development's (UNCTAD's) 2017 *Investor-State Dispute Settlement Review* shows that, within the 1887–2017 period, only three of the ten most frequent respondents are developed countries.[149] The same study for the same period also reveals that there are no developing countries among the ten most frequent home states of claimants.[150] Thus, although there may be different factors that explain

---

[144] Ibid.

[145] Ibid. p. 3.

[146] "(...) critics have long argued that there is ample evidence to show that these agreements to the contrary merely allow transnational corporations more freedom to exploit workers and to shape the national and global economies of the junior partners to suit the corporations' interests. Some commentators have argued at length that the self-interest mechanisms which are particular to the developed country sponsored bilateral agreements, such as the conditions relating to their enforcement and the control or check mechanisms for managing the agreements, are tantamount to an intrusion into domestic law making": Price (2017), p. 127.

[147] Mann (2013), p. 529.

[148] UNCTAD (2018).

[149] Ibid.

[150] Ibid.

these figures, it might be that, indeed, the system of investment treaty arbitration is designed to perpetuate the position of developed countries over developing countries.

The imperial pattern of treating environmental resources from developing countries as simple assets appears to be protected under BITs. In fact, investment treaty disputes originated from host state measures aiming to protect the environment are common. Viñuales' outstanding study, "Foreign Investment and the Environmental International Law," highlights this reality by presenting convincing evidence that in the last two decades, investment-treaty-environmental-related disputes have increased considerably and are set to become more frequent in the near future.[151] Disputes emerging from investors' operations "in environmental markets(. . .)"[152] or "(. . .) in other activities, where their impact on the environment or on certain minorities is part of the dispute,"[153] as well as disputes "where the application of domestic or international environmental law is at stake"[154] indicate that the economic interests of investors could rank higher or trump over environmental interests. These disputes are perceived with mistrust.[155] If the investment treaty law system is conceived as investor centered, embracing the collision between the investor's interests and environmental interests—including climate interest—within its scope leads to the commoditization of the latter.

In this vein, it has been suggested that these competing values (environmental/ climate protection and investment protection) are incompatible and do not fit in the same system; thus, investment treaty tribunals should reject jurisdiction when disputes require environmental or climate change considerations.[156] This approach is grounded on the belief that there is no proper translation for or understanding of nonmonetary values within an investor-centered treaty law system.[157] Since the investment treaty arbitration system is exclusively built upon the protection of foreign investment, there seems to be no place for weighing climate change as a global public interest within its rationale.[158]

---

[151] Viñuales (2015), pp. 17–23.

[152] Ibid. p. 17.

[153] Ibid.

[154] Ibid.

[155] "Challenging environmental regulation under international investment agreements has become normalised. There is a distinct distrust within the investment community of environmental protection measures and it is manifesting in investor-state arbitration. Initiating proceedings is now viewed by investors as an appropriate response to regulatory change and the threat of proceedings is used to lobby against the introduction of new regulation. The dismissive approach seen in awards such as Tecmed and Santa Elena to environmental regulation and to international environmental norms illustrates the primacy given to investor protection rules at the expense of competing norms and host state interests. It also indicates the methodology that can be expected from arbitral tribunals in any complaint made against climate change mitigation measures". Miles (2008), p. 26.

[156] Horst and Franzki (2016), p. 358.

[157] Ibid.

[158] Ibid. pp. 355–358.

Another approach toward the collision between an investor-centered treaty law system and noneconomic interests—such as climate change—is the modification to or expansion of the investment treaty law system to include public interests as a limit to the protection of investors' interests.[159] This conception involves a detachment from the idea of the investment treaty law as an instrument for dealing exclusively with collisions between private economic interests.[160] In this regard, this approach proposes a "distaste against itself,"[161] where the system tends to correct itself as a result of the identification of the shortcomings of its own rationale. In Horst and Franzki's words, "reflecting on its inherent rationality also bears the potential of an investment law that wants to be different, an investment law that internally opposes its own rationality."[162]

The findings of a number of critical studies underpin my argument that the climate change international legal framework is relevant to investment-treaty-climate-related disputes. If the investment treaty law system does not correct its inherent investor-centered conception by including public interests—such as climate interest—it would be perceived as a mere instrument of a transnational corporation that, on the one hand, perpetuates the exploitation relationship between developed and developing countries while, on the other hand, plays against the leveling of global warming. Discussions and reflections on the original disbalance of the investment treaty law must inform tribunals' decisions in order to impregnate the system's rationale with noneconomic considerations. Investment treaty arbitration tribunals have proven to be capable of challenging this postcolonial-neoliberal rationale by incorporating principles of international public law into their rulings as well as by adopting (creating) theories that allow global public interests to supersede investors' private economic interests.

The inclusion of climate interest in an investment arbitration dispute certainly requires an opening up of the analysis of the competing values at hand under the same key. Consequently, I reiterate—as argued above—that we need to abandon the conception of the conflict between the environment and the investor's interests as a conflict between the public interest of the host state against the private investor's economic interests. Framed in this way, the conflict cannot be dealt with without incurring contradictions since the rationale of the private interests and the public interests belongs to different dimensions that cannot be confronted without commoditizing the latter. As suggested by numerous critical studies, trying to analyze environmental interests under the original rationale of investment treaty arbitration will only lead to equating environmental protection to an economic interest, which is not acceptable. In contrast, if one frames the discussion as part of the conflict between climate as a global public interest against the global public values that lie beyond the protection of the investor—such as economic development or the

[159] Ibid.

[160] Ibid.

[161] Ibid p. 358.

[162] Ibid.

functioning of the global economic system– the competing values would be on the same plane, making it possible to weigh the interests against each other and to conclude which one trumps the other according to the context of the dispute.

4.2.3 The Survival of the Investment Treaty Law and Epistemic Communities: Concluding Remarks on the Sociopolitical Approach to the Relationship Between Investment Treaty Law and Climate Change

It has been shown in GAL studies that investment arbitration tribunals hold, to a great extent, the power to dictate the global (and domestic) climate change policy. In this regard, it would be counterintuitive to deprive this (*de facto*) climate change authority of the main instruments that regulate the international governance of the climate, such as the Paris Agreement or any other instrument that integrates the United Nations Framework Convention on Climate Change (UNFCCC). In such conditions, a tribunal would be like a toothless tiger.

No less important than proving the relevance of the climate change international legal framework is emphasizing that arbitration tribunals' climate change governance is at risk of being perverted by the influence of powerful industries' strategic use of litigation. In this context, the application of provisions protecting the global public interest in climate serves as an antidote to the (convenient) misuse of international investment arbitration. Furthermore, these provisions may counterbalance excessive investor protection, which critical studies have described as inherent to the investment treaty law and as prejudicial to the environment.

Epistemic communities that have been socialized under a postcolonial-neoliberal environment may resist the application of international climate change rules to international investment law. Arbitrators, consultants, academics, and practitioners representing the parties seem to belong to the same circle (if not acting randomly in one or the other role) that has modeled, since the late 1980s, the modern investment treaty law and that has profited from the large fees of international investment arbitration.[163] The natural consequence for a generation of lawyers who believe in

[163] "the community of international arbitrators is a small one. The inner circle has been described as a 'mafia' or the 'club', a community into which it is difficult to gain entrance (...) in order to succeed in the international market, arbitrators and potential arbitrators note the factors that are valued in the market place and lead to reappointments. They adjust their own conduct to display those characteristics and align themselves with 'successful' arbitrators, to certain degree, this inevitably involves demonstrating an affinity with the objectives of international commerce. This cultural dynamic is also reflected in the knowledge-creation process within international investment law – the public declaration of an arbitrator's knowledge, expertise, and cultural approach that is embodied in his or her awards. Through repetition of interpretations of the law, the informal pressures to emulate high-profile arbitrators, and the subtle reiteration of ideological structures, it

the maximization of economic growth over other global public interests would be the rejection of any relevance of noneconomic interests in investment treaty law.

However, more than a quarter of a century has passed since the boom of neoliberalism and the proliferation of BITs began. During this period, a new generation of lawyers has grown up aware of the dangers involved in the commoditization of all global public interests; while nongovernmental organizations (NGOs) have put pressure on the preservation of the environment, private-public partnerships that are concerned with sustainable development have been popularized, and networks of transitional companies have adopted corporate responsibility policies.[164] This evolution is reflected in the application of international law and the deployment (creation) of doctrines that represent the relevance of global public interests in international investment treaty law.[165] It is in this scenario that the climate change legal framework emerges as relevant in investment treaty law.

Although a complete change of the radical neoliberal narrative is not expected from the new generation of arbitrators and academics dealing with investment treaty arbitration, the message that not every foreign investment deserves the (same) protection from the system is gaining ground among international lawyers. Nowadays, questions on the quality of the investment, rather than the quantity of the investment, appear central to the discussions on the convenience of protecting foreign investment.[166] Thus, the concept of sustainable development is starting to be conceived as a defining feature in the analysis of the level of protection that investments deserve within the investment treaty law system.[167]

The change toward a permanent recognition of global public interests related to sustainable development—including climate—in investment treaty law is not only desirable in terms of achieving common global objectives but also necessary for the survival of the system. Developed and developing countries might be reluctant to sign new BITs and/or may be incentivized to cancel current agreements if the space for the recognition of sustainable development is restricted. Governments are aware that tribunals' embracing of theories promoting disproportionate protection of foreign investment and the commoditization of the environment might turn BITs into a threat to local stability. The latter manifestly opposes (sustainable) development and the prosperity that is expected from the adoption of BITs. Examples such as Venezuela, Ecuador, and Bolivia denouncing ICSID or Italy and Russia withdrawing from the ECT could be interpreted in this way. As Brillembourg firmly asserted

is a system that reinforces the prevailing conceptualizations that operate within the international investment law": Miles (2015), pp. 342–343.

[164] Ibid. pp. 339–341.

[165] Infra. See. Part II and Part III.

[166] "It requires a shift in thinking from all investment is good investment (and developing countries should be happy to have any at all!) to a recognition that states have a right, and indeed a duty, to seek to ensure that investments make a positive contribution to their sustainable development. It requires a shift in focus from looking at the quantity of investment as the only issue, to the quality of that investment as the key issue": Mann (2013), p. 534.

[167] Ibid; Alschner and Tuerk (2013), p. 217; Tienhaara (2018), pp. 229–230.

in a panel on "Global Public Interest in International Investment Law"[168]—including prominent academics in the discipline—"We need to find a way to better align these [global public] interests and commitments, otherwise state-users may ultimately choose to withdraw from the system or change it radically."[169]

References

Alschner W, Tuerk E (2013) The role of international investment agreements in fostering sustainable development. In: Baetens F (ed) Investment law within international law: integrationist perspectives. CUP

Asprilla R (2016) Demandas Internacionales contra el Estado, una Amenaza Latente. Congreso Visible http://www.congresovisible.org/agora/post/demandas-internacionales-contra-el-estado-una-amenaza-latente/8466/. Accessed 28 July 2022

Bankes N (2012) Decarbonising the economy and international investment law. J Energy Nat Resour Law 30(4):498. https://doi.org/10.1080/02646811.2012.11435309

Behn D, Fauchald O (2015) Governments under cross-fire? Renewable energy and international economic tribunals. Manchester J Int Econ Law 12(2):120

Bellantuono G (2016) Regulatory Stability in the Energy Sector: the Italian Experience. University of Trento, working paper http://ssrn.com/abstract=2790980. Accessed 28 July 2022

Cameron P (2010) International energy investment law: the pursuit of stability. Oxford University Press, pp 4–7

Caron D et al (2015) Global Public Interest in International Investment Law. ASIL. https://beluga.sub.uni-hamburg.de/vufind/PrimoRecord/gale_ofa450595289?rank=502. Accessed 28 July 2022

De Luca A (2015) Renewable energy in the EU, the energy charter treaty, and Italy's withdrawal therefrom. Bocconi Legal Studies Research Paper No. 2657395 https://ssrn.com/abstract=2657395. Accessed 28 July 2022

Draguiev D (2015) Bulgaria investment treaty arbitration against Bulgaria looming large in the energy sector. Young International Arbitration Group www.lcia.org/media/Download.aspx?MediaId=34. Accessed 28 July 2022

Erkan M (2011) International energy investment law: stability through contractual clause. Wolters Kluver

Franck S (2014) Using investor-state mediation rules to promote conflict management: an introductory guide. Washington & Lee Legal Studies Paper No. 2014–2013, p. 14 https://ssrn.com/abstract=2389763 or doi:https://doi.org/10.2139/ssrn.2389763. Accessed 28 July 2022

Geels F (2013) The Impact of the Financial-Economic Crisis on Sustainability Transitions: Financial Investment, Governance and Public Discourse. Working Paper no 39, European Commission, European Research Area

Gehring M, Kent A (2013) International investment agreements and the emerging green economy: rising to the challenge. In: Baetens F (ed) Investment law within international law: integrationist perspectives. CUP

Globate Climate Action Summit (2020) Press Release: National Governments Urged to Step Up Climate Action by 2020 at End of Landmark Summit. https://www.globalclimateactionsummit.org/step-up/. Accessed 28 July 2022

[168] Caron et al. (2015).

[169] Ibid.

Grewal D, Adkins C (2016) Democracy and legitimacy in investor-state arbitration. Faculty Scholarship Series, p. 5030 https://digitalcommons.law.yale.edu/fss_papers/5030. Accessed 28 July 2022

Horst J, Franzki H (2016) On the critical potential of law – and its limits: double fragmentation of law in Chevron v. Ecuador. In: Markard N et al (eds) Contested regime collisions. CUP, p 358

IA Reporter (2010) Parties announce settlement of dispute over German power plant https://www.iareporter.com/articles/parties-announce-settlement-of-dispute-over-german-power-plant/. Accessed 28 July 2022

International Energy Agency. Energy from Renewable Sources Act (ERSA). https://www.iea.org/policiesandmeasures/pams/bulgaria/name-42357-en.php. Accessed 28 July 2022

ISDS Platform. Case study: Vattenfall v. Germany I. https://isds.bilaterals.org/?case-study-vattenfall-v-germany-i&lang=en. Accessed 28 July 2022

Jetchev G (2013) Tackling boom and bust in the renewable energy sector: the case of Bulgaria. Master Thesis, *CEU-eTD Collection*, Central European University 2

Johnson L (2009) International investment agreements and climate change: the potential for investor-state conflicts and possible strategies for minimizing it. Eur Law Rev 39(12):11147, 11150

Johnson L et al (2018a) Costs and Benefits of Investment Treaties Practical Considerations for States. Columbia Center on Sustainable Investment. http://ccsi.columbia.edu/files/2018/04/07-Columbia-IIA-investor-policy-briefing-ENG-mr.pdf. Accessed 28 July 2022

Johnson L et al (2018b) Costs and Benefits of Investment Treaties. Policy Paper Columbia Center on Sustainable Investment, p. 11 http://ccsi.columbia.edu/files/2018/04/07-Columbia-IIA-investor-policy-briefing-ENG-mr.pdf. Accessed 28 July 2022

Kingsbury B, Schill S (2009) Investor-state arbitration as governance: fair and equitable treatment, proportionality and the emerging global administrative law. NYU School of Law, Public Law Research Paper No. 09-46. https://ssrn.com/abstract=1466980. Accessed 28 July 2022

Lax B (1983) Political risk in the international oil and gas industry. IHRDC, Boston, p 9

Manger M (2010) Common standards in a bilateral economy: a political-economy perspective on international investment law. Jap Yearb Int Law 53

Mann H (2013) Reconceptualizing international investment law: Its role in sustainable development. Lewis Clark Law Rev 17:521–544, 529, 532

Massari S (2015) The Italian Photovoltaic sector in two practical cases: how to create an unfavourable investment climate in renewables in lenders. Issue, OGEL, 13

Miles K (2008) International Investment Law and Climate Change: Issues in the Transition to a Low Carbon World. Society of International Economic Law (SIEL) Inaugural Conference, 13-17 https://ssrn.com/abstract=1154588 or https://doi.org/10.2139/ssrn.1154588. Accessed 28 July 2022

Miles K (2015) Origins of international investment law: empire, environment and the safeguarding of capital. CUP

Mohajeri E (2013) Overview of political and regulatory risks in international energy investment. OGEL, 2

OCDE (2019) Revenue Statistics in Latin America and the Caribbean 1990–2017. p 176 https://www.oecd.org/tax/tax-policy/revenue-statistics-latin-america-and-caribbean-2019-launch-version.pdf. Accessed 28 July 2022

Oeter S (2008) Toward a Richer Institutionalism for International Law and Policy. University of Illinois Law Review/U. of St. Gallen Law & Economics Working Paper No. 2007-15 https://ssrn.com/abstract=1000191. Accessed 28 July 2022

Perry S, Karadelis K (2014) Sun rises on Czech energy claims. GAR https://www.arblit.com/media/pdf/press/sun-rises-on-czech-energy-claims2014-02-19.pdf. Accessed 28 July 2022

Peterson L (2009) German media reveal details of Vattenfall claim v. Germany; NGOs raise enviro fears as two arbitrators named. IA Reporter. https://www.iareporter.com/articles/german-media-reveal-details-of-vattenfall-claim-v-germany-ngos-raise-enviro-fears-as-two-arbitrators-named/. Accessed 28 July 2022

Peterson L (2017a) In New Ruling, Contingency Fees can be Reimbursed – Including Slice of Award Owed to Lawyers – but Liability to Third-Party Funder Not Reimbursed. IA Reporter https://www.iareporter.com/articles/in-new-ruling-contingency-fees-can-be-reimbursed-including-slice-of-award-owed-to-lawyers-but-liability-to-third-party-funder-not-reimbursed/. Accessed 28 July 2022

Peterson L (2017b) Transcanada Says it has Permit for Keystone Pipeline, and ICSID Case is Swiftly Discontinued. IA Reporter. https://www.iareporter.com/articles/transcanada-says-it-has-permit-for-keystone-pipeline-and-icsid-case-is-swiftly-discontinued/. Accessed 28 July 2022

Peterson L (2017c) Transcanada's Threatened $15 Billion Claim Against the United States could Stoke Debate over Scope to Take "Politically-Motivated" Measures. IA Reporter https://www.iareporter.com/articles/transcanadas-threatened-15-billion-claim-against-the-united-states-could-stoke-debate-over-scope-to-take-politically-motivated-measures/. Accessed 28 July 2022

Price D (2017) Indonesia's bold strategy on bilateral investment treaties: seeking an equitable climate for investment?. Asian J Int Law 7(1):124, pp 126–127

Republic of Bulgaria, Ministry of Economy (2011) Energy and Tourism National Renewable Energy Action Plan drawn up in accordance with the template for national renewable energy action plans as set out in Directive 2009/28/EC of the European Parliament and of the Council http://pvtrin.eu/assets/media/PDF/EU_POLICIES/National%20Renewable%20Energy%20Action%20Plan/203.pdf. Accessed 28 July 2022

Restrepo T (2017) Modification of renewable energy support schemes under the energy charter treaty: Eiser and Charanne in the context of climate change. Goettingen J Int Law. 8:101-137. http://www.gojil.eu/issues/81/81_article_restrepo.pdf. Accessed 28 July 2022

Reuter A (2015) Retroactive Reduction of Support for Renewable Energy and Investment Treaty Protection from the Perspective of Shareholders and Lenders. OGEL 13(3). www.ogel.org/article.asp?key=3540. Accessed 28 July 2022

Schill S (2007) Do investment treaties chill unilateral state regulation to mitigate climate change? J Int Arbitr 24(5):469, p. 470

The New Climate Economy (2018) The 2018 Report of the Global Commission on the Economy and Climate, Unlocking the Inclusive Growth Story of the 21st Century: Accelerating Climate Action in Urgent Times. p 39. https://newclimateeconomy.report/2018/. Accessed 28 July 2022

Thurnell L (1977) Political risk in international business, investment behaviour of multinational corporations. Praeger Publisher, New York, p 4

Tienhaara K (2018) Regulatory chill in a warming world: the threat to climate policy posed by investor-state dispute settlement. Transnatl Environ Law 7(2):233

Timmons D et al (2014) The economics of renewable energy, global development and environment institute. Tufts University, Medford, p 32

Tirado J (2015) Renewable energy claims under the energy charter treaty: an overview. OGEL 13(3):8

UNCTAD (2018) Investor–state Dispute Settlement: Review of Developments in 2017. IIA Issues Note, Issue II, 3 https://unctad.org/en/PublicationsLibrary/diaepcbinf2018d2_en.pdf. Accessed 28 July 2022

Vadi V (2015) Beyond known worlds: climate change governance by arbitral tribunals? Vanderbilt J Transnatl Law 48(5):1286, 1292–1293

Vasani B et al (2015) Bulgaria: International Remedies for Foreign Investors in Bulgaria's Renewable Energy sector. Lexology. https://www.lexology.com/library/detail.aspx?g=b755a41f-d1b0-41a3-a485-6b4108a93505. Accessed 28 July 2022

Viñuales J (2015) Foreign investment and the environment in international law. CUP, p 23

Decisions

ICSID (2000a) Compañia del Desarrollo de Santa Elena S.A. v. Republic of Costa Rica, Case No. ARB/96/1 https://www.italaw.com/cases/3413. Accessed 28 July 2022

ICSID (2000b) Metalclad Corporation v. The United Mexican States, Case No. ARB(AF)/97/1 https://www.italaw.com/cases/671. Accessed 28 July 2022

ICSID (2003) Técnicas Medioambientales Tecmed, S.A. v. The United Mexican States, Case No. ARB (AF)/00/2 https://www.italaw.com/cases/1087. Accessed 28 July 2022

ICSID (2005) CMS Gas Transmission Company v. The Republic of Argentina, Case No. ARB/01/8 https://www.italaw.com/cases/288. Accessed 28 July 2022

ICSID (2006) Azurix Corp. v. The Argentine Republic, Case No. ARB/01/12 https://www.italaw.com/cases/118. Accessed 28 July 2022

ICSID (2011a) Libananco Holdings Co. Limited v. Republic of Turkey, Case No. ARB/06/8 https://www.italaw.com/cases/626. Accessed 28 July 2022

ICSID (2011b) Vattenfall AB et al. v. Federal Republic of Germany, Case No. ARB/09/6 https://www.italaw.com/cases/1654. Accessed 28 July 2022

ICSID (2013a) Antin Infrastructure Services Luxembourg S.À.R.L. and Antin Energia Termosolar B.V. v. Spain, Case No. Arb/13/31.

ICSID (2013b) EVN AG v. Bulgaria, Case No. ARB/13/17.

ICSID (2014a) Infrared Environmental Infrastructure Gp Ltd. et al v. Kingdom of Spain, Case No. Abr/14/12 https://www.italaw.com/cases/8049. Accessed 28 July 2022

ICSID (2014b) Masdar Solar & Wind Cooperatief Ua v. Kingdom of Spain, Case No. Abr/14/01 https://www.italaw.com/cases/6608. Accessed 28 July 2022

ICSID (2014c) Nextera Energy Global Holdings B.V. and Nextera Energy Spain Holdings B.V. v. Kingdom of Spain, Case No. Abr/14/11 https://www.italaw.com/cases/2585. Accessed 28 July 2022

ICSID (2014d) Renergy S.À.R.L. v. Spain, Case No. Abr/14/18 https://investmentpolicy.unctad.org/investment-dispute-settlement/cases/590/renergy-v-spain. Accessed 28 July 2022

ICSID (2014e) Rwe Innogy Gmbh And Rwe Innogy Aersa S.A.U., Case No. Arb/14/34 https://www.italaw.com/cases/7743. Accessed 28 July 2022

ICSID (2015a) Baywa R.E. Renewable Energy Gmbh and Baywa R.E. Asset Holding Gmbh, Case No., Arb/15/16.

ICSID (2015b) Belenergia S.A., v. Italian Republic, Case No. Arb/15/40 https://www.italaw.com/cases/7532. Accessed 28 July 2022

ICSID (2015c) Cavalum Sgps, S.A. v. Kingdom of Spain, Case No. Arb/15/34 https://www.italaw.com/cases/8480. Accessed 28 July 2022

ICSID (2015d) Cube Infrastructure Fund Sicav And Others v. Kingdom of Spain, Case No. Arb/15/20 https://www.italaw.com/cases/7477. Accessed 28 July 2022

ICSID (2015e) ENERGO-PRO v. Bulgaria, Case No. ARB/15/19 https://energycharter.org/what-we-do/dispute-settlement/investment-dispute-settlement-cases/. Accessed 28 July 2022

ICSID (2015f) Eskosol S.P.A. In Liquidazione v. Italian Republic, Case No. ABR/15/50 https://www.italaw.com/cases/5895. Accessed 28 July 2022

ICSID (2015g) E.On Se, E.On Finanzanlagen Gmbh And E.On Iberia Holding Gmbh. Case No. Arb/15/35 http://www.energycharter.org/what-we-do/dispute-settlement/investment-dispute-settlement-cases/78-eon-se-eon-finanzanlagen-gmbh-and-eon-iberia-holding-gmbh-v-spain/. Accessed 28 July 2022

ICSID (2015h) Hydro Energy 1 S.À R.L. and Hydroxana Sweden Ab v. Kingdom of Spain, Case No. Arb/15/42 https://www.italaw.com/cases/7980. Accessed 28 July 2022

ICSID (2015i) Jgc Corporation v. Kingdom of Spain, Case No. Arb/15/27 https://investmentpolicy.unctad.org/investment-dispute-settlement/cases/638/jgc-v-spain. Accessed 28 July 2022

ICSID (2015j) Ks Invest Gmbh and Tls Invest Gmbh v. Kingdom of Spain, Case No. Arb/15/25 https://www.italaw.com/cases/6637. Accessed 28 July 2022

ICSID (2015k) Matthias Kruck v. Kingdom of Spain, Case No. Arb/15/23 https://www.italaw.com/cases/6486. Accessed 28 July 2022

ICSID (2015l) Landesbank Baden-Württemberg and Others v. Kingdom of Spain, Case No. Arb/15/45 https://www.italaw.com/cases/7586. Accessed 28 July 2022

ICSID (2015m) Occidental Petroleum Corporation and Occidental Exploration and Production Company v. The Republic of Ecuador, Case No. ARB/06/11 https://www.italaw.com/cases/767. Accessed 28 July 2022.

ICSID (2015n) Operafund Eco-Invest Sicav Plc And Schwab Holding Ag. v. Kingdom of Spain, Case No. Arb/15/36 https://www.italaw.com/cases/3998. Accessed 28 July 2022

ICSID (2015o) Silver Ridge Power Bv, Case No. ABR/15/37.

ICSID (2015p) Stadtwerke München Gmbh, Rwe Innogy Gmbh et al. v. Kingdom of Spain, Case No. Arb/15/1 https://www.italaw.com/cases/7791. Accessed 28 July 2022

ICSID (2015q) Steag Gmbh v. Kingdom of Spain, Case No. Abr/15/4 https://www.italaw.com/cases/8512. Accessed 28 July 2022

ICSID (2015r) Soles Badajoz Gmbh, v. Kingdom of Spain, Case No. Arb/15/38 https://www.italaw.com/cases/7593. Accessed 28 July 2022

ICSID (2015s) *Watkins Holdings S.À R.L. and others v. Kingdom of Spain*, Case No. Arb/15/44, https://www.italaw.com/cases/3996. Accessed 28 July 2022

ICSID (2015t) *9ren Holding S.A.R.L*, Case No. Arb/15/15.

ICSID (2016a) Blusun S.A., Jean-Pierre Lecorcier and Michael Stein v. Italian Republic, Case No. ARB/14/3 https://energycharter.org/what-we-do/dispute-settlement/investment-dispute-settlement-cases/. Accessed 28 July 2022

ICSID (2016b) Espf Beteiligungs Gmbh, Espf Nr. 2 Austria Beteiligungs Gmbh, And Infraclass Energie 5 Gmbh & Co. Kg v. Italian Republic, Case No. Arb/16/5 https://www.italaw.com/cases/8445. Accessed 28 July 2022

ICSID (2016c) Eurus Energy Holdings Corporation And Eurus Energy Europe B.V. v. Kingdom of Spain, Case No. Arb/16/4 https://www.italaw.com/cases/8648. Accessed 28 July 2022

ICSID (2016d) Infracapital F1 S.À R.L. and *Infracapital Solar B.V. v. Kingdom of Spain*, Case No. Arb/16/18 https://investmentpolicy.unctad.org/investment-dispute-settlement/cases/717/infracapital-v-spain. Accessed 28 July 2022

ICSID (2016e) Pac Rim Cayman LLC v. Republic of El Salvador, Case No. ARB/09/12 https://www.italaw.com/cases/783, Accesses 28 July 2022.

ICSID (2016f) *Sevilla Beheer B.V.*, Case No. Arb/16/27.

ICSID (2016g) *Sun-Flower Olmeda Gmbh & Co Kg and Others*, Case No. Arb/16/17.

ICSID (2017) Eiser Infrastructure Limited and Energía Solar Luxembourg S.à r.l. v. Kingdom of Spain, Case No. ARB/13/36 https://www.italaw.com/cases/5721. Accessed 28 July 2022

ICISID (2018) RREEF Infrastructure (G.P.) Limited and RREEF Pan-European Infrastructure Two Lux S.à r.l. v. Kingdom of Spain, Case No. ARB/13/30 https://www.italaw.com/cases/2317. Accessed 28 July 2022

ICSID (2019) TransCanada Corporation and TransCanada PipeLines Limited v. The United States of America, Case No. ARB/16/21 https://www.italaw.com/cases/3823. Accessed 28 July 2022

SCC (2013) *Csp Equity Investment S.À.R.L v. Kingdom of Spain.* https://www.italaw.com/cases/2303. Accessed 28 July 2022

SCC (2015) *Alten Renewable Energy Developments Bv*.

SCC (2016a) *Charanne (The Netherlands) And Construction Investments (Luxembourg) v. Kingdom of Spain* https://www.italaw.com/cases/2082. Accessed 28 July 2022

SCC (2016b) Greentech Energy Systems And Novenergia v. Kingdom of Spain https://www.italaw.com/cases/7138. Accessed 28 July 2022

SCC (2016c) *Isolux Netherlands, BV v. Kingdom of Spain*, Case V2013/153 https://www.italaw.com/cases/5893. Accessed 28 July 2022

UNCITRAL (2011) *The Pv Investors v. Spain, Ad Hoc.*

UNCITRAL (2013a) *Antaris Solar and Dr. Michael Göde v. Czech Republic*, PCA administered.

UNCITRAL (2013b) *Natland Investment Group NV, Natland Group Limited v. Czech Republic, G.I.H.G. Limited, and Radiance Energy Holding S.A.R.L., Ad Hoc* https://www.italaw.com/cases/8351. Accessed 28 July 2022

UNCITRAL *(2013c) Photovoltaik Knopf Betriebs-GmbH v. Czech Republic; WA Investments-Europa Nova Limited Ad Hoc*.

UNCITRAL (2013d) *Voltaic Network GmbH v. Czech Republic, Ad Hoc* https://www.italaw.com/cases/2515. Accessed 28 July 2022

UNCITRAL (2014) Yukos Universal Limited (Isle of Man) v. The Russian Federation, Case No. AA 227 https://www.italaw.com/cases/1175. Accessed 28 July 2022

UNCITRAL (2016) Cosigo Resources Ltd. et al. v. Republic of Colombia, 'Notice of Demand for Arbitration and Statement of Claim https://www.italaw.com/cases/3961. Accessed 28 July 2022

International Legal Instruments

International Energy Charter. https://energycharter.org/what-we-do/dispute-settlement/investment-dispute-settlement-cases/. Accessed 07.01.2019.

Part II
Gravitational Power of the Climate Change International Legal Framework over the Bit Content-Interpretation: Orthodox Approach to Harmonisation

Chapter 5
The Orthodox Approach and the Sources of Law in Investment Treaty-Climate Change Disputes

The scholarship often proposes the adoption of climate change considerations in investment treaty law disputes through content interpretation of bilateral investment treaty (BIT) standards. The method deployed for this purpose consists of unearthing the balance of interests that is embedded in the text of the standards in connection with the general context of climate change. This approach corresponds to—as I refer to it in Part I—the private dimension of the tension of the investment treaty law[1] by analogy to the interpretation of law contracts, where the focus is to find the correct meaning of the parties' bargain. This layer of interpretation succeeds the pre(logical) interpretative phase, which evaluates whether the climate change international legal framework is materially and formally *relevant*, as described (proposed) in Part I. Also, and for practical reasons, the orthodox approach that characterizes content interpretation precedes the study of heterodox approaches—Part III—since it would not make sense to jump to the assessment of the *public law dimension* in cases where the private law dimension provides a satisfactory solution to protect the global climate interest. Besides, a decision based on the orthodox approach is impregnated with a higher degree of social acceptance and is closer to the ideals of certainty and unity than a decision grounded on the *public law dimension*.

In fact (paradoxically), one could say that resolving the (refocused) *global public interest challenge*—of harmonizing BITs with the global public interest in climate—through the private dimension layer could be perceived as a more refined system-based approach than doing so through the public dimension layer. Unlike the analysis of the public dimension, in content interpretation, the global public interest in climate is relevant as long as it serves to elucidate the standards' text. In this sense, the orthodox approach does not entail the direct application of the provisions of the climate change legal framework, nor does it involve conflicts of laws or tension of principles between the climate change legal framework and BITs. In content

[1] See. *supra*. 3.3.

T. Restrepo Rodríguez, *Investment Treaty Law and Climate Change*,
https://doi.org/10.1007/978-3-031-18655-4_5

interpretation, the gravitational power of the climate change legal framework influences the investment treaty law system, but it does not modify its internal structure.

However, mastering an interpretation of the standards' content that harmonizes with the global climate interest is not an easy task. It has been reiterated in this work that investment treaty law standards are open textured. If, on the one hand, this means that the standards accept a wide range of interpretations that could adapt to global interests—including the climate interest—on the other hand, a relative indeterminacy diminishes any aspiration of having one "correct" meaning. This has led investment arbitration tribunals to adopt divergent (incohesive) interpretative tactics, doctrines, and concepts that considerably change from one decision to another. Accordingly, the perfection of a convincing content interpretation that embraces the climate change international legal framework is a major challenge.

Part II is yet another attempt in the scholarship to interpret the content of the standards in accordance with the international climate change framework. However, from a reductionist legal positivism approach—which is the method I have chosen to adopt—this attempt must be made within the limits of the practice of tribunals. In this vein, the interpretation that I propose recognizes that the content of BITs is shaped not only by (nonbinding) prior decisions but also by the application of the principles of interpretation of the Vienna Convention on the Law of Treaties (VCLT), as well as through deference to the host state's authorities.

It would be imprecise to postulate that investment treaty law is a case law system. Considering that the vast majority of rulings expressly reject the idea of a precedent[2] is enough to confirm the inaccuracy of this postulation. In a strict sense, a tribunal's freedom to interpret the content of BIT standards is only subject to a BIT's authority[3] ("A tribunal is a creature of contract, tasked with resolving a particular dispute arising under a particular contract"[4]). However, it is no less true that what is commonly envisaged as investment treaty law is the product of international arbitration investment awards. On the one hand, tribunals openly accept that other arbitration decisions are "relevant for interpretation" [5] and provide "helpful guidance" [6] or, at least, "shed any useful light" [7] to solve the controversy at hand. [8] On

[2]Dolzer and Schreuer (2012), p. 33; Kadelbach (2014), p. 73. Bungenberg and Titi (2015) p. 1508; Grisel (2014), p. 224; Schefer (2013), p. 57.

[3]See. Supra. Ch. 3, 3.2.1.

[4]Ibid.

[5]*Liman Caspian Oil BV and NCL Dutch Investment BV v. Republic of Kazakhstan* (2010) para. 174; *RosInvestCo UK Ltd. v. The Russian Federation* (2013), para. 286; *Chevron Corporation (USA) and Texaco Petroleum Company (USA) v. The Republic of Ecuador* (2009) para. 165.

[6]*Mohammad Ammar Al-Bahloul v. The Republic of Tajikistan,* (2010), para. 44.

[7]*Liman Caspian Oil* (2010) para. 173; RosInvestCo (2013), para. 285; *Chevron Corporation* (2009), para. 164; *Caratube International Oil Company LLP v. The Republic of Kazakhstan* (2012), para. 235.

[8]*Churchill Mining PLC and Planet Mining Pty Ltd v. Republic of Indonesia* (2017), para. 253; *Saipem S.p.A. v. The People's Republic of Bangladesh* (2007), para. 67; *Burlington Resources v. Ecuador* (2012), para 187.

the other hand, the practice of tribunals shows that the social recognition of awards relies heavily on the support of prior rulings,[9] and decisions are even explicit regarding the need to preserve the unity of the investment treaty law system[10] (some authors refer to this practice as "*de facto* doctrine of precedent"[11]).

The social validation feature of prior decisions influences my study of the *orthodox approach* to the impact of the climate change international legal framework in investment-treaty-climate-change-related disputes since the content interpretation of the standards that I perform below is infused with social validity through the deployment of doctrines taken from the reasoning of tribunals' decisions—such as the sole-effects doctrine in the expropriation standard or the assessment of proportionality in the fair and equitable treatment (FET) standard. Nevertheless, this is not done without a critical perspective that profits from the identification of other rules that forge the interpretation of content, such as those embedded in the VCLT or the principle of deference to national authorities in investment treaty law.

The VCLT is central in almost every ruling, to the point that denying its binding character would be counterintuitive. If BITs are acknowledged as being the heart of the system, the VCLT is its backbone. The almost undisputed application of Articles 31 and 32 of the Vienna Convention on the Law of Treaties (VCLT)—concerning content interpretation—even to cases where the parties to the BIT are not parties to the Convention, seems to be the result of a sort of *opinio iuris* toward the practices of other tribunals, rather than the result of the recognition of states' practices. [12]

In this vein, the "ordinary meaning" of BIT standards in conjunction with the object, purpose, and context is crucial in my study to provide the "correct' meaning" of the text in investment treaty-climate change disputes. Particularly important for content interpretation in Part II is the fact that BITs' purposes and objects are to promote the flow of investment (in green energies or in other industries) but not to

[9]In *Von Pezold v. Zimbawe*, for example, having clarified that "in general, the standard of proof in international arbitration" is the *balance of probabilities*, the Tribunal added that in the concrete case there was not "any reason to depart from [this] standard practice": *Bernhard von Pezold and Others v. Republic of Zimbabwe* (2015), paras. 177–179. Similarly, in *Stans Energy Corp. v. The Kyrgyz Republic*—concerning the host-State authorities' refusal to re-execute a license for open-pit mining—the Tribunal expressed that its decision of rejecting that the denegation of a license constitutes an expropriation was "confirmed by the investment arbitration practice": *Stans Energy Corp. and Kutisay Mining LLC v. The Kyrgyz Republic (2015),* para. 91.

[10]"The ultimate integrity of the protections given to the many individuals made under Chapter 11 [NAFTA] is ensured by reference to a multitude of arbitral panels occupied by persons who are occasionally reappointed. The ultimate integrity of Chapter 11 system as a whole requires a modicum of awareness of each of these tribunals for each other and the system as a whole": *Glamis Gold, Ltd. v. The United States of America (2009)*, para. 5.

[11]See. McLachlan et al. (2007), pp. 72–76. Also, Wälde in *Thunderbird Gaming Corp. v. Mexico* (2006), para. 129; and Kadelbach (2014), p. 47.

[12]*Crystallex International Corporation v. Bolivarian Republic of Venezuela* (2016), para 537; *Romak S.A. (Switzerland) v. The Republic of Uzbekistan* (2009), para. 169; *Wintershall Aktiengesellschaft v. Argentine Republic* (2008), paras. 77, 26; *Merrill and Ring Forestry L.P. v. Canada* (2010), para. 183. See. Also Reinisch (2015), pp. 372–410.

the point of annulling the host state's power to establish its own (climate change or other) policies or take measures it considers more appropriate. Therefore, the difficult balance between the investor's and the host state's interests from a systemic point of view is at the optimal point where the conditions of the investment are still attractive while the autonomy of the state to govern itself is maximized.

No less important is the fact that the climate change international legal framework is part of the context that informs the understanding of content interpretation. In this regard, my elaboration on the concepts of legitimate expectations, proportionality, essentiality, and/or due diligence in the analysis of the FET standard as well as the temperance of the sole-effects doctrine in the study of expropriation embraces the gravitational force of the climate change context.

Apart from Article 31 VCLT—but not disconnected from it—deference to competent national authorities also plays a defining role in the harmonization of BIT contents with the climate change international legal framework. It is a general practice of investment tribunals to defer the definition of technical aspects to national authorities. This practice is grounded on the belief that specialized national institutions are better equipped to provide the meaning of technical terms in the national context and that the definition of certain aspects belongs to the host state's sovereignty.[13]

Very often, the definitional power of national authorities relates to questions on legality or validity under national law, which must be answered in order to establish the content of BIT obligations. In this particular scenario, national courts are considered to be better suited to deliver a more appropriate answer than international investment tribunals. In practice—unless there is a violation of due process[14]—the decisions of national courts are generally deemed to be satisfactory criteria to interpret the legal aspects of national law that are relevant to the international dispute.[15]

In addition to the interpretation of national law, there are other technical matters that are deferred to national authorities. Despite it being crucial for the definition of the content of BIT obligations, the definition of the public or national interest of the government's regulatory measures and the measure's suitability to achieve the results it pursues "is no part of [the] Tribunal's function [...]"[16] but rather "the respondent's sovereign right, as regulator."[17] Accordingly, the extension of the content of BIT obligations regarding the interpretation of "the protection of public health, safety, morals or welfare, as well as other functions related to taxation and

---

[13]Glamis Gold, Ltd. v. The United States of America (2009), para. 5.

[14]McLachlan (2007), p. 70.

[15]Hulley Enterprises Limited (Cyprus) v. The Russian Federation (2014), para. 52.

[16]Copper Mesa Mining Corporation v. Republic of Ecuador (2016), para. 6.64.

[17]Ibid.

police powers of states" [18] is a matter of national law. [19] Similarly, the scientific adequacy of the regulator's evaluation when approving or disapproving permits is not generally at odds in investment arbitration proceedings.[20]

Consequently, Part II takes special care not to invade the high degree of deference that corresponds to the host state's authorities when interpreting the content of the expropriation and FET standards in light of the climate change international legal framework. This involves avoiding the deployment of concepts such as necessity or suitability as an excuse for invading the host state's autonomy to dictate and implement its own policy—including the climate change policy.

Against this background, Chap. 6 focuses on the content interpretation of the expropriation standard, and Chap. 7 addresses the content interpretation of the FET standard, both in a climate change context. Chapter 6 posits that deploying the (dis)proportionality of the impact of the measure on the investor as the central element to define indirect expropriation leads to a balanced regulation of the parties' interests that is in line with climate interest. Chapter 7, in turn, postulates that harmonization with climate interest is possible by defining the (dis)proportionality and (un)reasonableness of the host state's measure by reference to its (non)-capriciousness or the (un)predictability and (in)essentiality of the changes to the legal framework it supposes.

Limiting the study to the content interpretation of the expropriation and FET standards is not meant to deny the influence of the climate change international legal framework in the understanding of other standards such as the most favored nation, national treatment, full protection and security, or qualifying requirements. Neither am I denying the relevance of general exceptions in shaping the content of the BIT in connection with climate interest. However, I have chosen to exclude the other standards from my study because the probability of climate-change-related disputes within their scope is still very low, or, in any case, there does not seem to be much discussion in the scholarship as to which the "correct" content interpretation would be when these disputes arise. Concerning BITs' general exceptions, instead of including their analysis in Part II, I found it convenient to study them in Part III in conjunction with the international defense of necessity—given the similar counterrational character of these notions and their interaction in investment treaty disputes.

[18] *Marion Unglaube v. Republic of Costa Rica* (2012), para. 246.

[19] *Saluka Investments B.V. v. The Czech Republic* (2006), para. 305–306; *Rusoro Mining Ltd. v. Bolivarian Republic of Venezuela* (2016), para. 385.

[20] This was the case in *Crystallex v. Venezuela*, where the Tribunal deferred the decision concerned with the technical adequacy of the reasoning that grounded the denegation of an environmental permit to the Ministry of Environment of Venezuela, because of governmental authorities' "[. . .] expertise and competence (which is assumed to be present in those institutions called to make the relevant decisions) and proximity with the situation under examination": *Crystallex v. Venezuela* (2016), para. 583.

References

Bungenberg M, Titi C (2015) Precedents in international investment law. In: Bungenberg M (ed) International investment law. C.H. Beck-Hart-Nomos, p 1508
Dolzer R, Schreuer C (2012) Principles of international investment law. 2nd edn. OUP
Grisel F (2014) The sources of foreign investment law. In: Douglas Z (ed) The foundations of international investment law. OUP, p 224
Kadelbach J (2014) Regimeübergreifende Konkretisierung im Internationalen Investitionsrecht. Nomos, p 73
McLachlan C et al (2007) International investment arbitration: substantive principles. OUP:72–76
Reinisch A (2015) The interpretation of international investment agreements. In: Bugenberg M (ed) International investment law. C.H. Bech/Hart/Nomos, pp 372–410.
Schefer K (2013) International investment law. Edward Elgar, p 57

Decisions

ICSID (2007) Saipem S.p.A. v. The People's Republic of Bangladesh, Case No. ARB/05/07 https://www.italaw.com/cases/951. Accessed 29 July 2022.
ICSID (2008) Wintershall Aktiengesellschaft v. Argentine Republic, Case No. ARB/04/14 https://www.italaw.com/cases/1171. Accessed 29 July 2022.
ICSID (2010a) Liman Caspian Oil BV and NCL Dutch Investment BV v. Republic of Kazakhstan, Case No. ARB/07/14 https://www.italaw.com/cases/1977#sthash.1BmlOYMa.dpuf. Accessed 29 July 2022.
ICSID (2010b) Merrill and Ring Forestry L.P. v. Canada, Case No. UNCT/07/1 https://www.italaw.com/cases/669#sthash.blsKTHTM.dpuf. Accessed 29 July 2022.
ICSID (2012a) Burlington Resources v. Ecuador, Case No. ARB/08/5 https://www.italaw.com/cases/181. Accessed 29 July 2022.
ICSID (2012b) Caratube International Oil Company LLP v. The Republic of Kazakhstan, Case No. ARB/08/12 https://www.italaw.com/cases/211#sthash.DjpvdB6B.dpuf. Accessed 29 July 2022.
ICSID (2012c) Marion Unglaube v. Republic of Costa Rica, Case No. ARB/08/1 https://www.italaw.com/cases/1134. Accessed 29 July 2022.
ICSID (2015) Bernhard von Pezold and Others v. Republic of Zimbabwe, Case No. ARB/10/15 https://www.italaw.com/cases/1472#sthash.9b8uV7nj.dpuf. Accessed 29 July 2022.
ICSID (2016a) Crystallex International Corporation v. Bolivarian Republic of Venezuela, Case No. ARB(AF)/11/2 https://www.italaw.com/cases/1530#sthash.FnlLIerj.dpuf. Accessed 29 July 2022.
ICSID (2016b) Rusoro Mining Ltd. v. Bolivarian Republic of Venezuela, Case No. ARB(AF)/12/5 https://www.italaw.com/cases/2048#sthash.VlFYPxlL.dpuf. Accessed 29 July 2022.
ICSID (2017) Churchill Mining PLC and Planet Mining Pty Ltd v. Republic of Indonesia, Case No. ARB/12/14 and 12/40 https://www.italaw.com/cases/1479#sthash.Psc1DedU.dpuf. Accessed 29 July 2022.
SCC (2010) Mohammad Ammar Al-Bahloul v. The Republic of Tajikistan, SCC Case No. 064/2008 https://www.italaw.com/cases/65#sthash.J9gKN91T.dpuf. Accessed 29 July 2022.
SCC (2013) RosInvestCo UK Ltd. v. The Russian Federation, SCC Case No. V079/2005 https://www.italaw.com/cases/923. Accessed 29 July 2022.
UNCITRAL (2006a) Saluka Investments B.V. v. The Czech Republic, ad hoc https://www.italaw.com/cases/961. Accessed 29 July 2022.

UNCITRAL (2006b) Thunderbird Gaming Corp. v. Mexico https://www.italaw.com/cases/571. Accessed 29 July 2022.

UNCITRAL (2009a) Chevron Corporation (USA) and Texaco Petroleum Company (USA) v. The Republic of Ecuador, Case No. 34877 https://www.italaw.com/cases/251#sthash.J8vcHWpC.dpuf. Accessed 29 July 2022.

UNCITRAL (2009b) Glamis Gold, Ltd. v. The United States of America, ad hoc http://www.italaw.com/cases/487#sthash.xCSTGU3d.dpuf. Accessed 29 July 2022.

UNCITRAL (2009c) Romak S.A. (Switzerland) v. The Republic of Uzbekistan, Case No. AA280 https://www.italaw.com/cases/documents/919 or https://www.italaw.com/cases/885. Accessed 29 July 2022.

UNCITRAL (2014) Hulley Enterprises Limited (Cyprus) v. The Russian Federation, Case No. AA 226, https://www.italaw.com/cases/544. Accessed 29 July 2022.

UNCITRAL (2015) Stans Energy Corp. and Kutisay Mining LLC v. The Kyrgyz Republic, Case No 2015-32. https://www.italaw.com/cases/7328. Accessed 29 July 2022.

UNCITRAL (2016) Copper Mesa Mining Corporation v. Republic of Ecuador, Case No. 2012-2 https://www.italaw.com/cases/4206#sthash.tqKhkan6.dpuf. Accessed 29 July 2022.

Chapter 6
Orthodox Approach to the Expropriation Standard in Investment Treaty-Climate Change Disputes: Determining the Breach Through the *Sacrificial-Lamb Doctrine*

It is an undisputed fact that the meaning of expropriation in international investment law is more comprehensive than the meaning that is usually attributed to it in national legal systems. The (ordinary) definition of BITs covers not only the legal or physical taking of the investor's assets by the host state (direct expropriation) but also the measures that are similar or have an equivalent or similar effect (indirect expropriation).[1] Direct expropriations appear not to be difficult to identify,[2] and they rarely happen.[3] On the contrary, even if the (ordinary) expropriation clause distinguishes indirect expropriation through its effects, its determination is a very conflictive issue. Claims in this regard are common practice in international investment litigation.[4]

[1]Isakoff (2013), p. 193: 'The expropriation provisions of many BITs and other international investment treaties reference measures "tantamount to" or "equivalent to" expropriation. This language has often been narrowly construed to avoid expanding the definition of expropriation; rather, it is simply meant to include expropriations that occur in substance but not in form. In fact, some arbitral tribunals have treated "measures tantamount to expropriation" as the functional equivalent of expropriation.'

[2]Ibid 192; *Anatolie Stati, Gabriel Stati, Ascom Group SA and Terra Raf Trans Traiding Ltd v. Kazakhstan* (2010): '"*Direct expropriation"* refers to the overt seizure of a foreign investor property or title such as property by the host state'(...) 'an outright seizure of physical assets, contractual rights, and legal title are textbook examples of "direct" expropriation'; McLachlan et al. (2007), p. 8.69: 'Arbitral tribunals have considered direct expropriation as being relatively easy to recognize: for example "governmental authorities get over a mine or factory, depriving the investor of all meaningful benefits of ownership and control", or there has been "a compulsory transfer of property rights"'.

[3]"In the area of environmental protection, direct expropriation does not seem to be a rule": Viñuales (2015), p. 23.

[4]Erkan's research found that indirect expropriation measures include: 'Excessive or discriminatory or abusive taxation; rejection of access to labor; Rejection of access to raw materials; Forced transfer of shares of an investment; Interference in the right of management; Revocation of Licenses or permits; Exchange control; Incidental interference that deprives owner of use of reasonable

T. Restrepo Rodríguez, *Investment Treaty Law and Climate Change*,
https://doi.org/10.1007/978-3-031-18655-4_6

In the current context of climate change, claims arising from indirect expropriation that are connected to climate change measures are expected to increase. Certainly, host states' measures that render useless or vacate the value of investments in the green economy, as well as measures promoting investments in the green economy that affect investments in other industries, have the potential to end up being the object of climate-change-related disputes. Accordingly—for example—it is equally conceivable for the host state to destroy the value of an investment in renewable energy by canceling the benefits that under a particular renewable energy support scheme were conceded to the investor, as envisaging the host state destroying the value of an investment in coal-energy production for the reason of implementing an energy policy that phases out coal in order to promote the leveling of global warming.

The reasoning in expropriatory climate-change-related disputes should be in tune with the climate change international legal framework but still adjusted within the limits of the pliability of the expropriation standard. The challenge of the *orthodox approach* to these disputes consists in finding—by interpreting the text—a solution that promotes the goal of mitigating, adapting, or preventing global warming.

The literature on the interpretation of the expropriatory standard concerning investment-treaty-climate-change-related disputes is inclined to recognize that the adoption of the police-power doctrine solves the dilemma of harmonizing climate interest with this obligation. Accordingly, it is believed that the (palpable) public interest of climate change measures impedes the perfection of indirect expropriation, or at least it does so under certain characteristics of the measure, such as proportionality, reasonability, necessity, etc. I argue that, unlike the sole-effect doctrine, the adoption of the police-power doctrine exceeds the pliability of the expropriation standard—by linking the breach of the obligation to the character of the measure rather than to its effects—and only partly harmonizes with the climate change international legal, as its application ignores the situation where the investors' interests run in parallel to the interest of leveling global warming (Sect. 6.1). Subsequently, I take the risk of proposing my own *orthodox approach* through the application of the sacrificial-lamb doctrine, which is a tempered version of the sole-effects doctrine that determines the breach of the expropriation standard by reference to the abnormality of the impact of the host state's measure (Sect. 6.2). Although this doctrine results—in most of the cases—in a balanced solution to the promotion of the climate interest without systemic contradictions, it still finds limitations in tailor-made clauses as well as restrictions stemming from the same particular nature of an *orthodox approach* (Sect. 6.3).

economic benefit of property (e.g., targeted trade restrictions, abuse of regulatory power), Environmental and human rights regulation': Erkan (2011), pp. 81–82.

6.1 Pliability of the Expropriation Standard in a Climate Change Context: Swinging Between the *Sole-Effects Doctrine* and the *Police-Power Doctrine*

Tribunals are consistent in identifying substantiality (severity) as the defining element of *indirect expropriation*.[5] However, it is not clear—neither in the doctrine nor in the literature—whether the verification of substantiality by itself is enough to uphold that the expropriation standard has been breached. Some decisions—which have been welcomed by a significant portion of the doctrine—establish that, in addition to the assessment of the impact of the measure on the investment, the evaluation of the character of the host state's measure is relevant to this end. The discrepancy in this regard is followed by theories intended to explain the elements that characterize the breach of the expropriation standard. As De Nanteuil puts it,

> Any attempt to elaborate a theory on indirect expropriation is intended to determine the elements that would identify the hypotheses through which the default in the payment of a financial compensation renders illegal the damage to the property and involves the liability of the state, and those through which the default in the payment could be justified[6] (my translation).

The key question would be, then, which climate change measures breach the expropriation standard and under which conditions? The answer is very tricky. First, there is no agreement—neither in the scholarship nor in case law—regarding the elements of indirect expropriation.[7] As a result, the definition of indirect expropriation is to a great extent modeled by each author and grounded on a very personal reading of the system's operation—this work is not an exception in this regard. Second, although the *orthodox approach* is intended to find an answer in the text, it is not devoid of systemic difficulties since it pendulates between a problem-based approach and a systemic approach.

The dissent on the definition of *indirect expropriation* is generally framed on the basis of two opposing theories: the sole-effects doctrine and the police-power doctrine. While the first defines indirect expropriation exclusively by reference to its effects, the second defines the concept by reference to the character of the measure. Therefore, while under the sole-effects doctrine the expropriation standard is breached when the impact on the investment is substantial, under the police-power doctrine, the breach requires not only the substantiality of the impact but also that of an illegitimate measure.

---

[5] For a comprehensive account of the deployment of substantiality in case law: See. *Anatolie Stati, Gabriel Stati, Ascom Group SA and Terra Raf Trans Traiding Ltd v. Kazakhstan (2013)*, para. 1111–1113.

[6] De Nanteuil (2014), p. 21.

[7] Waelde, Kolo (2001), p. 812. The confusion is so acute that the comment that Wälde made 20 years ago remains valid today: [this] "question.... has bedeviled governments, international tribunals, and international lawyers".

The police-power doctrine has served the scholarship to connect the climate change interest with the expropriation standard. Although tempting, I find this connection, on the one hand, insufficient to promote the objectives of the climate change international legal framework because it hinders investments in the green industry. On the other hand, it is contradictory with the text, purpose, and systematic coherence of the expropriation standard—as the police-power doctrine defines indirect expropriation by reference to the (il)legitimacy of a measure rather than by reference to its effects (Sect. 6.1.1). In contrast, the sole-effects doctrine seems to suit the pliability of the expropriatory standard, but its inclination toward favoring the investor's interest needs to be tempered, and in certain cases, it is important to bear in mind that some tailor-made expropriatory clauses exclude its application and swings back to the police-power doctrine (Sect. 6.1.2).

6.1.1 The Temptation to Adopt the Police-Power Doctrine in Expropriatory Climate Change Disputes and Its Drawbacks: Exposing the Failure of Methanex/Tecmed-Inspired Theories to Harmonize the Expropriation Standard with Climate Interest

The (irrefutable) public interest in climate change may tempt us to interpret the content of an expropriatory measure through the police-power doctrine. Apart from the scientific evidence on the current and future catastrophic effects of global warming or the considerations about the climate interest as a common concern of humankind; domestic laws, several international treaties, protocols, and soft-law instruments incorporate climate change provisions that together constitute a context that gives rise to the idea that the climate regulation should not be considered as expropriatory in investment treaty law. According to this logic, for example, the destruction of the value of investments in fossil fuels caused by the implementation of a renewable support energy scheme would not constitute an indirect expropriation and the investor would be called to internalize all the costs of the loss.

In its purest version the application of the police-power doctrine entails that a measure in the public interest of the climate cannot be regarded as expropriatory. In a more dynamic understanding of the police-power doctrine, a public measure concerning the climate is not considered expropriatory as long as it complies with certain characteristics, such as good faith, proportionality, reasonability, necessity, nondiscrimination, and/or respect for specific commitments. The understanding on which of these characteristics define the concept of indirect expropriation and their content is not settled.

In *Pope and Talbot*[8]—often cited as some of the seminal decisions concerning the police-power doctrine—the Tribunal affirmed that "a State is not liable for economic injury which is consequence of *a bona fide* regulation within accepted police power of the States."[9] This theory was subsequently followed by *SD Myers*, [10] *Saluka*,[11] and *Methanex* [12]—all regularly considered landmark decisions. The *Methanex* case—which is the most frequently cited in relation to the police-power doctrine—concerned a prohibition emanating from the state of California against the production of *methyl tertiary butyl ether*, which is an essential ingredient for the production of methanol. Although this prohibition rendered the investment useless—because the investor's business involved the production of methanol—the Tribunal found that the measure was not expropriatory because "as a matter of general international law, a non-discriminatory regulation for a public purpose (...) is not deemed expropriatory and compensatory (...)."[13]

There is no doubt that the police-power doctrine has been deployed in several decisions, and some even affirm that the trend is toward focusing on the character of the measure rather than on its effects.[14] In 2011, for example, the Tribunal in the case of *El Paso v. Argentina* established:

> In sum, a general regulation is a lawful act rather than an expropriation if it is non-discriminatory, made for public purpose and taken in conformity with the due process. In principle, general non-discriminatory regulatory measures, adopted in accordance with the rules of good faith and due process, do not entail a duty of compensation. [15]

It is important to stress the implausibility of encountering an absolute definition of the police-power doctrine in investment treaty decisions. Tribunals that rely on the police-power doctrine generally come up with exceptions to the principle that regulatory measures in the public interests do not constitute indirect expropriations. Even *Methanex*—usually described as an example of the purest version of the police-power doctrine—establishes that public interest measures are still expropriatory in the event that the host state dishonors specific commitments. Other decisions, in order to define whether a regulatory measure breaches the expropriatory standard by reference to its character, patchily refer to diverse concepts such as good faith, proportionality, reasonability, necessity, nondiscrimination, arbitrariness, predictability, legitimate expectations, specific commitments,

---

[8] Pope & Talbot Inc. v. The Government of Canada (2000), para. 96.

[9] Ibid.

[10] *S.D. Myers, Inc. v. Government of Canada (2002)*, para. 281–283.

[11] "[i]t is now established in international law that States are not liable to pay compensation to a foreign investor when, in the normal exercise of their regulatory powers, they adopt in a non-discriminatory manner bona fide regulations that are aimed at the general welfare": *Saluka Investments B.V. v. The Czech Republic (2006)*, para. 255.

[12] *Methanex Corporation v. United States of America (2002).*

[13] Ibid Part IV – para. 7.

[14] Schill (2007), p. 472.

[15] *El Paso Energy International Company v. The Argentine Republic (2011)*, para 240.

representations, deference, arbitrariness, commercial risk, and/or margin of appreciation. As a result—using a metaphor from the medical field—one may affirm that, as part of the bipolar disorder of case law that pendulates between the substantiality of the effects and the legitimacy of the measure as a key element to define indirect expropriation, the police-power doctrine could be defined more precisely as a category that roughly groups cyclothymic decisions that focus on the character of the measure to determine the breach of the expropriation standard rather than a category based on a refined, consistent, and healthy practice.

The promoters of the police-power doctrine concerning climate change measures range from proposing a *Methanex* approach to more restrictive conceptions that limit the regulatory space of the host state but still focus on the character of the measure. Baetens, for example—in her study called *Implementing the Kyoto Protocol Through Private Investment*[16]—supports the application of the *Methanex* approach; i.e., a climate measure would not be regarded as expropriatory unless it ignores specific commitments. She points out that this posture represents a closer approximation to the Kyoto Protocol in comparison to the theses that rely on the analysis of reasonability, proportionality and nondiscrimination.[17] In her view, these criteria—besides being imprecise—may result in "many legitimate, potentially effective and useful [climate change] measures to be struck down or to have their scope of application restricted to an extent that their achievements would be rendered negligible."[18] In the same vein, the International Institute for Sustainable Development (IISD) study on climate change and international investment agreements labels *Methanex* and *Saluka* as the "most progressive of approaches,"[19] and Condon, on the basis of *Methanex*, affirms that "[climate change] regulation based on the precautionary principle can survive a challenge by foreign investors."[20]

I admit that it might appear simplistic to categorize in, the preceding paragraph, conceptions that limit the regulatory space of the host state but still focus on the character of the measures, by opposition to *Methanex*, in order to define a branch of the police-power doctrine that inspires an interpretation of the expropriatory standard in line with the climate interest. However, in my defense, the lack of consistency in case law with regard to the exceptions or limitations to the police powers of the state provokes a high degree of uncertainty and subjectivity in the scholarship. Nevertheless, there seems to be a certain degree of agreement in the literature on the interpretation of the expropriatory standard that the application of the police-power doctrine in absolute terms might be lean excessively toward the host state. Therefore, the idea of balance between the investor's interests and the host state emerges (again) as an ideal reading of the expropriatory standard.

[16]Baetens (2010).

[17]Ibid. pp. 22–23.

[18]Ibid. p. 22.

[19]Marshal (2010).

[20]Condon (2015), p. 337.

The ruling generally considered an authority when it comes to limiting the police power of the state is *Tecmed*. In this case, the Mexican national environmental agency did not extend the investor's permit to operate a hazardous waste landfill apparently due to *Tecmed*'s noncompliance with some environmental regulations. The host state was found liable to pay damages for the breach of the expropriation standard because the nonextension of the permit was considered to run counter to the investor's legitimate expectations and nonproportional. The investor argued that it had the legitimate expectation to operate long term, recover the investment, and obtain returns. With regard to proportionality, the Tribunal clarified that "[t]here must be a reasonable relationship of proportionality between the charge or weight imposed to the foreign investor and the aim sought to be realized by any expropriatory measure." [21] Based on this ruling, tribunals and academics refer to reasonableness, proportionality, legitimate expectations, and other characteristics that could be attributed to regulatory measures as concepts that might limit the state's power to expropriate.

For instance, Vadi, who advocates in favor of the application of the police-power doctrine in expropriatory treaty (climate-change-related) disputes, subject to some restrictions, is of the opinion that

> [c]oncepts such as **reasonableness and proportionality** may help the arbitrators to assess whether the modalities of state regulation are suitable and appropriate to achieve the state objective of climate change mitigation and do not constitute a camouflaged indirect expropriation of the given foreign investment[22] (emphasis added)

Similarly, Schill posits that regulatory limits to gas emissions do not constitute indirect expropriation "as long as the measures are **non-discriminatory and impose proportionate burdens** to the investor"[23] (emphasis added). Adding more restrictions to the expropriatory power of the host state in investment treaty law, Mitchel and Henckels, based on their observation of case law, do not discard the application of the concept of *necessity*—independently or as an element of *proportionality*—to expropriatory disputes.[24] Furthermore, Viñuales—based on his very comprehensive analysis of the relationship between environmental law and investment law, contained in Chapter 14 of his book[25]—concludes that, apart from being relevant to defining the proportionality of the measure, the purpose of the regulatory measure serves to assess whether it involves a commercial risk—in contrast to a political risk—that the investor should bear.[26]

Against this backdrop of concepts, one might become confused as to how they operate and interact. For example, one may think that the "reasonableness" and "proportionality" of a measure are synonymous or, inversely, are different concepts,

---

[21] *Técnicas Medioambientales Tecmed, S.A. v. The United Mexican States (2003)*, para. 122.

[22] Vadi (2015), pp. 1285, 1328.

[23] Schill (2007), p. 472.

[24] Mitchell and Henckels (2013), pp. 122–124.

[25] Viñuales (2015), pp. 293–315.

[26] Ibid. p. 311.

or reasonableness may be understood as merely an element of proportionality, among others, such as the necessity or public interest of a measure, or one may also think exactly the opposite, i.e., that proportionality is a characteristic of reasonableness. To make things more complex, one could think that the functions of both reasonableness and proportionality can serve as a basis for defining whether a measure goes against the legitimate expectations of the investor since a measure that is unreasonable and disproportional is certainly not expected by the investor. Conversely, one could think that the purpose of assessing whether a measure goes against legitimate expectations is to determine its proportionality and/or reasonableness since a measure that is against the legitimate expectations of the investor might appear to be disproportional and unreasonable. The same analysis could be applied to the concept of necessity. Is necessity an element to establish whether measures are proportional, reasonable, and legitimately expected by the investor? Or does a disproportional, unreasonable, and unexpected measure show itself as unnecessary?

The riddle in the paragraph above is not a mere play on words. Any person who embarks on the study of the investment treaty law would find that decisions regarding the limitations on the police-power doctrine are imprecise and disconnected. As a result, academics who support this approach are compelled to build their own coherent version of the system's operation by cherry-picking the pieces of tribunals' decisions that favor their position. But leaving aside the lack of conceptual precision that navigates in the framing of the character of the measure and its (il)legitimacy—which will be revisited in Chapter 7 (fair and equitable treatment (FET))—the police-power doctrine, in its *Methanex*- or *Tecmed*-inspired version, does not suit the pliability of the (ordinary) expropriation standard and does not appear to be a satisfactory solution to the harmonization of the expropriation standard with the objectives of the climate change international legal framework.

Studies that support the application of the police-power doctrine to expropriatory (climate-change-related) disputes are generally concerned with investors challenging the power of the host state to implement measures aiming to mitigate, prevent, or adapt to global climate change. This concern is commonly buttressed by the finding that the investment treaty law system favors the position of the investor and that it is grounded on the chilling effects that expropriatory decisions might have on the host state's capability to promote the leveling of global warming, as explained in Part I.[27] The adoption of the police-power doctrine "regards, in consequence, a total swap in the viewpoint towards a perspective that takes into account the concerns related to the general interest, but which totally excludes those related to the investor's interest."[28] Undoubtedly, the adoption of the principle that legitimate regulation cannot be expropriatory—subject or not to exceptions—is appealing because it boosts measures that protect the climate by encouraging states to take measures against climate change but, on the other hand, is insufficient to protect climate interest when the host state's measures are against it (Sect. 6.1.1.1). Moreover, this

[27] Supra, Chap. 4, 4.1.2 and 4.2.2.

[28] De Nanteuil (2014), p. 26.

principle does not adapt to the text, purpose, and coherence of the system (Sect. 6.1.1.2).

6.1.1.1 The Insufficiency of the Police-Power Doctrine to Promote Climate Interest: Accusing the Police-Power Doctrine of Disincentivizing Investments in the Green Industry

Although the deployment of the police-power doctrine promotes the goals of the climate change international legal framework when the interest of the host state matches the global climate interest, it is insufficient for—and even contrary to—these goals when it is the interest of the foreign investor that matches the climate interest. This is so because under such a doctrine, the public interest inherent in the host state measures, in principle, outweighs the investor's individual interest. If one applies the police-power doctrine to RESS investment treaty disputes, regulatory measures that are grounded on any other public interest would still hinder this possibility. The promoters of the application of the police-power doctrine to climate change measures either dismiss this fact or undermine its relevance.

It could be argued against this conclusion that, through the limits of reasonableness and proportionality, the police-power (*Tecmed*-style) doctrine is appropriate to deal with cases where the investor's interest, rather than the host state's interest, matches the climate interest. However, this argument is still not satisfactory since the analysis of the proportionality or reasonableness of a measure might often yield a result wherein foreign investors in green energy industries are required to bear the costs of the public interest. Taking, once again, as an example the renewable energy support schemes in the case of Spain, it is highly probable that host governments' measures of cutting benefits for producers of photovoltaic energy are regarded as reasonable in light of and proportional to the objective of stabilizing the energy deficit of the national system in a context of a financial crisis. This reasonability and proportionality would even be achieved when the damage to the investor in green energies is catastrophic and transfers the costs of the public interest measure from the community to itself. In a context such as this, investments in renewable energy remain very risky, and the expropriation standard would provide poor protection to them.

Admittedly, most of the studies on the connection between the expropriation standard and climate change are oriented particularly toward solving the problem of how to define indirect expropriation in a manner that the regulatory power to pursue the goal of tackling global warming is maximized. Thus, it could be considered that cases where the regulation destroys investments in the green industry are outside the scope that these studies aim to cover, and consequently, concerns about the interpretation of the expropriatory standard in this regard might appear irrelevant to them. Yet if the purpose of these studies is to harmonize the climate change international legal framework with the investment treaty law system, the application of the police-power doctrine is an unsatisfactory solution that misses half of the picture. While the police-power doctrine in its purest (*Methanex*-inspired) version neglects any protection to foreign investors in green industries through the expropriation standard—because it posits that legitimate measures cannot be expropriatory—the police-power doctrine in its renewed (*Tecmed*-style) version neglects the protection to investors in green energies in any event where "the public interest to be pursued is very high."[29] Under any of these alternatives, the police-power doctrine discourages the promotion and protection of climate interest.

6.1.1.2 Inconsistency of the Police-Power Doctrine with the Text, the Purpose, and the Coherence of the Expropriation Standard

While, on the one hand, the (ordinary) expropriatory standard establishes as a principle that direct and indirect expropriations are forbidden, on the other, it permits expropriation when it is perfected for a public purpose, in good faith, and in a nondiscriminatory manner and is supported by a compensation:

> **Investments of investors of either Contracting Party shall not be nationalized, expropriated or subjected to measures having the effect of dispossession, direct or indirect, or having** effect equivalent to nationalization or expropriation (hereinafter referred to as "expropriation") in the territory of the other Contracting Party **except for a public purpose related to the internal needs of that Party on a non-discriminatory basis under due process of law not being contrary to any contractual obligation undertaken by a Contracting Party in favour of an investor and against prompt adequate and effective compensation**[30] (emphasis added).

At first glance, one can spot the contradiction between the text and the police-power doctrine. While the text allows expropriation for legitimate purposes (public purpose, good faith, due process, etc. [...]) against the payment of compensation (legal expropriation), the police-power doctrine rejects the compatibility between a legitimate regulation and the breach of the expropriation standard. This means whereas according to the text legitimate expropriatory measures require compensation in order to be legal, according to the police-power doctrine, the legitimate nature of the regulatory measures justify the absence of compensation.[31]

---

[29] Ranjan (2014), pp. 853, 970.

[30] Art. 6, Agreement Between the Government of The United Arab Emirates and the Government of the United Kingdom of Great Britain and Northern Ireland for the Promotion and Protection of Investments; See also, Art.6, Agreement between the Republic of Colombia and the Swiss Confederation on the Promotion and Reciprocal Protection of Investments.

[31] 'The argument made by the S.D. Myers tribunal is somehow contradictory [*police-power doctrine*]. According to it, the BIT would require that investments not be expropriated except for a public purpose and that there be compensation if such expropriation takes place and, at the same time, regulatory measures that may be tantamount to expropriation would not give rise to a claim for compensation if taken for a public purpose': *Azurix Corp. v. The Argentine Republic (2006)*, para. 311. 'On the one hand, it admits that the public interest is a condition for the legality of the expropriation, but this does not involve the nullification of the obligation to compensate the proprietor that has been disposed; on the other hand, the theory of the police powers posits that

Certainly, the police-power doctrine has no contradictions with the concept of direct expropriations—since the latter can be differentiated from the concept of regulatory measures that have an equivalent effect—but still opposes the concept of indirect expropriation. According to the text, measures with an equivalent effect are not in breach of the expropriation standard if they are legitimate measures and are supported by a compensation to the investor. *Contrario sensu*, measures that are illegitimate **or** are not supported by compensation constitute a breach of the expropriation standard. Nevertheless, the text does not require a verification of both requirements—illegitimacy **and** noncompensation—for the standard to be breached; i.e., if the measure is legitimate but is not perfected against the payment of compensation, the measure is still in breach of the expropriation standard. As noncompensation is inherent in the concept of indirect expropriation, the legitimacy of the regulatory measure loses its relevance; the host state is in breach when it does not pay compensation to the investor for equivalent measures, be they legitimate or illegitimate. The application of the police-power doctrine, then, seems inconsistent with the text because it defines the concept of indirect expropriation—the breach of the standard—by reference to the legitimacy of the measure.

Having demonstrated that the police-power doctrine is not grounded on the text, one may think that there is a justification for its application rooted in the system or in the purpose of the measure. The contention of the supporters of the police-power doctrine in this regard consists of the argument that in international law, it is generally accepted that legitimate regulatory measures do not entitle the victim to compensation, so one has to understand that indirect expropriation only covers measures whose character is illegitimate (arbitrary, discriminatory, disproportional, etc. [...]). This reasoning might be misleading. Firstly, it is not consistent with the special character of the provision; secondly, it does match the purpose of the expropriatory standard as it neglects the centrality of this provision as regards the promotion of foreign investment; and, thirdly, it is unsystematic because it erases the boundaries with the FET.

Although the claim that legitimate regulation does not require compensation under international law could be true, the application of this reasoning to the (ordinary) expropriation standard undermines the fact that the latter regards a more specific situation. Whereas the expropriation standard's hypothesis addresses the breach of the obligation to not expropriate without compensation, the international rule of noncompensation for legitimate regulation refers, in general, to the liability of the state for damages originated in regulatory measures. Therefore, although both rules share the characteristic of referring to the liability of the state for regulatory measures, the expropriatory standard adds an additional ingredient that makes it *lex specialis*: the destruction of the investment.

Connected to its special character, the purpose of the provision is not unnecessarily repeating the international law principle dictating that measures within the

certain measures escape from this obligation, precisely, because its public interest": Ranjan (2014), p. 869.

police power of the states are not expropriatory but rather providing a particular protection to the investor in the event that the state—through a formal process or indirectly—annihilates the value of or renders useless the investment. Actually, expropriation represents the most aggressive attack to an investment. The reason why bilateral investment treaties (BITs) contain an exclusive standard for expropriation that defines indirect expropriation by reference to the result is the need to provide investors with the certainty that they would be protected against the most acute risk that they face. Expropriation—relying again on the use of the metaphorical language—without compensation is the capital sin of the investment treaty law; thus, it is reasonable that the breach of the expropriation standard is, in principle, regarded as inexcusable in sovereignty arguments or public interest arguments.

The purpose of protecting the investor against legitimate measures that have an equivalent effect also makes sense from the perspective of the role that the expropriation standard plays in the whole investment treaty law system. It must be reiterated that the protection of the private interest of investors is an instrumental goal to achieve the broader objective of promoting the flow of investment between states—the functioning of a market economy is based on the premise that property will not be taken by the state without fair compensation, in the same way that the flow of transnational investment in a globalized economy is based on the premise that a foreign investment will not be expropriated by the state without fair compensation. In this vein, providing a high degree of protection to the investor against the risk of suffering a substantial loss that can obliterate its investment is in line with the rationale behind BITs.

Identifying indirect expropriation through the *effects* rather than the character of the regulatory measure also brings coherence to the interaction between the expropriation standard and FET. Indeed, it would not be acceptable to give the same meaning to both the concepts of indirect expropriation and FET. Yet claiming that the character of the measure is relevant to establish indirect expropriation eradicates the difference between the two concepts and, unavoidably, leads to the overlapping of the operation of the two standards. In fact, if one applies the police-power doctrine to expropriation, one would be centering the analysis of the breach on notions such as discrimination, due process, public interest, proportionality of measure, etc., which are precisely the same elements that one analyzes to define the breach of the FET standard. The existence of both standards in the same instrument would only make sense if they have a different functional and operational definition. While the operation of indirect expropriation is characterized by the verification of the measure's effect, which is equivalent to direct expropriation (substantiality and disproportionality of the impact), the breach of the FET standard is characterized by the verification of the illegitimacy of the measure (arbitrariness, discrimination, disproportionality of the measure, etc. [...]), and while the function of indirect expropriation is to compensate the destruction of the investment (absolute/strict liability/*obligation de resultat*), the function of FET is to compensate the damages that are derived from illegitimate measures (fault liability). The police-power doctrine ignores this distinction.

To sum up what has been demonstrated in this section, although the police-power doctrine appears to be an appealing interpretation of the concept of indirect expropriation concerning climate change measures, its deployment might end up disincentivizing investments in the green industries and outdoing the pliability of the expropriation standard (i). Given that it postulates that legitimate regulatory measures promoting climate change would be almost unbeatable against investment treaty claims, this theory certainly incentivizes climate action from the host state (ii). Yet—by flipping the coin—the systematic application of the police-power doctrine entails an imbalance that places the investor in green industries under a higher risk than the one it might be willing to assume (iii).

Moreover, the analysis of the police-power doctrine against the scope of the expropriation standard reveals that the definition of indirect expropriation based on the character of the measure both ignores the *lex specialis* principle and runs counter to its text, purpose, and coherence with the system (iv). The (ordinary) text of the expropriation standard would be simply contradictory if it, simultaneously, establishes the legality of expropriations supported by compensation—as it does—and the rejection of compensations for indirect expropriations deriving from legitimate measures—as the police-power doctrine appears to posit (v). Similarly, it would be contradictory to exclude the host state's legitimate measures from the scope of indirect expropriation since the standard aims to provide maximum protection to foreign investors against the maximum risk they face (expropriation) in order to promote the flow of this type of investment (vi). Finally, identifying indirect expropriation through the (il)legitimacy of the measure would be incoherent with the system since it cancels the functional and operational differences between indirect expropriation and the FET standard (vii).

Conversely, the sole-effects doctrine, by defining expropriation based on its effects, harmonizes with the (ordinary) text and purpose of the standard, as well as with the other BIT provisions and the investment treaty law system. Nevertheless, as this theory is recognized as clearly pro-investor and entails chilling effects on the host state's initiative to take action against climate change, I argue that the sole-effects doctrine needs to be tempered through the sacrificial-lamb doctrine in order to preserve a balanced relationship between the investor's interest and the host state's interest, which, incidentally, matches the promotion of climate interest. However, before jumping into mentioning the virtues of the sacrificial-lamb doctrine, the next section will deal with the sole-effects doctrine and its limitations in the context of climate change measures.

6.1.2 Adaptation of the Sole-Effects Doctrine to the Expropriation Standard in Expropriatory (Climate-Change-Related) Disputes and Its Limitations: Metalclad as Inspiration to Unravel Content Interpretation of the Expropriatory Standard

The sole-effects doctrine is a restatement of the expropriation standard's text. The textual interpretation that the equivalence of the effects is the defining feature of indirect expropriation also matches the purpose (incentivizing the transnational investment) and reinforces the independence of the standard from other standards that focus the breach on the (il)legitimacy of the measures, such as the FET. For these reasons—without undermining the problems that could arise from its pro-investor proclivity—the sole-effects doctrine has the virtue of adapting to the pliability of the standard without much struggle. Besides, its application promotes the goals of the climate change international legal framework by protecting investors in the green economy from the opportunistic behavior of the host state of shifting the burden of public interest to the investor. This reinforced protection of the investor is translated into a major flow of foreign green investments that favors climate interest.

The sole-effects doctrine contends that the expropriation standard is automatically breached in the event that the investment loses all of its value or becomes useless as a consequence of the host state's measures. The deprivation of the investment as such—which is the sole result—would be enough to establish that the measure is—indirectly—expropriatory independently of its public interest or legitimacy. Article 13 of the Energy Charter Treaty (ECT) exemplifies the definition that is usually included in BITs:

> Investments of investors of a Contracting Party in the area of any other Contracting Party shall not be nationalized, expropriated or subjected to a *measure or measures* **having effect equivalent** *to nationalization or expropriation* (hereinafter referred to as "expropriation")[32] [emphasis added].

Another example of the typical expropriation clause is Article VI of the BIT between Colombia and the United Kingdom, which provides:

> Investments of investors of a Contracting Party in the territory of the other Contracting Party shall not be the subject of nationalisation, *direct or indirect expropriation, or any measure* **having similar effects** *(hereinafter "expropriation")*[33] [emphasis added].

Metalclad[34] is often cited as the seminal decision deploying the sole-effects doctrine. This dispute concerns the impossibility for the investor to operate a hazardous waste deposit in *San Luis de Potosí* (Mexico) because of the lack of a

[32] Energy Charter Secretariat (2014).

[33] Bilateral Agreement for the Promotion and Protection of Investments between the Government of the United Kingdom of Great Britain and Northern Ireland and Republic of Colombia (2010).

[34] *Metalclad Corporation v. The United Mexican States (2000).*

construction permit and the restrictions stemming from an ecological decree declaring the area as a natural reserve for the protection of cacti. The Tribunal was categorical in establishing that

> expropriation under NAFTA includes not only open, deliberate and acknowledged takings of property, such as outright seizure or formal or obligatory transfer of title in favor of the host State, but also cover incidental interference with the use of property which has the **effect** *of depriving the owner, in whole or in significant part, of the use of reasonably-to-be-expected economic benefit of property* **even if not necessarily to the obvious benefit of the host State**[35] [emphasis added].

In the same vein as *Metalclad*, it is also common to find references to *Asurix* [36] and *Santa Elena* in the literature. [37] In the former decision, the adoption of the sole-effects doctrine was even more patent: "The issue is not so much whether the [expropriatory] measure concerned is legitimate and serves to a public purpose, but whether it is a measure that, being legitimate and serving a public purpose, should give rise to a compensation claim." [38] As regards the latter—*Santa Elena*—the Tribunal highlighted the irrelevance to expropriatory disputes of the host state's environmental legitimate purpose through the application of the sole-effect doctrine:

> *Expropriatory* **environmental measures**—no matter how laudable and beneficial to society as a whole—are, in this respect, **similar** *to any other expropriatory measures that a state may take* in order to implement its policies: where property is expropriated, even for environmental purposes, whether domestic or international, **the state's obligation to pay compensation remains**. [39]

The application of the sole-effects doctrine is common in the practice of tribunals, and some authors even claim that it is the majoritarian position in case law.[40] In *Oxy v. Ecuador*, for example, the Tribunal concluded that the termination of an oil participation contract was "tantamount to expropriation" under Article III.1 of the North American Free Trade Agreement (NAFTA) [41] based on the elaboration of *Metalclad*.[42] Similarly, in *Unglaube v. Costa Rica*, the Tribunal—citing *Santa Elena* as authority—considered that a decree declaring part of the Pacific coast of Costa Rica as a national park for the protection of leatherback turtles was expropriatory, [43] even if "there can be no question concerning the right of the government of Costa Rica to expropriate property for a bona fide public purpose (...)."[44]

---

[35] Ibid para. 103.

[36] *Azurix Corp. v. The Argentine Republic (2006).*

[37] *Compañia del Desarrollo de Santa Elena S.A. v. Republic of Costa Rica (2000).*

[38] *Azurix Corp. v. The Argentine Republic (2006),* para. 310.

[39] *Compañia del Desarrollo de Santa Elena S.A. v. Republic of Costa Rica (2000),* para. 72.

[40] De Nanteuil (2014), p. 21.

[41] *Occidental Petroleum Corporation and Occidental Exploration and Production Company v. The Republic of Ecuador (2012),* paras. 443, 456.

[42] Ibid. para. 455.

[43] *Marion Unglaube v. Republic of Costa Rica (2012),* para. 257.

[44] Ibid. para 205.

Since the relationship between climate change law and investment treaty law has been analyzed by the scholarship mainly from the perspective of the host state's interest in proffering climate-friendly regulation, the importance of providing investors in the green economy with the certainty that they would not be defrauded through the activation of the state's police powers has been overlooked. Consequently, the relevance of the sole-effects doctrine as a plausible interpretation in the context of climate change appears as unpopular. Boute, for example, although expressing concern that "Withdrawing support could lead to the destruction of low-carbon investors' rights and expectations (...)"[45] and, accordingly, could "place most financial costs of renewable energy and GHG emission reduction policies on the private sector,"[46] he is confident that tribunals' decisions would find that "the destruction of investors' rights of support would fail the proportionality test,"[47] rather than relying on the sole-effects doctrine to argue that the measure is expropriatory.

However, if one analyzes the relationship between climate change and investment treaty law from the investor's perspective, or from a broader perspective that takes into account the tension between investors in the green industry and the power of the state to regulate in favor or against the global climate interest, one might reach a conclusion that the deployment of the sole-effects doctrine is positive in terms of meeting the international climate change framework's goals. In this regard, in a previous work that focused on the modification of the renewable energy support schemes under the ECT, I argued that

> [f]rom our pragmatic approach of reinforced stability of RESs, it is appropriate to adopt the sole-effect doctrine rather than the police-power doctrine. The fact that the host government is aware that the good faith or the legitimate public interest of the measure would not serve as defense for excluding liability, would incentivize the authorities to implement the modification of the RESs in a subtle way that does not deprive the investment from its whole value and, therefore, would protect as much as possible the production of low carbon emissions from dramatic changes in the regulation.[48]

Even though this comment is specific to the ECT and partly stems from the understanding that this treaty gives preponderant value to climate interest, it summarizes the advantages of the minoritarian opinion that expropriatory (climate-change-related) disputes concerning restrictions to investments in the green industry should be resolved by reference to the effects of the measure rather than to the character of the measure. The alignment between the rationale of the investment treaty law and the climate change international legal framework's objectives in this respect is obvious: the higher is the level of protection to the investor through the expropriation standard, the higher is the level of predictability of the risk and

[45] Boute (2012), p. 613, 651.

[46] Ibid.

[47] Ibid.

[48] Restrepo (2017), pp. 101, 130.

profitability of the investment and, accordingly, the higher is the volume of investments flowing to the host state's green industry.

Yet the application of the sole-effects doctrine concerning climate change measures has limits. The first, and most obvious, is the case where the text of the BIT could be distinguished from the ordinary standard by referring to the character of the measure as a basis for assessing indirect expropriation, thus swinging back to the police-power doctrine (Sect. 6.1.2.1). The second limitation is derived from the same climate change context, which comes up against the undeniable fact that the sole-effects doctrine's disbalance in favor of the investor needs to be tempered in order to avoid chilling effects on the government's capability to act in favor of the climate change international legal framework's goals (Sect. 6.1.2.2).

6.1.2.1 Tailor-Made Clauses That Limit (or Exclude) the Application of the Sole-Effects Doctrine in Investment Treaty-Climate Change Disputes: Swinging Back to the Police-Power Doctrine

Apart from the *general exceptions* that are common to all standards, the suspicion of the investment treaty law system constraining the host states' capability to regulate public interest matters, such as climate change issues, has led to the—more or less—recent trend of drafting the expropriation standard by referring to the character of the measure as the defining feature of indirect expropriations. In this context, climate interest plays a significant role in legitimizing the severe impact of the host state's measures on the investor's interest. The drafting techniques in this regard are characterized either by adopting a *Methanex*-inspired approach or a *Tecmed*-inspired approach, with some references to legitimately backed expectations.

Methanex-Inspired Standard

> Non-discriminatory regulatory measures by a Party or measures and awards by judicial bodies of a Party that are designed and applied in pursuit of public policy to achieve legitimate public interest or public welfare objectives, such as the protection of public health, safety, and the environment, do not constitute expropriation of the type referred to in subparagraph 2(b) of this Article [i.e., indirect expropriation].[49]

This Annex to the investment agreement between the Association of Southeast Asian Nations and Hong Kong provides a fair illustration of the *Methanex*-inspired approach,[50] where regulatory measures within the police power of states are not

[49] Agreement on Investment among the Governments of the Hong Kong Special Administrative Region of the People's Republic of China and the Member States of the Association of Southeast Asian nations (2017) Annex 2(4).

[50] See. for example, Art. 6(2), Agreement Between the Government of the Republic of Rwanda and the Government of the Republic of Turkey Concerning Tile Reciprocal Promotion and Protection of Investments (2016).

subject to compensation. According to the text, the confirmation that the host state is not in breach simply requires verification that the measure is not discriminatory and that it is adopted for a legitimate purpose. Furthermore, the list of public interests that are categorized as welfare objectives is not restrictive. Thus, given the irrefutability and importance of global warming concerns, there is no doubt that climate change measures suit these legitimate interests. Under this clause, the space for breach of the expropriation standard is almost inexistent for climate change measures since it is limited to discriminatory or arbitrary measures.

Tecmed-Inspired Standard[51]

The distinctive feature of this standard is the language restricting indirect expropriation to exceptional or rare circumstances. In this event, unlike the *Methanex*-inspired standard, there is room for host states' climate change measures different from discriminatory and arbitrary measures to be regarded as expropriatory. Nevertheless, depending on the specific text, the threshold to reach the level of indirect expropriation could be more or less stringent, in the sense that it allows more or less discretion to the tribunal to establish whether the measure is within exceptional circumstances or rare ones. In this range, there are clauses that merely establish that

> [e]xcept in rare circumstances, non-discriminatory regulatory actions by a Party that are designed and applied to protect legitimate public welfare objectives, such as public health, safety, and the environment, do not constitute indirect expropriations.[52]

And, on the other hand, there are BITs that provide what should be understood as rare circumstances in greater detail:

> For greater certainty, except in the rare circumstance **when the impact of a measure or series of measures is so severe in light of its purpose that it appears manifestly excessive**, non-discriminatory measures of a Party that are designed and applied to protect legitimate public welfare objectives, such as health, safety and the environment, do not constitute indirect expropriations[53] [emphasis added].

This second formula suggests the appropriateness of concepts such as proportionality and/or reasonableness for assessing whether the host state's action breaches the expropriation standard. In the case of climate change measures, the tribunal or

[51] See. for example, Art.8(3)(c) Agreement on Investment under the Framework Agreement on Comprehensive Economic Cooperation between the Association of Southeast Asian Nations and the Republic of India (2014); See. for example, Annex I(3), Investment Promotion and Protection Agreement Between the Government of the Federal Republic of Nigeria and The Government of The Republic of Singapore (2016).

[52] Central America-Dominican Republic-United States Free Trade Agreement (2004).

[53] Comprehensive Economic and Trade Agreement (CETA) between Canada, on the one Part, and the European Union and its Member States, 3 Annex 8-A (2016). See also Mexico-EU FTA Annex on Expropriation 3 (2018); Article 9 (3)(c) Agreement Between the Belgium-Luxembourg Economic Union, on the one hand, and the Republic of Colombia, on the other hand, on The Reciprocal Promotion and Protection of Investments (2009).

interpreter would need to weigh the impact of the measure on the investor against the purpose of promoting or restricting climate interests and conclude whether the effect was too severe. This provision, despite restraining the tribunal's discretion by defining a methodology to find out to what extent the circumstances can be considered rare or exceptional, still perceives the potentiality of nondiscriminatory climate change measures as breaching the expropriatory standard.

Testing *the proportionality or reasonableness* of a measure as a basis for determining a breach of the standard under the second *Tecmed*-inspired model deserves the same criticism as that formulated below *à propos* the proportionality test in the FET[54] standard concerning ambiguity and lack of deference. Conversely, under the sacrificial-lamb theory, a balanced decision between the investor's interests and the host state's interest could be reached without invading the regulatory autonomy of the latter by weighing the proportionality of the impact on the investor against the burden that the public in general assumes for the same purpose rather than against the character of the measure, as alleged below.[55]

Investment-Backed Expectations

The concept of investment-backed expectations operates as a limitation to the application of the police-power doctrine in expropriatory treaty disputes. It refers to the investor's expectation—based on the government's representations—that the investment would not be expropriated. As was mentioned above, even *Methanex*—which is usually described as an example of the purest version of the police-power doctrine—establishes that legitimate measures are still expropriatory in the event that the host state dishonors specific commitments. Similarly, in *Tecmed*, the examination of investment-backed expectations was significant to determine the breach of the expropriation standard:[56]

> (...) it was widely known that the investor expected its investments in the landfill to last for a long term and that it took this into account to estimate the time and business required to recover such investment and obtain the expected return upon making its tender offer for the acquisition of the assets related to the Landfill. To evaluate if the actions attributable to the respondent -as well as the Resolution- violate the agreement, such expectations should be considered legitimate and should be evaluated in the light of the agreement and of the international law. Based on this and the fact that the Resolution was not proportionate to the infringements by *Tecmed,* the Tribunal found that the Resolution and its effects amounted to expropriation.[57]

In the same vein, most of the treaties that adopt the police-power doctrine condition its effectivity on the alignment of measures with the investment-backed

[54] Supra. Chap. 7, 7.3.1.

[55] See. Sect. 6.2.

[56] *Técnicas Medioambientales Tecmed, S.A. v. The United Mexican States* (2003).

[57] Ibid., para. 150.

expectations of the investor. Accordingly, BITs include different model expressions that involve different levels of protection to investment-backed expectations, depending on the particular choice of language. For instance, a common trend in the last generation of treaties is that

> [t]he determination of whether a measure (...) constitutes an indirect expropriation requires (...) a case-by-case, fact-based inquiry that considers, among other factors (...) the extent to which the government measure **interferes with the distinct and reasonable expectations of the investor arising out of the investment**.[58]

The expressions "distinct" and "reasonable" link the relevance of the expectations to the subjective convincement and the objective basis to believe that the investment would not be expropriated. In other words, the investment-backed expectation that the investment will not be expropriated needs to be buttressed, not only by the particular perspective of the investor but also by the reasonableness of the expectation from a contextual perspective.

By contrast, some BITs—such as the Trans-Pacific Partnership (TPP)—provide a detailed definition of the concept of investment-backed expectations concerning the expropriation standard:

> For greater certainty, whether an investor's investment-backed expectations are reasonable depends, to the extent relevant, on factors such as whether the government provided the investor with binding written assurances and the nature and extent of governmental regulation or the potential for government regulation in the relevant sector.[59]

Other BITs, instead of referring to investment-backed expectations, establish that public interest regulatory measures are expropriatory when they are contrary to the investor's legitimate expectations[60]—which is a more general concept commonly used in the interpretation of FET. In contrast, other models of the expropriation standard, instead of referring to legitimately backed expectations or legitimate expectations, determine that there is an indirect expropriation when regulatory measures are against a "specific commitment,"[61] a "particular commitment,"[62] or a "prior binding written commitment to the investor whether by contract, licence[sic]

---

[58] See. 2(C)Annex 8-A 14. Comprehensive Economic and Trade Agreement (CETA) between Canada, of the one Part, and The European Union, of the other part (2014); Annex I (3) (ii), Investment Promotion and Protection Agreement between The Government of the Federal Republic of Nigeria and The Government Of The Republic Of Singapore (2016); Mexico-EU FTA (2018) Annex on Expropriation 4 (IV); Agreement Between the Belgium-Luxembourg Economic Union, on the one hand, and the Republic of Colombia, on the other hand, on The Reciprocal Promotion and Protection of Investments (2009) article 9, (3)(b) 3(c).

[59] Trans-Pacific Partnership (2018) Annex 9B-Expropriation.

[60] Art. 6, Acuerdo Entre El Gobierno de La Republica de Colombia y El Gobierno de la Republica Francesa Sobre el Fomento y Protección Recíprocos De Inversiones (2014).

[61] Art 5, Agreement between the Government of the Republic of France and The Government of the Republic of Uganda on the Reciprocal Promotion and Protection of Investments (2003).

[62] Art. 5, Agreement between the Government of The Federal Republic of Ethiopia and The Government of The Republic of France for The Reciprocal Promotion and Protection of Investments (2003).

or other legal document."[63] All three terms further reduce the radio of interpretation, in the sense that they involve a more detailed description of the host state's representations capable of engaging its liability.

Overall, when the BIT is tailor-made by mirroring the police-power doctrine, it is still possible to consider that the host state is liable for indirect expropriation in investment treaty-climate change disputes when the measure frustrates legitimately backed expectations (i). In this regard, it is crucial to identify whether the investor had a legitimately backed expectation that measures whether promoting climate interest would not destroy its investment (i); in the same way, it is crucial to determine whether the investor in green industries had a legitimately backed expectation that regulatory measures would not empty the value of its investment (iii). Under some treaties, only written assurances in this respect will ensure that the state is liable for expropriation (iv); other treaties will leave to the tribunals more room for discretion, either to define which representations are sufficient to create legitimately backed expectations or to establish the representations that constitute particular or specific commitments capable of engaging state responsibility (v).

In a climate change context, when the tribunal has the discretion to establish which representations create legitimately backed expectations or specific commitments, investors seem not to have many chances to succeed in their claim that the regulatory measure constitutes indirect expropriation (vi). In the absence of a clearly binding statement from the host state, it might be very difficult to argue that the investor in green energies had the legitimate expectation that its investment would not be expropriated because of changes in the regulatory framework. The uncertainty in terms of the outcome of the operation of green technologies from a macroeconomic perspective as well as in terms of market efficiency suggests that government regulation might require a wide range of elasticity to adjust these technologies to the interest of their community.

Similarly, investors in other industries could envisage that the adaptation of the host state to the context of climate change might require the enactment of a regulation against their interests. Therefore, these investors in other industries would need to rely on truly clear representations of the host state to succeed in the argument that the investments' loss of value is unexpected (vii).

6.1.2.2 The Climate Change Context as a Limitation to the Sole-Effects Doctrine: Revealing the Need for Rebalancing the Investor's Supremacy over the Host State's Interest and Reassessing the Investor's Commercial Risk

To this point, we have demonstrated that the classic understanding of the sole-effects doctrine suits the (ordinary) standard and the investment treaty law system in

[63] ASEAN-India BIT (2009) Art. 8.3 (b). See. also. Annex II 3(b) Hong-Kong-ASEAN BIT (2017) Annex II 3(b).

general. However, the gravitation of the climate change international legal framework as part of the context of any investment treaty (climate-change-related) dispute calls for an interpretation of the expropriation standard that harmonizes with the objectives of preventing, mitigating, and adapting to the effects of global warming. The climate change context compels us to rethink the application of the *sole*-effects doctrine in terms of rebalancing the investor-host state relationship and reassessing the commercial risk that the investor should assume in disputes related to global warming.

Although the police-power doctrine has the virtue of unearthing the need for rebalancing the investor-state relationship, it does not mold itself to the pliability of the (ordinary) expropriation standard and it is insufficient to provide a satisfactory answer to climate change concerns in investment treaty law. However, a classic understanding of the sole-effects doctrine is also unsatisfactory for this purpose since it clearly leans in favor of the investor.[64] In fact, it is reiterated throughout this chapter that, although the sole-effects doctrine harmonizes with the objectives of the climate change international legal framework by establishing the verification of substantiality as the only requirement for the indirect expropriation of green investments, the sole-effect doctrine disincentivizes the host states from taking climate action, given the investor's significant chances to succeed by only demonstrating the destruction of the investment.

The chilling effects on the host state's capability to regulate in favor of climate interest suggest that the application of the sole-effects doctrine encounters limitations on incentivizing the leveling of global warming. In other words, an interpretation of the expropriation standard in context with the climate change international legal framework cannot ignore the need for rebalancing the tension between the host state's interest and the investor's interest.

Additionally, the application of *sole-effects doctrine ignores* that the foreign investor assumes—or should assume—a commercial risk in a market that is sensitive to climate change regulations.

Certainly, an (expert) investor should be aware that the climate change context involves a level of uncertainty that is translated into inconsistent and fluctuant regulations. Against this observation, the classic understanding of the sole-effects doctrine does not provide any weight on the due diligence of the investor nor on the existence of certain risks that are natural to its business. Indeed, by recognizing that the investor expects to be compensated on any occasion where the loss is substantial, the sole-effects doctrine transfers the whole commercial risk of destruction of the investment to the host state, disregarding the fact that instability in climate change

[64] "The *Metalclad* award set out an approach that was completely closed to integrative arguments by interpreting the expression 'tantamount to expropriation' in NAFTA Art. 1110 in very broad terms. The Tribunal reference to 'incidental interference', 'significant part' and 'reasonably-to-be-expected benefit led to an excessive dilatation of the category of indirect expropriation, according to a model of rigid teleological interpretation that was completely dependent on the treaty aim of protecting investor interests, and which ultimately resulted in a self-contained approach to the matter at hand": Di Benedetto (2013), p. 121.

regulation is a normal part of the business. An interpretation of the expropriation standard in context with the climate change international legal framework suggests that the sole-effects doctrine should recognize that the investor has the duty to absorb the loss of the investment in proportion to the commercial risk it takes when investing.

The need for rebalancing the host state-investor relationship and identifying the commercial risk that the investor should absorb in a climate change context might find its legal translation in the sacrificial-lamb doctrine. This theory is a renewed version of the sole-effects doctrine that defines indirect expropriation not only through the substantiality but also through the (dis)proportionality of the impact of loss on the investor by reference to the community at large. The operation of the sacrificial lamb and its advantages in a climate change context is dealt with in the next section.

6.2 Orthodox Approach to the (Ordinary) Expropriation Standard in a Climate Change Context: Tempering the Sole-Effects Doctrine Through the Sacrificial-Lamb Doctrine

The sacrificial-lamb theory evokes the idea of a foreign investor victimized for the common good. It does not contradict the tenet that investors should bear the risks that are *normal* within their commercial activity, nor does it ignore their responsibility concerning due diligence. While acknowledging that limitations to investments are normal under a dynamic regulatory environment and a changing reality, the sacrificial-lamb theory recognizes that abnormal limitations that destroy investment constitute indirect expropriatory measures. This (ab)normality is defined in terms of the (blatant) (dis)proportionality between the negative externalities of the measure that the investor internalizes and the burden that the public at large experiences for the public purpose that is pursued through the measure. If the result of the assessment shows that the damage to the investor was especially detrimental to it, then the measure constitutes an indirect expropriation. What lies beyond the concept is the unfairness of shifting the costs of public interest from the community at large to the foreign investor without proper compensation.[65] In brief—back to the metaphor—the investor is not supposed to be the lamb that is sacrificed for the

[65] "The duty to pay a compensation because of finding of '"regulatory taking"' is here part of the normal function of the constitutional guarantee: to protect a minority's rights against the majority, to make majority pause and consider the cost of its action —rather than shift the cost to the minority. This applies particularly so where there is often no well-reflected and supported action by a true majority, but rather capturing of the government machinery by well organized special interest groups. The constitutional (or treaty-based) duty to pay compensation means that the costs of such capture of the machinery of government should be made transparent": Waelde and Kolo (2001), p. 846.

wealth of the community; thus, the expropriatory standard is breached when this sacrifice takes place without compensation.

The sacrificial-lamb doctrine is not a brand-new discovery in investment treaty law. Actually, it is an adaptation to the investment treaty law of the French theory of *préjudice anormal et special* (abnormal and special prejudice)[66]—which establishes that no one has the duty to suffer an abnormal burden for a public purpose that is meant to be assumed by the whole community—or it can be grasped as an adjustment of the *Sonderopfer* (special sacrifice) doctrine, which the German case law has endorsed in the past as a basis for expropriation.[67] Comparative law studies also show that this theory has been deployed by the US Supreme Court and implemented by the European Court of Human Rights.[68] Its translation to the domain of expropriation in investment treaty law was (bashfully) suggested by Dolzer in a paper that dates back to 1987[69] and openly defended by Wälde in 2001, the latter concerning environmental measures.[70]

This approach—as evident—steps away from any theory that bases the determination of indirect expropriation on the character of the measure—such as the police-power doctrine. It is a renewed version of the *sole-effects doctrine* since the definition of indirect expropriation is still centered on the equivalent effects on the investor.

Indeed, the sacrificial-lamb doctrine entails a satisfactory orthodox interpretation of the expropriation standard in investment treaty (climate-change-related) disputes. Its operation does not only match the trend of tribunals' decisions concerning expropriatory environmental matters (Sect. 6.2.1); it also promotes a balanced understanding of the indirect expropriation standard that is influenced by the gravitation of the climate change international legal framework (Sect. 6.2.2) without distorting the pliability of the expropriation standard.

6.2.1 (Dis)proportionality of the Impact on the Investor as a Common Basis for Expropriation: Demonstrating the Conformity of the Sacrificial-Lamb Doctrine to the Trend of Expropriatory (Environmental) Awards

Not all climate change measures are the same. This may sound obvious, but it serves to simplify the understanding of climate measures through their categorization in order to facilitate the content interpretation of the expropriation standard in a climate

[66] De Nanteuil (2014), pp. 326–329.

[67] Perkams (2010), pp. 131–133.

[68] Ibid. pp. 112–121.

[69] Dolzer (1986), p. 61.

[70] Waelde and Kolo (2001), pp. 845–846.

change context. Although this can be done through different criteria,[71] I consider that Viñuales' differentiation between regulatory measures and targeted measures[72] is useful for making the point I would like to stress in this section, that is, that the (dis)proportionality of the climate measure's effect on the investor in relation to the effects on other stakeholders from society at large defines whether an indirect expropriation occurred.

Certainly, it is possible to distinguish government measures that primarily address the public at large (regulatory measures)—such as measures implementing a Kyoto Protocol style emission trading mechanism—from those that address a particular investor or group of investors (targeted measures)—such as the cancelation of a permit to operate a coal plant. Concerning indirect expropriation disputes, Viñuales observes that while targeted measures annihilating the value of investments are prone to be labeled as a breach, regulatory measures are not.[73] He adds that even if it is very difficult to deduct legal consequences from this comparison, in general terms, the most likely reason for justifying the trend is that the regulatory measures should not, as a general rule, violate the standard.[74]

It is doubtful whether Viñuales' guessing that "the trend is that the regulatory measures should not, as a general rule, violate the standard" is correct. Indeed, he himself seems unsure.[75] First, this explanation appears to be more of a repetition of the trend than a justification of the trend, and, second, it conflicts with my claim that the text of the expropriation standard, by referring to the equivalent effects of the measure (sole-effects doctrine), is the key to determining whether a measure is expropriatory rather than a direct application of the general rules of international law.[76] However, the pattern of case law that he reveals—where environmental targeted measures are generally expropriatory by contrast to environmental regulatory measures—urges an explanation of the concept of indirect expropriation that, simultaneously, fits the pattern and accommodates itself to the fact that the pliability of the standard is determined by the effects of the measure.

In fact, the same Viñuales—when discussing possible additional reasons for justifying the pattern—proposes that the reason for explaining that environmental targeted measures are more prone to degenerating into indirect expropriation could be the "similar effects of direct takings and targeted measures (by contrast with those regulatory measures)."[77] I value as principal the reasons that explain the existence of the pattern that he regards as secondary. Grounding this trend among similar effects

---

[71] For example, is possible to distinguish between "disguised expropriations (to borrow the term from the ELSI case) and those substantial interferences with property rights that take place in breach of due process": Paparinskis (2011), p. 305.

[72] Viñuales (2015), pp. 293–315.

[73] Ibid. pp. 297–298.

[74] Ibid. pp. 298–299.

[75] Ibid. p. 298, footnote 16.

[76] See. *supra.* Sect. 6.1.2.

[77] Viñuales (2015), p. 298.

is certainly congruent with the definition of indirect expropriation by reference to its effects. Yet it would need to be complemented with a description of the content of these effects in order to provide a concrete basis for explaining why environmental regulatory (and climate) measures are rarely expropriations but are, at the same time, capable of constituting expropriation in the cases they do.

The operation of the sacrificial-lamb doctrine is fairly clear when it concerns direct expropriations. An everyday example is direct land expropriations for the development of energy projects where the property is taken for installing grids, turbines, or solar panels, among others. The sacrifice of the investor for the community in the event of direct expropriations is evident, and no one would argue against the duty to compensate investors.

Targeted measures that deprive an investment of its value are not in essence different from direct expropriations. Both measures expressly address a particular investor or a particular group of investors, annihilate the value of the investment, and shift the costs of pursuing a public interest to the investor. In *Cooper v. Ecuador*, for example, the termination of the *Junin* concession for the extraction of copper was grounded on the execution of a "2009 Mining Law," which was addressed to "certain defined mining concessions" and which established that those investors who did not submit the EIS (Environmental Impact Study) to the regulators would face termination of the concession. [78] If one compares direct expropriations in energy projects with this issue, one should come to the conclusion that the only difference stems from the fact that in the latter, the formal resolution of expropriation is absent.[79] In the end, in both cases, investors are sacrificed for a community interest: climate[80] (among other values).

Although the deprivation that is a product of targeted measures leads to compensation on most occasions, I argue that in exceptional cases—such as systemic financial crises or energy crises—the targeted measure may not be viewed as disproportional because it pertains to a general package of government policies that affect different segments of society equally. This reasoning suits Viñuales' observation that targeted measures causing a substantial loss almost always breach the expropriation standard, but he also acknowledges that the breach of the expropriation standard is defined through the result rather than by the character of the host state's measure.

Concerning environmental regulatory measures—keeping in line with Viñuales' findings—the application of the sacrificial-lamb theory also explains why these measures might turn into expropriation only under exceptional circumstances. It is highly unlikely that climate change measures whose recipient is undetermined

[78] *Copper Mesa Mining Corporation v. Republic of Ecuador* (2016), p. 19.

[79] Sharing the same opinion: Paparinskis (2011), p. 306.

[80] The effects that were expected from the *Junin* concession covered: "massive deforestation will give rise to an increase in dry conditions (known as "Desertification"), effects on the local climate and alteration of the vegetation around the boundary of the split area [...]": *Copper Mesa Mining Corporation v. Republic of Ecuador* (2016), Part 4, page 9 of 101, section 4.25.

destroy the value of a particular investor in a particular way by comparison to the public at large, precisely because the measure addresses the whole community. The costs of adopting a renewable electricity support scheme (RESS), for example, are generally spread among diverse agents and different segments of society in a more or less proportional fashion: consumers paying higher prices for energies, polluting industries affording to adopt new technologies, power generators being obliged to produce at higher costs, transmission companies being required to extend grids, retailers decreasing profit margin, etc.

The occasions on which regulatory measures cause severe impacts that appear disproportionate are connected to the possibility of identifying a particular investor or a group of investors that internalize the biggest share of the costs for satisfying public purpose. In the recent cases of modification of renewable support schemes in Spain, for instance, it could be argued that the regulatory measures that established the calculation of the regulated tariff according to model costs instead of real production costs resulted in a substantial and disproportional loss to the particular group of photovoltaic energy producers, and the loss could, consequently, amount to indirect expropriation.[81]

While direct expropriations and targeted measures resulting in substantial losses are usually disproportional as there is quite a clear direct causal link between the measure and the specialty of the impact; regulatory measures resulting in substantial losses are usually not. This is because the causal link between the measure and the specialty of the impact is difficult to establish when the measure aims to affect the public at large on a nondiscriminatory basis. For instance, the disproportionality of the loss in the example of the cancelation of renewable energy support schemes presented above might be difficult to establish since the investor would be required to prove that the substantiality of the damage puts it in a particular detrimental situation in comparison with other investors in the photovoltaic industry and with other agents that were subject to the same package of measures. It is highly probable that, concerning regulatory measures, the substantial loss that the investor suffered is not so different from what other members of society experienced for the same purpose, or it may be that the substantiality of the loss is the result, totally or in part, of the investors' incompetence. This last hypothesis would be within the domain of the analysis of contributory negligence.

Against this background, the operation of the sacrificial-lamb doctrine concerning climate change mitigation measures matches the observation of Viñuales regarding tribunals' decisions on investment treaty (environmental-related) measures: any climate measure whose effect is substantial and disproportionate is an expropriation, but regulatory climate change measures are less likely to breach the standard than targeted climate change measures.

[81] Supra. Part I, Chap. 4, 4.1.1.2.

6.2.2 Consistency of the Sacrificial-Lamb Doctrine with the Climate Change International Legal Framework's Context: Bridging the Gaps of the Sole-Effects Doctrine Through the Concept of (Dis)proportionality of the Impact on the Investor

The adoption of the sacrificial-lamb theory concerning climate change measures seems to be in harmony with the (ordinary) text, purpose, and context of the BIT expropriation standard and, in general, the investment treaty law system. Thus, it is adjusted to the principles of interpretation of Article 31 of the Vienna Convention on the Law of Treaties (VCLT). This harmony is grounded, to a great extent, on the characterization of the sacrificial-lamb theory as a version of the sole-effects doctrine. As explained in detail above,[82] the identification of indirect expropriation through its effects rather than through the character of the measure adapts to the pliability of the (ordinary) expropriation standard. However, the sacrificial-lamb doctrine pushes the sole-effect doctrine one step further, tempering the latter's inclination to favor the investor's interest. Accordingly, it defines indirect expropriation in a manner that embraces the need for rebalancing the investor-state relationship as well as acknowledging the investor's duty to assume the commercial risk. Moreover, an interpretation in these terms harmonizes with the international climate change legal framework's context in which the expropriation standard is embedded when dealing with investment treaty (climate-change-related) disputes.

Firstly, the sacrificial-lamb doctrine promotes the flow of foreign investments in the green economy by providing the highest possible degree of protection to the investor against the highest risk that it faces (expropriation). An investor's greatest fear is for the investment to lose all of its value because of the government's opportunistic action. By using the term opportunistic in this context, I intend not to formulate a negative moral judgment on the government's action but to express that the investor is always subject to the risk of the host state finding the limitation on the investor's position as an opportunity to achieve legitimate and—in the worst-case scenario—illegitimate purposes. If the objective of the BIT is to protect investments against opportunistic behavior, defining the breach of the expropriation standard based on the impact of the measures—as the sacrificial-lamb doctrine does—appears to be more suited to this goal than the thesis that promotes the assessment of expropriation through the character of the measure, such as the police-power doctrine.

Secondly, the sacrificial-lamb doctrine preserves—as far as possible—the host state's right to regulate in favor of climate interest by restricting indirect expropriation only to regulatory measures that cause the investor to make an abnormal sacrifice. As a result, the regulatory chill effects that stem from the expropriation standard are considerably reduced. If the host state is aware that only substantial loss

[82] See. *Supra.* Sect. 6.1.2.

that is disproportionate in relation to the community at large breaches the standard, it is incentivized to take climate change measures. Indeed, the host state's room for maneuver in terms of taking climate change measures is significantly expanded in comparison to the space it has under the sole-effects doctrine. Therefore, the application of the sacrificial-lamb doctrine overcomes the deficiencies of the sole-effects doctrine concerning the lack of promotion of the host state's climate action.

Thirdly, the sacrificial-lamb doctrine compels the investor to absorb the commercial risk that is normal in a climate change context. While the sole-effects doctrine shifts all the commercial risk to the host state, the sacrificial-lamb doctrine recognizes that the investor should bear the substantial loss as long as it is not a disproportional burden in comparison to the burden that the community at large experiences for the public interest.

Transferring the commercial risk to the investor has two crucial consequences. On the one hand, investors—either in green energies or in other industries—have to put all their efforts into protecting themselves from the risk of a constant fluctuation of climate change regulation as a natural part of the business. Otherwise, they can attribute the loss as a matter of contributory negligence. The other crucial consequence is that the investor's expectation to be compensated is limited to the case of (dis)proportionality of the impact, unless there are representations of the host state that broaden these expectations. In fact, if representations of the host state induce the investor to reasonably expect the commercial risk to be shifted to the former, a verification of substantiality is enough to breach the expropriation standard even if the impact on the investor is not abnormal.[83] The interplay between due diligence, expectations, and representations will determine on a case-by-case basis whether (dis)proportionality is an element of indirect expropriation.

Fourthly, the concept of (dis)proportionality of the measure's impact on the investor in expropriatory climate-change-related disputes acknowledges that climate action is everyone's responsibility—not just of states. In this vein, the Paris Agreement establishes that the "[e]arth's climate and its adverse effects are a common concern for humankind"[84] and recognizes "the importance of the engagements of all levels of government and various actors (...) in addressing climate change."[85] The sacrificial-lamb theory, in contrast to the sole-effects doctrine, requires the investor to assume its responsibility of bearing part of the costs of climate action by establishing that proportional losses, even if substantial, would not lead to compensation.

Fifthly, the sacrificial-lamb doctrine is also in line with the premise of the climate change international legal framework that responses to climate change should be coordinated with social and economic development. This is particularly important in cases where measures that affect investors in green energies are the response to

---

[83] For a deeper analysis of expectations concerning the expropriation standard, See. De Luca (2015), pp. 58–75.

[84] United Nations Paris Agreement (2015).

[85] Ibid.

general economic crises in the host state. Certainly, if the host state complies with the requirement of proportionality—by spreading the negative effects of pursuing public interest over community interest at large—the expectations of investors in the green industries would not be frustrated unless there are representations that indicate that, even in the case of crisis, the loss of value entitles them to compensation. Under these conditions, climate change interest is not seen as a barrier to economic or social development.

In brief, the sacrificial-lamb doctrine promotes the flow of foreign investments in the green economy (i), preserves the host state's right to regulate in favor of climate interest as much as possible (ii), compels the investor to absorb the commercial risk that is normal in a climate change context (iii), and acknowledges that climate action is also the responsibility of investors (iv) and that responses to climate change should be coordinated with social and economic development (v). All these characteristics suggest that the application of the sacrificial-lamb doctrine to investment treaty-climate change disputes bridges the gaps that the sole-effects doctrine may present, in terms of coping with an interpretation of indirect expropriation that is adjusted toward a climate change context, and conforms to the pliability of the expropriation standard by defining breach by virtue of the effects rather than by means of the character of the measure (vi).

6.3 The Limits of the *Orthodox Approach* to Promote the Global Climate Interest: Concluding Remarks on the Interpretation of the Expropriation Standard in Investment Treaty-Climate Change Disputes

As I mentioned at the beginning of the chapter, studies on expropriatory climate-change-related disputes tend to find a solution to the tension between the expropriatory standard and the climate change international legal framework in the police-power doctrine. Against this trend, I argued that the police-power doctrine is unfit to provide a balanced answer that harmonizes with the climate change interest because it exceeds the pliability of the expropriation standard and only promotes climate interest when it matches the host state's interests (i). Instead, I proposed to deploy the sacrificial-lamb doctrine, which is a tempered version of the sole-effects doctrine that determines the breach by reference to the abnormality of the investor's loss in comparison to the burden that the community bears as a result of the measure (ii). This approach leads, simultaneously, to a balance between the parties' interests and the promotion of the objectives of the climate change international legal framework when climate interest matches the investor's interest and the host state's interest (iii).

Concerning the incompatibility of the police-power doctrine with the expropriation standard, I found that the former runs counter to the text, purpose, and systemic coherence of the latter (iv). It goes against the (ordinary) text because while the text requires compensation even in cases where expropriation is for a legitimate purpose, the police-power doctrine rejects the duty to pay compensation when the host state's measure is taken for a legitimate public objective (v); it is against the purpose because its application leaves the investor defenseless against the most aggressive action from the host state and, thus, discourages foreign investments (vi); and it runs counter to its systemic coherence since it erases any boundaries between the FET standard and the expropriatory standard by focusing on the character of the measure (vii).

Regarding the inability to harmonize with the global climate interest, it was claimed that the police-power doctrine results in undermining the role of the investor in green industries (viii). Conceiving that regulatory action does not lead to a breach of the expropriatory standard (*Methanex* version) or that public interest legitimizes expropriating the investor without granting compensation (*Tecmed* version) locates green investors in a no-win situation that persuades them to channel their investment somewhere else (ix), and, thus, it clashes with the promotion of the objectives of the climate change international legal framework.

Given these drawbacks of the police-power doctrine, I recommended the adoption of the *sacrificial-lamb doctrine*. From a systemic perspective, on the one hand, it has the virtue of matching the (ordinary) text of the standard by associating the determination of the expropriation with the effects of the regulatory measure—as in the sole-effect doctrine—(xi), and, on the other hand, it recognizes considerable regulatory flexibility to the host state by limiting the breach to cases of abnormality of the loss (xii).

This systemic coherence leads to a harmonized content interpretation of the expropriation standard with the climate change international framework. First, under the sacrificial-lamb theory, the host state would be motivated to take climate action (xiii) since its power to regulate is only limited by the extraordinary situation where the substantial impact of the measure is abnormal to the investor in comparison with the community. Second, the investor in green industries would be persuaded to invest (xiv) as it knows that any substantial damage will be compensated, unless it is evident that the costs of the public interest pursued with the regulatory measure are shared fairly with the community. Third, this approach acknowledges the caveat that climate action is everyone's responsibility (xv) as it involves the investor bearing the commercial risks that are natural in a climate change context (xvi) and entails the community's duty to take on a share of the costs of climate change policies (xviii).

Although the systemic coherence of the sacrificial-lamb doctrine—and the fact that its operation provides a feasible explanation to the unlikeliness of environmental-investment awards to order compensation for regulatory-expropriations by contrast to targeted-expropriations (xix)—it is limited in terms of providing an interpretation that fosters the climate interest in all cases. The most evident limitations are those arising from the adoption in the BIT of tailor-made clauses that are inspired by the police-power doctrine (xx). Nevertheless, these clauses often offer protection to the investor in the green industry through the recognition of its legitimately backed expectations (xxi).

Other limitations come from the nature of the orthodox approach. Indeed, the interpretation of the expropriation standard through the sacrificial-lamb doctrine may still conflict with the protection or promotion of climate interest on exceptional occasions. The power of the circumstances that surround green policies may compel the host state to trigger a substantial loss for the investor that appears abnormal, and it will still have the duty to compensate. For instance, the host state may need to expropriate particular investors whose operation is located in areas—such as tropical forests—that are strategic for the protection of the climate interest. Similarly, the host state may cause substantial damage to investors in the green industry in order to pursue other policy objectives without causing disproportional damage and, then, not being required to compensate, i.e., when carrying out a plan that involves affecting different industries to trigger an economic crisis.

In these cases, the balance between the investor and the host state, which is embedded in the expropriation standard and interpreted under the authority of Article 31 VCLT, is not enough to promote and protect the climate interest. In the end, just as in physics the gravitational power shapes an object without changing its internal structure, the gravitational power of the international climate change framework shapes the investment treaty law system without modifying the balance between the parties that is entrenched in the text. As a result, if the interpreter wants to go beyond these limits in order to protect the climate interest, he is compelled to find a harmonized answer through the direct application of other sources of international law. In other words, the interpreter needs to shift perspectives from an orthodox approach to a heterodox one that promotes the protection of the climate change interest through the application of principles that are different from the doctrine of *pacta sunt servanda* (public dimension), this being the topic for PART III.

References

Baetens F (2010) Foreign Investment Law and Climate Change: Legal Conflicts Arising from Implementing the Kyoto Protocol Through Private Investment. The Center for International Sustainable Development Law & The International Development Law Organization, Working Paper Series https://papers.ssrn.com/sol3/papers.cfm?abstract_id=2117950. Accessed 30 July 2022

Boute A (2012) Combating climate change through investment arbitration. Fordham Int Law J 35: 613, 651 https://ir.lawnet.fordham.edu/ilj/vol35/iss3/7. Accessed 30 July 2022

Condon B (2015) Climate change and international investment agreements. Chinese J Int Law 8, 2 (14), p 337 https://doi.org/10.1093/chinesejil/jmv023. Accessed 30 July 2022

De Luca A (2015) Renewable energy in the EU, the energy charter treaty, and Italy's withdrawal therefrom. Bocconi legal studies research paper no. 2657395. https://ssrn.com/abstract=2657395. Accessed 28 Jul 2022

De Nanteuil A (2014) L'Expropriation Indirecte en Droit International de l'Investissement. Editions A. Pedone, p. 21

Di Benedetto S (2013) International investment law and the environment. Edward Elgar, Cheltenham

Dolzer R (1986) Indirect expropriation of Alien Property. ICSID-Rev 1:41, 61

Erkan M (2011) International energy investment law: stability through contractual clause. Wolters Kluwer, pp 81–82

Isakoff P (2013) Defining the scope of indirect expropriation for international investments. Global Bus Law Rev 3(2)

Marshal F (2010) Climate Change and International Investment Agreements: Obstacles or opportunities? IISD Trade, Investment and Climate Change Series 48 http://www.fao.org/fileadmin/user_upload/rome2007/docs/Climate%20Change%20and%20International%20Investment%20Agreements.pdf. Accessed 30 July 2022

McLachlan C et al (2007) International investment arbitration: substantive principles, OUP

Mitchell A, Henckels C (2013) Variations on a theme: comparing the concept of "Necessity" in international investment law and WTO law. Chicago J Int Law 1(14):93, 122–124. https://chicagounbound.uchicago.edu/cjil/vol14/iss1/5. Accessed 30 July 2022

Paparinskis M (2011) Regulatory expropriation and sustainable development. In: Cordonier M et al (eds) Sustainable development in world investment law. Wolters Kluwer/IDLO, p 305

Perkams M (2010) International investment law and comparative public law —an introduction. In: Schill S (ed) International investment law and comparative public law. OUP

Ranjan P (2014) Using the public law concept of proportionality to balance investment protection with regulation in international investment law: a critical appraisal. Cambridge J Int Comp Law 3: 853, 970

Restrepo T (2017) Modification of renewable energy support schemes under the energy charter treaty: Eiser and Charanne in the context of climate change. Goettingen J Int Law 1(8)

Schill (2007) Do investment treaties chill unilateral state regulation to mitigate climate change? J Int Arbitr 24:472

Vadi V (2015) Beyond known worlds: climate change governance by arbitral tribunals. Vand J Transnatl Law 48(1285):1328

Viñuales J (2015) Foreign investment and the environment in international law. CUP, p 23

Waelde T, Kolo A (2001) Environmental regulation, investment protection and regulatory taking in international law. Int Comp Law Q 50(4):812. http://www.jstor.org/stable/761516. Accessed 30 July 2022

Decisions

ICSID (2000a) Compañia del Desarrollo de Santa Elena S.A. v. Republic of Costa Rica, Case No. ARB/96/1 https://www.italaw.com/cases/3413. Accessed 30 July 2022

ICSID (2000b) *Metalclad Corporation v. The United Mexican States,* Case No. ARB(AF)/97/1 http://www.italaw.com/cases/671#sthash.2TPwidW3.dpuf. Accessed 30 July 2022

ICSID (2003) *Técnicas Medioambientales Tecmed, S.A. v. The United Mexican States*, Case No. ARB (AF)/00/2) http://www.italaw.com/cases/1087#sthash.9Pg82zdi.dpuf. Accessed 30 July 2022

ICSID (2006) Azurix Corp. v. The Argentine Republic, Case No. ARB/01/12 *http://www.italaw.com/cases/118#sthash.pz1dK8UU.dpuf.* Accessed 30 July 2022

ICSID (2012) *Marion Unglaube v. Republic of Costa Rica*, Case No. ARB/08/1 https://www.italaw.com/cases/1134. Accessed 30 July 2022

ICSID (2015) Occidental Petroleum Corporation and Occidental Exploration and Production Company v. The Republic of Ecuador, Case No. ARB/06/11 https://www.italaw.com/cases/767. Accessed 30 July 2022

SCC (2013) *Anatolie Stati, Gabriel Stati, Ascom Group SA and Terra Raf Trans Traiding Ltd v. Kazakhstan*, Case No. v. 116/2010

UNCITRAL (2000) Pope & Talbot Inc. v. The Government of Canada, Interim Award http://www.italaw.com/cases/863#sthash.4lXLuYCz.dpuf. Accessed 30 July 2022

UNCITRAL (2002a) *Methanex Corporation v. United States of America* http://www.italaw.com/cases/683#sthash.JisiDQv5.dpuf. Accessed 30 July 2022

UNCITRAL (2002b) S.D. Myers, Inc. v. Government of Canada http://www.italaw.com/cases/969S. Accessed 30 July 2022

UNCITRAL (2006) *Saluka Investments B.V. v. The Czech Republic* http://www.italaw.com/cases/documents/963#sthash.gEUKlO8O.dpuf. Accessed 30 July 2022

UNCITRAL (2011) El Paso Energy International Company v. The Argentine Republic https://www.italaw.com/sites/default/files/case-documents/ita0270pdf. Accessed 30 July 2022

UNCITRAL (2016) *Copper Mesa Mining Corporation v. Republic of Ecuador, Case No. 2012-2* https://www.italaw.com/sites/default/files/case-documents/italaw7443.pdf. Accessed 30 July 2022

International Legal Instruments

Acuerdo Entre El Gobierno de La Republica de Colombia y El Gobierno de la Republica Francesa Sobre el Fomento y Protección Recíprocos De Inversiones (2014). dn.ramajudicial.gov.co/SIDN/NORMATIVA/TEXTOS_COMPLETOS/7_LEYES/LEYES 2017 (1822-)/Ley 1840 de 2017 (Aprueba el Acuerdo entre Colombia y Francia sobre inversiones recíprocas).pdf Accessed 30 July 2022.

Agreement Between the Belgium-Luxembourg Economic Union, on the one hand, and the Republic of Colombia, on the other hand, on The Reciprocal Promotion and Protection of Investments (2009). https://www.fdfa.be/sites/default/files/atoms/files/822_Certificate%20in%20English.pdf. Accessed 30 July 2022

Agreement between the Government of The Federal Republic of Ethiopia and The Government of The Republic of France for The Reciprocal Promotion and Protection of Investments (2003). https://investmentpolicy.unctad.org/international-investment-agreements/treaties/bit/1475/ethiopia%2D%2D-france-bit-2003-. Accessed 30 July 2022

Agreement between the Government of the Republic of France and The Government of the Republic of Uganda on the Reciprocal Promotion and Protection of Investments (2003). https://investmentpolicy.unctad.org/international-investment-agreements/treaty-files/1289/download. Accessed 30 July 2022

Agreement Between the Government of the Republic of Rwanda and the Government of the Republic of Turkey Concerning Tile Reciprocal Promotion and Protection of Investments (2016). https://investmentpolicy.unctad.org/international-investment-agreements/treaty-files/5415/download. Accessed 30 July 2022

Agreement Between the Government of The United Arab Emirates and the Government of the United Kingdom of Great Britain and Northern Ireland for the Promotion and Protection of Investments (1992). https://investmentpolicyhub.unctad.org/Download/TreatyFile/2367. Accessed 30 July 2022

Agreement between the Republic of Colombia and the Swiss Confederation on the Promotion and Reciprocal Protection of Investments (2006). https://wipolex.wipo.int/en/text/249638. Accessed 30 July 2022

Agreement on Investment among the Governments of the Hong Kong Special Administrative Region of the People's Republic of China and the Member States of the Association of Southeast Asian nations (2017). https://investmentpolicy.unctad.org/international-investment-agreements/treaty-files/5655/download Accessed 30 July 2022

Agreement on Investment under the Framework Agreement on Comprehensive Economic Cooperation between the Association of Southeast Asian Nations and the Republic of India (2014). https://investmentpolicy.unctad.org/international-investment-agreements/treatyfiles/3337/downloada. Accessed 30 July 2022

Bilateral Agreement for the Promotion and Protection of Investments between the Government of the United Kingdom of Great Britain and Northern Ireland and Republic of Colombia (2010). available at: https://assets.publishing.service.gov.uk/government/uploads/system/uploads/attachment_data/file/317686/40694_Cm_8887_print_ready.pdf. Accessed 30 July 2022

Central America-Dominican Republic-United States Free Trade Agreement (2004). https://ustr.gov/archive/Trade_Agreements/Regional/CAFTA/CAFTA-DR_Final_Texts/Section_Index.html. Accessed 30 July 2022

Comprehensive Economic and Trade Agreement (CETA) between Canada, of the one Part, and The European Union, of the other part http://trade.ec.europa.eu/doclib/docs/2014/september/tradoc_152806.pdf. Accessed 30 July 2022

Comprehensive Economic and Trade Agreement (CETA) between Canada, on the one Part, and the European Union and its Member States (2016). http://trade.ec.europa.eu/doclib/docs/2016/february/tradoc_154329.pdf Accessed 30 July 2022

Energy Charter Secretariat (2014) The energy charter treaty (with incorporated trade amendment) and related documents. https://www.europarl.europa.eu/meetdocs/2014_2019/documents/itre/dv/energy_charter_/energy_charter_en.pdf. Accessed 06 Dec 2023

Energy Charter Treaty (ECT), (1994). https://www.europarl.europa.eu/meetdocs/2014_2019/documents/itre/dv/energy_charter_/energy_charter_en.pdf. Accessed 30 July 2022

Investment Promotion and Protection Agreement Between the Government of the Federal Republic of Nigeria and The Government of The Republic of Singapore (2016a). https://investmentpolicy.unctad.org/international-investment-agreements/treaty-files/5410/download. Accessed 30 July 2022

Investment Promotion and Protection Agreement between The Government of the Federal Republic of Nigeria and The Government Of The Republic Of Singapore (2016b). https://investmentpolicy.unctad.org/international-investment-agreements/treaty-files/5410/download. Accessed 30 July 2022

Mexico-EU FTA Annex on Expropriation 3 (2018). https://trade.ec.europa.eu/doclib/docs/2018/april/tradoc_156791.pdf. Accessed 30 July 2022

Trans-Pacific Partnership (2018) Annex 9B-Expropriation https://ustr.gov/sites/default/files/TPP-Final-Text-Investment.pdf. Accessed 30 July 2022

United Nations Paris Agreement (2015) FCCC/CP/2015/L.9/Rev.1 Preamble https://unfccc.int/resource/docs/2015/cop21/eng/l09r01.pdf. Accessed 30 July 2022

Chapter 7
Orthodox Approach to the Fair and Equitable Treatment Standard (FET) in Investment Treaty-Climate Change Disputes: Determining the Breach Through the (Dis)Proportionality/(Un)Reasonableness of the Measure

The (ordinary) FET standard—as almost all BIT standards—is open-textured. At first sight, the expression *fair and equitable treatment* may lead us to think in terms of fairness and/or equity. In this regard, for example, Dugan associates the FET with the ordinary business notion of fairness.[1] Yet, in the practice of investment treaty arbitration tribunals and in the literature the standard is generally connected to the stability of the investment. Naturally, many questions arise about the level of stability granted to the investor and the flexibility of the host state to regulate under the FET.

Claims concerning the breach of the FET standard for climate change measures are expected to increase in a climate change context. An analogous dilemma to the one addressed in the chapter of expropriation requires attention in this respect: how to provide stability to investors in the green industry and at the same time encourage host states to take measures to prevent, mitigate and adapt to the risk of global warming in the context of the climate change international legal framework. Although defining the breach of the FET as the result of any change in the regulation that affects the investor would incentivise investments in the green industries, it would also prevent the host state from adopting climate change measures.[2] Similarly, if it is argued that any legitimate modification of the regulation in favour of the climate does not breach the FET, the host state will be encouraged to fight against climate change, but investors in the green industry would be disincentivised and might invest somewhere else. Again, the idea of balance between the investor's interests and the interest of the host state in connection to the gravitation of the climate change interest appears to be the nucleus of the inquiry.

An *orthodox approach* to the FET standard in investment climate change-related disputes is faced with the challenge of discovering a solution that promotes the goal of mitigating, adapting or preventing global warming by interpreting the text. Most

---

[1]Dugan et al. (2011), p. 504.

[2]Johnson (2009), pp. 11147–11153.

T. Restrepo Rodríguez, *Investment Treaty Law and Climate Change*,
https://doi.org/10.1007/978-3-031-18655-4_7

of the studies in this regard point out the concepts of proportionality and/or reasonableness as a satisfactory channel to solve the puzzle. This trend is also visible in the investment treaty disputes concerning RESSs. However, the natural difficulties associated to defining these concepts in legal terms have led to different approaches that characterise the notions of proportionality and reasonableness through the interaction of a confusing hotchpotch of elements that, for instance, not only overlap but also end up in superficiality and/or invading the radio of deference that is natural to the host state.

Schill, for example, based on the premise that proportional measures are not against the *legitimate expectations* of the investor under the FET standard, concludes that the imposition of emission standards 'under normal circumstances'[3] does not breach the latter.[4] In his perspective, the proportionality demands, on the one hand, 'that a **reasonable balance** be struck between the purpose of the regulatory change, the benefit reported from it, and the burden imposed on investors'[5] and, on the other hand, preventing any '**unreasonably changing** [in] emissions standards or **unforeseeable changing** [in] production standards'.[6] Therefore, he considers that if measures are adopted—for instance—through a foreseeable transition period, 'it is hard to conceive that fair and equitable treatment would prevent the introduction of new regulations that aim at the reduction of greenhouse gases'.[7]

In the same vein—but embracing a broader perspective that covers not only the analysis of the restrictions to emissions but also the need to provide stability to investments in the green industry, Isotalo finds that through proportionality '[. . .] tribunals can ensure that climate change considerations are given due weight in the arbitral analysis (. . .)'.[8] She contends that the legitimate expectations of the investor could be outweighed by the legitimate public interest[9] of the host state as the result of a test of proportionality that includes the assessment of the **legitimate purpose, suitability, necessity, and proportionality (*strictu sensu*)** of the climate change measure.[10]

Other authors, instead of relying directly on the notion of proportionality, emphasise that the reasonableness of the measure is the key element to defining whether legitimate public measures in favour of climate supersede the investor's legitimate expectations. VanDuzer, apart from recognising that the consistency of any climate change regulation with the FET standard requires for measures to be adopted 'transparently, follow the due process, and refrain from taking arbitrary or

[3] Schill (2007), pp. 469–476.

[4] *Ibid.*

[5] *Ibid.* (emphasis added)

[6] *Ibid.* (emphasis added)

[7] *Ibid.*

[8] Isotalo (2016), p. 32.

[9] *Ibid.* p. 13.

[10] *Ibid.* pp. 29–32. (emphasis added)

discriminatory measures',[11] asserts that the risk of breaching the FET standard is mitigated by '**acting reasonably**';[12] e.g. 'by consulting an investor in advance of changes that would affect it, giving adequate notice of their implementation and time to implement any new requirement'. [13] Similarly, Vadi, when addressing the conflict between the right to health and the development of nuclear energy in investment arbitration—a tension that is closely related to the climate law-investment law tension—suggests that even if there is a level of deference that must be granted to the host state, 'arbitrators should assess the **reasonableness** of the measure adopted by the host state'.[14]

The complications to define the concepts of reasonableness and proportionality as the key elements to determine the breach of the FET call for an effort to provide consistency to the determination of its elements in line with the climate change international legal framework. Therefore, I suggest precising the notion of proportionality/reasonableness in investment treaty-climate change disputes, by finding a coherent pattern in the recent decisions on RESSs (Sect. 7.1), as well as extending this pattern to the cases where the host state takes measures to promote climate change by affecting investors in non-green industries (Sect. 7.2).

As a result, it is found that the determination of the (un)reasonableness/(dis)-proportionality of regulatory measures designed to address climate change—and, accordingly, the breach of the FET—through the analysis of *(non)capriciousness*, *(non)essentiality* and *(un)predictability* not only conforms to the investment treaty law system but also preserves the balance between the interests of the parties to the dispute and acknowledges the gravitational power of the climate change context.

I conclude that the scholarship's tendency of applying a public law-modeled (interest-weighing) proportionality test to FET disputes is unsystematic and that—although generally successful—the application of an *orthodox approach* is still insufficient in terms of harmonising the FET standard with climate change international legal framework in all cases (Sect. 7.3).

7.1 (Un)Reasonableness/(Dis)Proportionality in RESS Disputes: Discovering (Non)Capriciousness, (Un)Predictability and (In)Essentiality as a Pattern

Although the concepts of reasonableness and proportionality in RESS disputes could be regarded as confusing—given their polysemic deployment—(Sect. 7.1.1) it is possible to trace a pattern that moves from a vague understanding of these concepts to a more technical definition of their elements, namely (non)capriciousness,

[11] VanDuzer (2016), p. 442.

[12] *Ibid*. p. 443. (emphasis added)

[13] *Ibid.*

[14] Vadi (2015–2016), p. 1069, 1136. (emphasis added)

(un)predictability and (in)essentiality (Sect. 7.1.2). The observation of the operation of these elements in RESS disputes leads me to conclude that the differences in the reasoning among RESS awards do not alter the pattern, that the legitimate expectation to a reasonable return could stem from de-formalised representations, and that the examination of essentiality requires a material and a formal test (Sect. 7.1.3).

7.1.1 Polysemic Character of the Concepts of Proportionality and Reasonableness in RESS Disputes

One of the trends present in decisions on RESSs is the use of the expression reasonableness as equivalent to fairness or equity. In *Antin*, for example, the tribunal asserts that the host state 'cannot alter the regulatory framework applicable to the Claimants investments in ways that would be unfair, **unreasonable** and inequitable'[15] [emphasis added]. Similarly, in *RREEF* and *EISER* it is explained that, under the FET standard, host states are not compelled to avoid changes in the regulation as long as regulatory powers are not exercised in an 'unfairly, unreasonably or unnecessary'[16] manner. This assimilation of reasonableness with justice, fairness, equity or any other similar expression does not seem to be provided in the decisions with any pragmatic consequences different from describing, in general terms, the purpose of the FET. Therefore, it is compelling to further the analysis of FET and define parameters to establish whether the host state's behaviour runs counter to this general conception of fairness/necessity/reasonability/equity.

Accordingly, a second trend consists of including reasonableness and proportionality within a hotchpotch of elements to assess whether the FET standard has been breached. In this regard, in *Charanne,* the tribunal found that the claimant did not prove that the 2010s limitation to the period that the investor benefits from the regulated tariff was neither '**unreasonable**, arbitrary, contrary to public interest, or disproportionate...nor that these norms were unfair or inconsistent'[17] [emphasis added]. However, this trend does not delimit a clear distinction between the meaning of these concepts, nor does it establish how they interrelate with other elements.

In *RREEF*, for example, the tribunal first highlighted that 'the main criterion to be applied for the interpretation of the FET standard is that of *reasonable*ness'.[18] Then, the same tribunal established that finding out whether host state's measures are 'unfairly, unreasonably or inequitably'[19] deployed, depends not only on the analysis

[15] Antin Infrastructure Services Luxembourg S.à.r.l. and Antin Energia Termosolar B.V. v. Kingdom of Spain (2018), para 364.

[16] RREEF Infrastructure (G.P.) Limited and RREEF Pan-European Infrastructure Two Lux S.à r.l. v. Kingdom of Spain (2018), para. 324; Eiser Infrastructure Limited and Energía Solar Luxembourg S.à r.l. v. Kingdom of Spain (2017), para. 382.

[17] Charanne and Construction Investments v. Spain (2016), para. 539.

[18] RREEF Infrastructure (G.P.) Limited and RREEF Pan-European Infrastructure Two Lux S.à r.l. v. Kingdom of Spain (2018), para. 263.

[19] *Ibid.* para. 324.

of the creation of legitimate expectations but also 'on whether or not the changes can be held as being reasonable and proportionate'. Subsequently, this ruling determined 'that *proportionality* is the main test for *reasonable*ness'.[20] In sum, the tribunal suggested that proportionality represents, at the same time, an independent element and a part of the definition of reasonableness, as well as *reasonable*ness the solution to the problems that derive from the interpretation of the FET standard.

Conversely, in *Charanne*, reasonableness does not prevail over proportionality nor proportionality over reasonableness as prevalent element. These concepts are, apparently, treated in this ruling as independent objects. According to the tribunal, while a proportional measure involves that 'changes are not *capricious* or unnecessary and do not amount to suddenly and unpredictably eliminate the *essential* characteristics of the existing regulatory framework'[21] [italics added], a reasonable measure is characterised by the fact that it suits the purpose that the host state is willing to achieve—for example 'the requirement to cover voltage sags is reasonable because it aims at preventing the technical collapse of the system and contributes to ensure better security and better management'.[22] Yet, this definition of reasonableness could be confused with the requirement that conditions proportionality to the necessity and non-capriciousness of regulatory changes.[23] In fact, if a measure is suitable to pursue a policy goal, it could be said that it is neither capricious nor unnecessary. In this vein, the reasoning of *Charanne* might be interpreted as if reasonableness were an element for assessing proportionality, instead of interpreting proportionality as an element for assessing reasonableness, which is exactly the opposite to the *RREEF*'s approach described in the paragraph above.

A third trend is conceiving the determination of (dis)proportionality through a 'balancing or weighing exercise'. [24] The Tribunal in *Novenergia II* asserted that this balancing exercise consists of pondering 'the state's regulatory interests (...) against the investors' *legitimate expectations*'.[25] However, this statement is only repeating what is expected from an interpretation of the BIT in general, which is preserving the host state's right to regulate as much as possible while protecting the investor's interest as much as possible. The recognition of the need to weight or balance the parties' interest does not provide a methodology in itself or any element to determine what is (dis)proportional. Indeed, in this case, it was found that the host state breached the standard because the changes in the regulation were radical and unexpected without giving any specific weight to the competing interests and/or pondering them against each other.

---

[20] *Ibid*. para. 463.

[21] Charanne and Construction Investments v. Spain (2016), para. 517.

[22] *Ibid*. para. 538.

[23] *Ibid*. para. 517.

[24] *Novenergia II - Energy & Environment (SCA) (Grand Duchy of Luxembourg), SICAR v. The Kingdom of Spain* (2018), para. 657.

[25] *Ibid*. para. 650.

The *RREEF* award invokes *Novenergia II* to support the statement that the assessment of (dis)proportionality is a weighing exercise[26] but is more precise than the latter in defining what it means by this. The tribunal in *RREEF* held that the aim of the proportionality is to reach a 'fair balance between competing interests and/or principles affected by the regulation, taking into account all relevant *circumstances*'.[27] Nevertheless, despite the explicit effort that the tribunal makes to clarify the interrelationship between proportionality, reasonableness and other elements, these concepts are still blurry in the award. After asserting that the concept of proportionality is at the heart of the notion of reasonableness, the tribunal characterises reasonableness by reference to elements that the scholarship often uses to assess proportionality in investment treaty law:

- *Legitimacy of purpose, inasmuch as it represents interests of the society as a whole and does not alter the substance of the rights affected by the regulation*
- *Necessity, which implies the existence of a pressing social need. The threshold for 'necessary' is more demanding than the one for 'useful' or 'desirable'*
- *Suitability, in that it must make it possible to achieve the legitimate objective pursued*[28]

In conclusion, there is little clarity in the language of the RESS awards of the differences between (dis)proportionality and (un)reasonableness, nor is there any certainty about which elements—necessity, suitability, legitimacy, public interest, non-capriciousness, essentiality, non-radicality of the changes, predictability—are encapsulated within one or the other notion. However, I argue that it is possible to trace a pattern in the decisions of RESSs of the elements that define whether a host state's measures are (un)reasonable and/or (dis)proportional—regardless of the discussion on the proper meaning of these terms, their differences or similarities. (Un)Predictability, (in)essentiality and (non)capriciousness appear to be a common basis to assess (un)reasonableness and/or (dis)proportionality.

7.1.2 Technical Definition of (Un)Reasonableness/(Dis)Proportionality: Unfolding the Concepts of (Non)Capriciousness, (In)Predictability and (In)Essentiality in RESS Disputes

Although the language of RESS disputes is polysemic and ambiguous on the notion of proportionality and/or reasonableness, the *ratio decidendi* of the awards shares common features linked to these notions. I propose to focus on the common

[26] RREEF Infrastructure (G.P.) Limited and RREEF Pan-European Infrastructure Two Lux S.à r.l. v. Kingdom of Spain (2018), para. 466.

[27] *Ibid.* para. 465.

[28] *Ibid.* para. 464.

elements of the tribunal's reasoning that characterise proportionality and/or reasonableness rather than attempting to find a correct definition or criteria to distinguish the two notions. As the *RREEF* tribunal highlighted: '[a]lthough intellectually different, both concepts (...) are closely linked, because the breach of one of them normally entails the breach of the other'. [29] Therefore, it makes sense to deploy these terms—if not as synonyms—as compounded by the same elements to assess whether the FET standard has been breached.

The assessment of (dis)proportionality/(un)reasonableness was determinant for defining the breach of the FET standard in all the RESS awards. Indeed, the decisions that have been issued until now, in general terms, follow the same pattern and complement each other. In this respect, tribunals follow the basic reasoning of *Charanne,* differing in minor details rather than in the substance.

Charanne establishes that the proportionality/reasonableness of the measure is verified when the regulation is not '[...] *capricious* or *unnecessary* [...]' and do not '[...] suddenly and *unpredictably* eliminate the *essential* characteristics of the existing regulatory framework'[30] [italics added]. These elements—non-capriciousness, predictability and non-essentiality of the host state's measures—are reiterated in all the following rulings; often the replication of *Charanne's* approach in this respect is explicit. For example, *Eiser* both refers to *Charanne* in this regard as a 'relevant insight',[31] and adopts the same formula: 'Article 10(1) of the ECT [FET] entitled them [the investors] to expect that Spain would not *drastically and abruptly* revise the regime, on which their investment depended, in a way that destroyed its value' [32] [italics added]. Likewise, in *RREEF,* it is considered that regulatory changes are *proportional/reasonable* as long as they are not '**random, unnecessary or arbitrary**, that is, provided that **they do not significantly modify the legal framework applicable to the investors**'[33] [emphasis added]—and the award invokes as references in the footnotes *Charanne*, *Eiser* and *Novenergia*. Similarly, *Antin* endorses the same formula of *proportionality/reasonableness*—'**suddenly and unexpectedly** eliminate the ***essential*** features of the regulatory framework in place'[34] [emphasis added]—and makes express reference to *Charanne* and *Eiser* in this respect.

In other cases, the adoption of *Charanne's* elements to assess the proportionality/reasonableness is not explicit. However, the *ratio decidendi* shows the pattern of considering non-capriciousness, predictability and non-essentiality as the basis of

[29] *Ibid.* para. 463.

[30] Charanne and Construction Investments v. Spain (2016), para. 517.

[31] Eiser Infrastructure Limited and Energía Solar Luxembourg S.à r.l. v. Kingdom of Spain (2017), para. 370.

[32] *Ibid.* para. 387.

[33] RREEF Infrastructure (G.P.) Limited and RREEF Pan-European Infrastructure Two Lux S.à r.l. v. Kingdom of Spain (2018), para. 460. (emphasis added)

[34] Antin Infrastructure Services Luxembourg S.à.r.l. and Antin Energia Termosolar B.V. v. Kingdom of Spain (2018), para. 531. (emphasis added)

the FET. *Novenergia,* for instance, after embracing *Eiser's* thesis *that* 'fair and equitable treatment necessarily embraces an obligation to provide **fundamental stability in the *essential* characteristics of the legal regime',** [35] concludes that the changes to the regime of RD 661/2007 through RDL/2013—in conjunction with other modifications adopted in 2013 and 2014—were '**drastic and unexpected** in a manner that is contrary to the Kingdom of Spain's obligation to provide FET to investors'.[36] In the same vein, the whole argumentation of the Tribunal that is addressed to contend that the FET was not breached in *Isolux* is based on the fact that the modifications to the regime still allowed the investor a 'reasonable return', which was the basic characteristic of the framework that induced it to invest (non-*essential*ity); that there were technical and economic needs that justified the modifications (non-capriciousness); and, that the changes to the regime were predictable (predictability), although the award seems to link this more to the absence of legitimate expectations than to the determination of proportionality/reasonableness.[37]

Whether explicit or not in the analysis of the (dis)proportionality/(un)-reasonableness of the measure, the elements that were depicted in *Charanne* and developed through the subsequent rulings on RESSs require further explanation. Accordingly, this section unfolds the operation of (dis)proportionality/(un)-reasonableness through (non)capriciousness (Sect. 7.1.2.1), (in)essentiality and (un)-predictability (Sect. 7.1.2.2).

7.1.2.1 (Non)Capriciousness

The element of (non)capriciousness requires host state measures to be justified in objective reasons to be proportional/reasonable. A capricious, illegitimate, arbitrary and/or unnecessary measure is regarded as unreasonable/disproportional and, accordingly, as contrary to the FET standard. The non-capriciousness could be unfolded in the non-capriciousness of the policy objectives that the governmental activity pursues with the regulatory measure and the non-capriciousness of the concrete regulatory measures to achieve those policy objectives. While in the former, the interpreter should analyse whether the regulation emerges as a result of public interest needs or sensible public concerns (necessity/legitimate public interest), in the latter, it should assess whether it is possible for the measures adopted to achieve the policy goal (suitability). However, it must be emphasised that this double-analysis is not oriented towards unearthing whether the measures were appropriate or the best-possible to tackle the public interests that they were pursuing. It would be against the purpose of BITs to give tribunals the competence to dictate

---

[35] *Novenergia II - Energy & Environment (SCA) (Grand Duchy of Luxembourg), SICAR v. The Kingdom of Spain* (2018), para. 645. (emphasis added)

[36] *Ibid*. para. 695. (emphasis added)

[37] *Isolux Netherlands, BV v. Kingdom of Spain (2016)*, para. 773–784.

the host sates' policy agenda and impose the measures to follow this agenda. The high level of deference given to host states in choosing, designing and applying regulation impedes tribunals from invading the regulatory space that is reserved for the host state as an expression of its sovereign powers.

There is consensus in the REESs decisions that changes to RESS regulation obeyed a legitimate public interest purpose. In fact, as *Charanne* accurately described that the increase of the price to consumers, and the considerable rise in the payment of the regulatory tariff to the investors, evinced a tariff deficit in the energy system that required adjustments. This certainly constitutes a legitimate reason to enact legislation rather than 'arbitrary, irrational or contrary to public interest'.[38] Likewise, the Tribunal in *Eiser* clarified that since the host state's measures were addressing the 'legitimate public policy problem'[39] of the energy system's tariff deficit, the 'appropriateness of Spanish authorities adopting *reasonable* measures to address the situation'[40] was uncontestable. In the same line of reasoning, *Antin* remarks that '[i]t is undisputed that the Tariff Deficit poses a legitimate public policy problem for Spain';[41] *RREEF* states that the regulator measures were part of 'a comprehensive plan' to overcome a problem of a 'structural nature';[42] *Isolux* establishes that the frequency of the changes made to the RESS obeys 'the technical and economic needs of the moment'; [43] and, *Novenergia* accepts that not only the 'changes to the Special Regime were motivated by the Kingdom of Spain's need to address the so-called tariff deficit(....) but also that the Kingdom of Spain had a regulatory right to do so'. [44]

The assessment of the plausibility of the concrete measures to tackle the deficit varies in each case according to the regulatory measures that are the object of the dispute. The tribunal in *Charanne*—which concerns the package of measures from 2007 to 2010—determined that the regulatory measures in the specific case were not arbitrary because they were grounded on objective criteria.[45] The tribunal found that the limitation of operation hours as well as the limitation of the period that the investor would benefit from the regulated tariff was justified because of 'the climate zone in which the plant is located and the technology used', as well as the 'lifetime of a photovoltaic power plant'—respectively. In contrast, in the *Antin* case—which deals with a broader range of measures than *Charanne* that ended up in 2013/2014

[38] Charanne and Construction Investments v. Spain (2016), para. 535–536.

[39] Eiser Infrastructure Limited and Energía Solar Luxembourg S.à r.l. v. Kingdom of Spain (2017), para. 371.

[40] *Ibid.*

[41] Antin Infrastructure Services Luxembourg S.à.r.l. and Antin Energia Termosolar B.V. v. Kingdom of Spain (2018), para. 570.

[42] RREEF Infrastructure (G.P.) Limited and RREEF Pan-European Infrastructure Two Lux S.à r.l. v. Kingdom of Spain (2018), para. 469.

[43] *Isolux Netherlands, BV v. Kingdom of Spain (2016),* para 788.

[44] *Novenergia II - Energy & Environment (SCA) (Grand Duchy of Luxembourg), SICAR v. The Kingdom of Spain* (2018), para. 688.

[45] Charanne and Construction Investments v. Spain (2016), para. 534.

dismantling the priority access and canceling the special benefits of the 2006/2007 regulated tariff scheme (FIT)—the tribunal observed that the measures were not suitable to address the tariff deficit since in its perspective it was not visible that the 'FIT [regulated tariff] for CSP plants played a significant role in the accumulation of the Tariff Deficit'.[46]

In *RREEF*, after the tribunal defined suitability as an element of reasonableness that tests whether the measure makes it **'possible to achieve the legitimate objective pursued'**,[47] surprisingly, there is not any insight on the suitability of the concrete measures needed to tackle the tariff deficit in the case at hand. This absence could be explained by two plausible reasons. The first is that the tribunal confers considerable deference to the host state in the sense that the former considers that it is not its task to 'substitute (. . .) [the latter's] views either on the appropriateness of the measures at stake or on the characterization of the situation which prompted them', [48] specifically, not 'taking any position on the issue of the existence of other or more appropriate possible measures to face this situation'. [49] The second is that deciding on suitability does not seem necessary to support the tribunal's reasoning. In fact, it appears that the tribunal equates—not without contradiction—arbitrariness (non-capriciousness) to the alteration of the essential characteristics of the RESS under RD 1614/2010,[50] which in its view was the 'reasonable return' [51] to the investor through the regulated tariff. Therefore, the tribunal's finding that the amendment to the tariff represents a return below the threshold of reasonable return, leads to the conclusion that the measure is arbitrary and unreasonable—without the need to determine whether the measure could effectively achieve the goal that it is supposed to achieve.

Similarly, it appears that in *Isolux* it was not necessary to discuss the plausibility of the measures that were adopted from 2012 onwards to stabilise the energy system,

[46] Antin Infrastructure Services Luxembourg S.à.r.l. and Antin Energia Termosolar B.V. v. Kingdom of Spain (2018), para. 571–572: Moreover, Brattle showed that, '"Spain provided cumulative financial support of €0.6 billion to CSP from 2000 to 2011, which was less than 0.6% of €118 billion in cumulative costs of regulated activities other than generation". Accordingly, the Tribunal is not persuaded that the FIT for CSP plants played a significant role in the accumulation of the Tariff Deficit'.

[47] RREEF Infrastructure (G.P.) Limited and RREEF Pan-European Infrastructure Two Lux S.à r.l. v. Kingdom of Spain (2018), para. 464. (emphasis added)

[48] *Ibid.* para. 468.

[49] *Ibid.*

[50] RREEF Infrastructure (G.P.) Limited and RREEF Pan-European Infrastructure Two Lux S.à r.l. v. Kingdom of Spain (2018), para. 460–461: 'The Arbitral Tribunal considers that the *proportionality* requirement is fulfilled inasmuch as the modifications are not random, unnecessary or arbitrary, that is, provided that they do not significantly modify the legal framework applicable to the investors'.

[51] *Ibid.* para. 470: Considering the Respondent's margin of appreciation, the Tribunal considers that, for appreciating the *reasonable*ness of the disputed measures, it is necessary to start by recalling that the only established *legitimate expectation* of the Claimants is the guarantee of a *reasonable* return of their investments.

not only because it was possible to anticipate the modification of the RESS as of the date of the investment, [52] but also because the investor was obtaining higher returns than the returns it was obtaining before the modifications, [53] i.e. although the changes to the retribution systems were significant, the investor was earning a '*reasonable* return', [54] thus the examination of the suitability of the measures was irrelevant to solve the dispute.

Nevertheless, when *Isolux* addressed the investor's claim about the allegedly discriminatory and exorbitant nature of the measures that were adopted from 2012 onwards—as an independent violation from the FET standard—it was considered that even if other alternative measures were found to 'have been preferable and more favorable to the claimant [...] this wouldn't be enough to conclude that the adopted measures were 'exorbitant' or not *reasonable*'. [55] In the perspective of the tribunal, the amendment to the solar energy regulation 'was a *reasonable* policy, which was, like it or not, to protect the consumer'. [56]

Similarly to *RREEF, Novenergia*, despite recognising that measures should suit the purpose they are meant to serve as part of the FET, ignores in the *ratio decidendi* the analysis of the suitability of the measure. Certainly, the tribunal acknowledges that the possibility to achieve the result is enough to fulfill the exam of suitability by supporting the award with the transcription and adoption of *AES Summit* and *Electrabel* reasoning. While the fragment of the first award that is adopted establishes that a 'sub-optional regulation' can still be in accordance with 'the acceptable range of legislative and regulatory behavior without breaching the FET standard', [57] the fragment from the second award establishes that a 'State can thus be mistaken without being *unreasonable*'. [58] However, the tribunal did not need to make this analysis, since it concludes that the fundamental changes to the regulation from 2013 to 2014 that altered essential features of the RESS' framework under which the claimant invested in 2007 were sufficient to verify unreasonableness/ disproportionality. [59] In the perspective of the tribunal, measuring the retribution to the investor through model costs instead of real costs was a radical and unexpected change in the conditions of the investment. [60] This conclusion was also reached by the Tribunal in *Eiser*,[61] which might be the reason why this ruling also fails to examine the suitability of the measure.

---

[52] *Isolux Netherlands, BV v. Kingdom of Spain (2016)*, para. 804.

[53] *Ibid.*

[54] *Ibid.* paras. 804–811.

[55] *Ibid.* para. 823.

[56] *Ibid.*

[57] *Novenergia II - Energy & Environment (SCA) (Grand Duchy of Luxembourg), SICAR v. The Kingdom of Spain* (2018), para. 655.

[58] *Ibid.* para. 655.

[59] *Ibid.* paras. 691, 695, 697.

[60] *Ibid.* para. 695.

[61] Eiser Infrastructure Limited and Energía Solar Luxembourg S.à r.l. v. Kingdom of Spain (2017), paras. 387, 389, 391, 398.

Some conclusions can be extracted from the analysis of case law on RESSs concerning the requirement of non-capriciousness of the regulatory measures. First, all of the cases confirmed that the objective of addressing the systemic energy tariff deficit was a legitimate public purpose that needed to be addressed (i). Moreover, this confirmation was not the result of a strict examination but rather the reinstatement of a fact that was evident (ii). Moreover, it appears that the host state has a considerable discretion—the right—to choose which policy objectives to pursue without breaching the FET standard on the grounds of disproportionality/unreasonableness (iii). On the other hand, the decisions on RESSs also coincide in the deduction that reasonableness/proportionality involves measures that are equipped to address the policy objective of stabilising the energy system (suitability) (iv). However, while some cases actually examined the suitability of the regulatory measures (e.g., *Antin*, *Charanne* and *Isolux*), other cases only incorporated the concept of suitability in theoretical terms, omitting an examination (e.g., *RREFF* and *Novenergia*). In the case of *Eiser,* the concept was not incorporated at all (v). The most likely reason for omitting the test in these cases is the irrelevance of its deployment when the regulatory measures were proven to unexpectedly change the essential characteristics of the legal framework that induced claimants to invest, which suffices to verify that the measures were unreasonable/disproportional (vi). Therefore, it appears that to assess reasonableness/proportionality, the non-capriciousness could be analysed independently of the analysis of predictability and non-essentiality, i.e. the verification of either capriciousness or unpredictable essential changes to the framework is considered to be unreasonable/disproportional (vii). Finally, although in *Antin,* the regulatory changes were considered unsuitable to achieve the purpose, tribunals in general appear to be mindful in not invading the regulatory space and the sovereignty that belongs to the host state by not proposing alternative measures, better measures or correct measures in comparison to those that were adopted. In other words, when analysing the suitability of the measures in RESS investment disputes, tribunals recognise a high level of deference to the host state measures (viii).

7.1.2.2 (In)Essentiality and (Un)Predictability

RESS awards are consistent in recognising that changes to the regulation are reasonable/proportional as long as they do not unpredictably modify the essential characteristics of the legal framework that motivated the investment. Accordingly, unexpected modifications to the essential characteristics are regarded as unreasonable/disproportional and, consequently, they breach the FET standard. The determination of *(un)predictability* and the *essential* characteristics of the legal framework is particular to each case, and heavily depends on the host state's representations that can generate legitimate expectations as well as on the context that surrounds the investment. Although we could say—as a generalisation—that the investors in RESSs legitimately expected a reasonable return, it is convenient to describe the circumstances of each decision to plainly understand the reasoning of the investment treaty tribunals concerning the preservation of the essential characteristics of the framework as a requirement of reasonableness/proportionality. Grouping the awards stemming from the range of years that the regulatory modifications cover facilitates their analysis.

Charanne (2007–2010)

The investments in *Charanne* were perfected under the conditions of *Real Decreto 661 de 2007* and *Real Decreto 1578/2008*. The investor challenged the changes in the regulation of the photovoltaic energy sector of 2010 for breaching the FET; mainly focusing on the negative effects of *Real Decreto 1614/2010* on the investment. According to this, the modifications reduced the investor's profit by 10% on PV installations in relation to the conditions of *Real Decreto 1578/2008* and by 8.5% in relation to the conditions of *Real Decreto 661/2007*.[62]

The enactment of *Real Decreto 661 de 2007* (the 2007 Decree) was the result of an aggressive campaign by the government of Spain to promote the development of renewable energies that included road shows and the circulation of official communications promoting the stability of the system as well as predicting considerable returns to investors. [63] The 2007 Decree—in accordance with the *Ley 54/1997*[64]—promoted the special regime for PV installations which was characterised by the right to connect and priority access to distribution and transmission grids[65] as well as by the possibility for investors to sell the electricity directly to consumers and, most importantly, to calculate the retribution to the investors through a regulated tariff or premium (FIT).[66]

The impact of *Real Decreto 661/2007* attracted significant investments in PV installations and considerably increased the production of PV energy in Spain.[67] The appeal of the legal framework was grounded on the FIT applicable for a 25-year period (Arts. 35-37 and 44; Annexes 7-8 and 12), with the possibility for the investor to get an extension for a second period (Art. 36). A year later, the scheme was modified through *Real Decreto 1578 de 2008* (the 2008 Decree) by limiting the application of the FIT to the first period of 25 years (Art. 11.5.) as well as requiring investors to participate in tender processes as a condition to be entitled to the FIT (Arts. 4 and 5).

In 2010, the Spanish government adopted several instruments that reduced the profitability of the investments (the 2010 Decrees). *Real Decreto* 1565/2010 extended the application of the modifications of the 2008 Decree to the investments

[62] Charanne and Construction Investments v. Spain (2016), para. 284.

[63] *Ibid*. paras. 95, 102–130; Eiser (2017), para. 358.

[64] Charanne and Construction Investments v. Spain (2016), para. 84.

[65] *Ibid*.

[66] *Ibid*. para. 118.

[67] *See* Comisión Nacional de Energía (2008), pp. 10–12.

in PV Installations that were completed under the 2007 Decree (Art. 29.2). This means that the possibility of extending the benefits further from the 25-year period was cancelled. *Real Decreto Ley 14/2010'* levied a 0.5 euro/MW fee to the investors in PV energy for accessing the grid, *Real Decreto 1614/2010* reduced the operation hours that the investor would benefit from the FIT (Art. 2) and *Real Decreto 1565/ 2010* required the investor to implement some technical measures to facilitate the control of tension gaps.[68]

The tribunal in *Charanne* found that the 2010 Decrees did not unpredictably change the essential characteristics of the framework because the investor could have reasonably expected modifications to the legal framework, in addition to the fact that the changes did not alter the basic structure of the FIT for the whole life of the PV Plants, nor did it cancel priority access.

The tribunal reached the conclusion that changes should have been reasonably expected by the investor based on the absence of specific commitments and grounded on the legal and technical circumstances of the investment. Concerning specific commitments, the tribunal denied any theoretical possibility for the legal framework by itself to constitute a commitment capable of generating the expectation that the legislation will remain untouched in the future. [69] Under this perspective, in the absence of a stability clause or a similar representation, even if the regulation addresses an identifiable group of investors, the host state does not acquire the specific commitment to keep the regulation unaltered. [70] A different conclusion 'would constitute an excessive limitation on the power of states to regulate'. [71] Therefore, the tribunal was categorical in establishing that *Real Decreto 661 de 2007* and *Real Decreto 1578/2008* did not create the expectation that the FIT would not be modified. [72] Neither the tribunal considered that the representations of Spain during the promotional campaign of the scheme—which comprehended presentations and documentation directed to attract investment—were enough to create the legitimate expectation that the framework would not change. [73]

On the technical and legal context, the award highlights the burden placed of the investor to be diligent in analysing the conditions of the investment. Certainly, in the perspective of the tribunal, an appropriate legal expert study would have shown that the decisions of the Spanish supreme court about the validity of the modifications to the electricity system—although they were not specifically concerned about the operation of PV Plants—'could have **easily foreseen** possible adjustments to the regulatory framework as those introduced by the rules of 2010'. [74] Besides, the tribunal highlights that, as any investment in the energy sector, the PV energy

[68] Charanne and Construction Investments v. Spain (2016), paras. 490, 152.

[69] *Ibid.* para. 490.

[70] *Ibid.* para. 493.

[71] *Ibid.* para. 493.

[72] *Ibid.* paras. 490–494.

[73] *Ibid.* paras. 496–497.

[74] *Ibid.* para. 505. (emphasis added)

operation is a highly regulated market that requires permanent adjustments, thus the expectations on the legal framework's immutability would be unreasonable. [75]

Furthermore, concerning the technical aspects, the reasoning of the decision, points out that the investor could not reasonably expect the benefits of the RESS to last more than the lifespan of a PV plant; 25 years average. This conclusion was taken from the expert's report on the duration of PV plants, and buttressed by the promotional documents that the host state circulated to attract investment as well as from the cash flow projection that the same investors anticipated. [76] Consequently, the tribunal dismissed the claim of the investor that the modification which cancelled the possibility of extending the benefit for a second period beyond 30 years (*Real Decreto* 1565/2010) was against an FET. [77] Similarly, it is established in the award that the modification of the operation hours of the FIT according to the climate zone and the technology was predictable from the text of both decrees with special reference to the annex of the 2007 Decree. [78] Accordingly, the tribunal concludes that the modification of the operation hours is not disproportional/unreasonable or against legitimate expectations. [79]

Apart from asserting that the changes to the regulation were predictable, the tribunal reveals that the changes did not affect the essential characteristics that motivated the investor to invest. The *essential*s of the legal framework according to the award were the FIT and the priority access. [80] As these elements continued to exist under the 2010 Decrees, the investor's claim in this regard was dismissed. [81]

Importantly, the tribunal emphasised that the investor had the legitimate expectation of obtaining a reasonable return in connection to the FIT. [82] This is a thesis that is reiterated throughout investment treaty decisions on RESSs. In fact, in contrast to the modifications of the 2010 Decrees, the modifications that followed in Spain appeared to affect this reasonable return. Mainly for this reason, the disputes that are described in the next section, although consistent with *Charanne* in general terms, reached the opposite conclusion, which is that the host state's behaviour was disproportional/unreasonable.

Eiser, Novenergia, Antin (2007–2014)

The investments in *Eiser, Novenergia, Antin* were perfected under the conditions of *Real Decreto 661 de 2007*. [83] The investors in each of the cases challenged the changes in the regulation of the photovoltaic energy sector that commenced in 2010

[75] *Ibid.* para. 507.

[76] *Ibid.* para. 527–528.

[77] *Ibid.* para. 529.

[78] *Ibid.* para. 531.

[79] *Ibid.* para. 532.

[80] *Ibid.* para. 533.

[81] *Ibid.* para. 533.

[82] *Ibid.* para. 518.

[83] *Novenergia II - Energy & Environment (SCA) (Grand Duchy of Luxembourg), SICAR v. The Kingdom of Spain* (2018), para. 531–541; Antin Infrastructure Services Luxembourg S.à.r.l. and

and finished in 2014—with the cancellation of any benefits from the special regime to investors—for breaching the FET. While in the *Eiser's* dispute, the investor claims damages amounting to €256 million for loss of profit and €13 million for historical loss,[84] in *Antin,* the investor claims €25 million for historical loss and €155 million for loss of profit.[85] In *Novenergia*, the investor calculates the total loss as €61 million as a result of the breach of the FET standard.[86]

Subsequent to the modifications to the regulation of REESs that were described in the case of *Charanne* (immediately above), the Spanish government openly communicated its intentions to take new measures addressing the tariff deficit, which at the time was estimated at €22.000 million.[87] Accordingly, several measures were adopted (The 2010-onwards regulation). *Ley 15/2012* levied a tax of 7% on access to the grid (Art. 8), and *Real Decreto 2/2013* eliminated the alternative for investors to opt for a regulated premium, reducing their possibilities to calculate retributions by reference to either the market price or the regulated tariff (Art. 3). Furthermore, this decree detached the methodology used to assess the adjustment of retribution from the factor of inflation (Art. 1).

Next, the host state opted to disarticulate the RESS through several measures. *Real Decreto Ley 9/2013* replaced *Real Decreto 661/2007* and set a completely different system for determining retribution, whose main factor was the calculation of model costs rather than real costs (Art. 1). These changes were later complemented through the *Ley 24/2013*, which formally abolished the special regime of *Ley 54/1997 (Disposición derogatoria única)*, and the *Real Decreto 413/2014*, which established as a reference to calculate the model operational costs of a hypothetical efficient plant (Art. 13). Finally, *Orden Ministerial IET/1045/2014* specified the technical terms for establishing retribution and set the rate of return at 7.398%.

In the three awards, the arbitrators concluded that the 2010-onwards regulation was disproportional/unreasonable because it unpredictably changed the essential characteristics of the framework, mainly, by altering the basic structure of the FIT. However, the rulings might appear to have some discrepancies on the nature of an essential change and the notion of specific commitments—as is explained below.

Just as in *Charanne*, the analysis of (in)essentiality and (un)predictability is connected to the creation of legitimate expectations as well as to the context of the

---

Antin Energia Termosolar B.V. v. Kingdom of Spain (2018), para. 509; Eiser Infrastructure Limited and Energía Solar Luxembourg S.à r.l. v. Kingdom of Spain (2017), paras. 126, 135.

[84] Eiser Infrastructure Limited and Energía Solar Luxembourg S.à r.l. v. Kingdom of Spain (2017), paras. 460, 457.

[85] Antin Infrastructure Services Luxembourg S.à.r.l. and Antin Energia Termosolar B.V. v. Kingdom of Spain (2018), para. 584–586.

[86] *Novenergia II - Energy & Environment (SCA) (Grand Duchy of Luxembourg), SICAR v. The Kingdom of Spain* (2018), para. 768.

[87] Eiser Infrastructure Limited and Energía Solar Luxembourg S.à r.l. v. Kingdom of Spain (2017) paras. 137, 150.

investment. On the question of whether the legal framework is capable by itself of creating the legitimate expectation of the crystallisation of the regulation, the three cases coincide in a negative answer. *Eiser* emphasised that in the absence of a stability clause or any similar commitment, the investor should envisage that the regulation might change.[88] Likewise, *Novenergia* acknowledges that '"*the fair and equitable treatment standard does not give a right to regulatory stability per se*" (...) and investors must expect that legislation may and will change' [89] [italics added]. In the same vein, *Antin* establishes that '[t]he requirement of stability under the Treaty does not equate to the immutability of the legal framework'.[90] However, it was clarified in all these cases that the absence of the legitimate expectation of general immutability of the framework, does not mean that the host state has unlimited discretion for changing the regulation. It is limited by the assurances that were given to the investor and the stability of the basic features of the legal framework.[91]

Consequently, it is plausible to affirm that there are two different types of legitimate expectations applicable to RESS disputes according to how they are created. While, on the one hand, some expectations would stem from unambiguous formal-instrumentalised representations—such as stability clauses, stability contracts, etc.—on the other hand, the expectation that the essential features of the legal framework would not be changed stems from the basic obligation to provide stability that characterises the FET.[92] In this regard, *Novenergia* states that '[w]hile specific assurances given by the host State may reinforce the investor's expectations,

[88] Eiser Infrastructure Limited and Energía Solar Luxembourg S.à r.l. v. Kingdom of Spain (2017), para. 362.

[89] *Novenergia II - Energy & Environment (SCA) (Grand Duchy of Luxembourg), SICAR v. The Kingdom of Spain* (2018), para. 654.

[90] Antin Infrastructure Services Luxembourg S.à.r.l. and Antin Energia Termosolar B.V. v. Kingdom of Spain (2018), para. 555.

[91] *Ibid.*: 'The State is certainly entitled to exercise its sovereign power to amend its regulations to respond to changing *circumstances* in the public interest. However, any such changes must be consistent with the assurances on stability of the regulatory framework provided by the State and required by the ECT.' *Novenergia II - Energy & Environment (SCA) (Grand Duchy of Luxembourg), SICAR v. The Kingdom of Spain* (2018), para. 654: 'The FET standard does, nevertheless, protect investors from a radical or fundamental change to legislation or other relevant assurances by a state that do not adequately consider the interests of existing investments already made on the basis of such legislation'. Eiser Infrastructure Limited and Energía Solar Luxembourg S.à r.l. v. Kingdom of Spain (2017), para. 382: '[T]he Tribunal concludes that Article 10(1)'s obligation to accord fair and equitable treatment necessarily embraces an obligation to provide fundamental stability in the *essential* characteristics of the legal regime relied upon by investors in making long-term investments. This does not mean that regulatory regimes cannot evolve. Surely they can'.

[92] Eiser Infrastructure Limited and Energía Solar Luxembourg S.à r.l. v. Kingdom of Spain (2017), para 382. 'Taking account of the *context* and of the ECT's object and purpose, the Tribunal concludes that Article 10(1)'s obligation to accord fair and equitable treatment necessarily embraces an obligation to provide fundamental stability in the *essential* characteristics of the legal regime relied upon by investors in making long-term investments'.

such an assurance is not always indispensable'.[93] This citation reflects the general trend in these cases where, without undermining the force of specific assurances, the tribunals admit that the legal framework in a given context generates the expectation that the essential aspects of the RESS that motivated the investor are reasonable and legitimate expectations.

The tribunals' reading of the facts in the three cases concur in inferring that the text of *Real Decreto 661/2007* is clearly intended to attract investments in the solar energy industry to meet emission targets and that the adoption of the special regime was the instrument for achieving this objective.[94] Apart from this, the examination of legitimate expectations in the three decisions includes a set of legislation and promotional communications and official programmes revealing representations from the host state incentivising investors by ensuring the stability of the system and promising attractive returns.[95] For example, '[that the investor would] *obtain reasonable rates of return*',[96] 'a durable, objective, and transparent framework',[97] '[t]he return on the investment is reasonable and can sometimes reach up to 15%',[98] '[that the investor would obtain] significant financing of the investment',[99] 'the proper functioning of these mechanisms must be guaranteed [...] to maintain investor's confidence'.[100]

In this context and bearing in mind the general obligation to provide stability under the FET, the tribunals found that the investor had objective grounds to expect that the particular features of the special regime for solar energy production would be maintained during the operation of the project. Therefore, the radical alteration of these special features was regarded as disproportional/unreasonable. The description of the new regime in comparison to the 2007 RESSs made in *Antin,* provides a clear picture of the disproportionality/reasonability of the measures:

> 564. First, the New Regime fixes the "applicable rate for *reasonable* return" at 7.398% before taxes. Such rate is based on the average yield in the secondary market of 10-year Spanish government bonds, plus a differential of 3%. The evidence in the record does not point to an identifiable basis for determining such adjustment(...)
>
> 565. Second, the Special Payment is calculated considering the costs of a standard installation (...)There is no evidence on the record as to which parameters were considered in determining what is a standard installation (...) the Respondent provided no evidence as to which studies—technical, financial or otherwise—were made or considered to determine the criteria for grouping the installations and to determine which installations shared the same technical characteristics and the same year of commissioning.

---

[93] *Novenergia II - Energy & Environment (SCA) (Grand Duchy of Luxembourg), SICAR v. The Kingdom of Spain* (2018), para. 650.

[94] *Ibid.* para. 665.

[95] *Ibid.* para. 666–669.

[96] *Ibid.* para. 666.

[97] *Ibid.*

[98] *Ibid.*

[99] *Ibid.* para. 668.

[100] *Ibid.*

> 566. Third, the "*reasonable* rate of return" is subject to a mandatory revision every three years and a possible revision every six years. At the three-year point, the forecast of the market price is adjusted to the actual market price, which could go up or down. After six years, Spain could revise the figure for the "*reasonable* return" and apply a different adjustment (...) There is thus no evidence in the record that shows that the Government has established an identifiable set of criteria for the revision of the remuneration for RE installations.
>
> 567. Fourth, (...) the already reduced "*reasonable* rate return" [7.398%] may be lowered considering previous payments to the relevant installation. Moreover, if the revenues obtained by a group of installations under the same standard installation exceeds the 7.398% rate of return, future compensation under the New Regime will be reduced accordingly so that the rate of return does not exceed the aforesaid 7.398%. This means that the rate of return under the New Regime may apply to periods preceding the establishment of the New Regime.
>
> 568. Based on the evidence on the record of this arbitration, the Tribunal concludes that the methodology for determining the "reasonable rate of return" under the New Regime is not based on any identifiable criteria. On the contrary, what Spain labels a "reasonable rate of return" seemingly depends on governmental discretion. This is in plain contrast with the relative precision of the Original Regime—in force when the Claimants made their investment—which provided for objective and identifiable criteria for determining the remuneration due to CSP plants, which were expressly specified in the regulation and were dependent on the market.[101]

These tribunals did not elaborate much on the concept of predictability. This could be explained by the fact that before 2007—when the investment was made—it was clear that there were no signs that the special regime could be modified much in the near future, or because the Respondent did not insist much on the predictability of the measure as a defence against the breach—its main argument was that the investor could predict changes as a general rule of any regulatory system but did not really focus on the statements of the government prior to the investment anticipating changes. Besides, the conclusion that the stability of the main features of the RESSS was objectively and reasonably expected implies—to a great extent—that changes were unpredictable. By contrast to these cases, the analysis of predictability was one of the central elements in *Isolux* and *RREEF*, as is described below.

Isolux and RREEF 2011–2014

The investments in *RREEF* were perfected between February and August 2011;[102] and in *Isolux,* in October 2012. Therefore, Although the legal framework applicable to these investments was still (mainly) the 2007 or 2008 Decrees, the comprehensive modifications of the 2010 Decrees were also a significant part of them. The date of

[101] Antin Infrastructure Services Luxembourg S.à.r.l. and Antin Energia Termosolar B.V. v. Kingdom of Spain (2018); A similar reasoning on the changes: Eiser Infrastructure Limited and Energía Solar Luxembourg S.à r.l. v. Kingdom of Spain (2017), para. 382–400.

[102] RREEF Infrastructure (G.P.) Limited and RREEF Pan-European Infrastructure Two Lux S.à r.l. v. Kingdom of Spain (2018), para. 392.

investment differentiates these cases from the two group of cases described above. However, similar to *Eiser, Novenergia* and *Antin*, the claimants in these controversies challenged the 2010-onwards regulation that cancelled any benefits from the special regime for investors, for breaching the FET. While in the *RREEF* dispute the investor claims that damages amount to €252 million for loss of profit and €48.3 million for historical loss, [103] in *Isolux* the investor claims €68.9 million in total. [104]

In *Isolux* the tribunal concluded, on the one hand, that the host state did not breach the FET because changes of regulation could have been anticipated in the particular case [105] and, on the other hand, that the investor's retribution under the new regime still ensured a reasonable return. [106] In contrast, in *RREEF,* it was found that the respondent breached the standard since the new regime did not guarantee a reasonable return to the investor, which was the basic feature of the scheme and expectation from the investor.[107]

This group of cases also draws the (in)essentiality and (un)predictability liaised closer to the analysis of legitimate expectations and the context of the investment. The tribunal in *Isolux* expressed that the question of whether the legal framework is capable by itself of creating the legitimate expectation of the crystallisation of the regulation was not relevant, since the claimant did not suggest the existence of such an expectation. [108] However, the tribunal followed the same line of reasoning of the two group of cases described above that consists in acknowledging that legitimate expectation could originate from unambiguous formal-instrumentalised representations—such as stability contracts or stability clauses—as well as from the regulation in conjunction with the circumstances and other representations of the host state. [109]

RREEF, by contrast, does answer to the question about the legitimate expectation of crystallisation of the regulation by (categorically) stating that 'Stability is not an absolute concept; absent a clear stabilization clause, it does not equate with immutability'.[110] Furthermore, also along the same lines as other decisions on RESSs, the tribunal of this award asserted that 'the obligation to create a stable environment certainly excludes any unpredictable radical transformation in the conditions of the investments'. [111]

[103] *Ibid.* para. 493–496.

[104] *Isolux Netherlands, BV v. Kingdom of Spain (2016),* para. 478.

[105] *Ibid.* para. 802–804.

[106] *Ibid.* para. 810–811.

[107] RREEF Infrastructure (G.P.) Limited and RREEF Pan-European Infrastructure Two Lux S.à r.l. v. Kingdom of Spain (2018), para. 589.

[108] *Isolux Netherlands, BV v. Kingdom of Spain (2016),* para. 774.

[109] *Ibid.* para. 775–778.

[110] RREEF Infrastructure (G.P.) Limited and RREEF Pan-European Infrastructure Two Lux S.à r.l. v. Kingdom of Spain (2018), para. 315.

[111] *Ibid.*

Clearer in *RREEF* than in *Isolux,* the adoption of a division between legitimate expectations arising from unambiguous-instrumentalised representations and the legitimate expectation of the preservation of the legal framework's essentials as an expression of the obligation of stability, is present in these decisions. Naturally, these tribunals also adopt the approach that the determination of the essentials of the framework depends on the circumstances of the investment and the representations of the host state in the particular case. [112] In this regard, both cases coincide that the investor legitimately expected to obtain a reasonable return. [113]

Isolux reached the conclusion that the investor had the legitimate expectation of obtaining a reasonable return based on Article 30.4 of *Ley 54/1997* and the Spanish Supreme Tribunal's interpretation of this article. [114] The Tribunal revealed that before the investment was perfected, it was clear in the ruling of the Spanish Supreme Tribunal that changes to the FIT were regarded as valid as long as they maintained the principle of a reasonable return established in *Ley 54/1997.* [115] Accordingly, in the tribunal's perspective the only expectation that the investor could reasonably have at the time of the investment was to obtain a reasonable return. [116] Moreover, the tribunal added that according to *Ley 54/1997* and the reasoning of the Spanish Supreme Tribunal, the reasonable return should be assessed by reference to the cost of money in the capital market. [117] Similarly, grounded on the texts of both *Ley 54/1997* and *Real Decreto 661/2007* as well as in the official energy sector development programme, *RREEFF* is firm in expressing that 'a *reasonable* return or profitability was, no doubt, part of the guarantees given to the Claimants - and, more generally, to all investors in the Spain's RE'.[118] Furthermore, the tribunal in this case stated that the '"*reasonable return*" or "*reasonable profitability*" was the main *specific commitment* of Spain *vis-à-vis* the investors in the Special Regime'. [119] Additionally, like *Isolux*, this ruling contends that a reasonable return must be determined by reference to the cost of money in the capital market. [120] Yet, apart from this factor, the tribunal highlights that the determination of a *reasonable return* also has to be determined against the host state's 'concern about the cost of electricity and the competitiveness with other means of production of energy'. [121]

---

[112] *Isolux Netherlands, BV v. Kingdom of Spain (2016)* paras. 794–798, 804–801.

[113] *Ibid.* para. 786.

[114] *Ibid.* para. 804–807.

[115] *Ibid.* paras. 794–798, 804–807.

[116] *Ibid.* paras. 794–798, 810–811.

[117] *Ibid.* para. 806–808.

[118] RREEF Infrastructure (G.P.) Limited and RREEF Pan-European Infrastructure Two Lux S.à r.l. v. Kingdom of Spain (2018), para. 381.

[119] *Ibid.* para. 384.

[120] *Ibid.* para. 387.

[121] *Ibid.* para. 385.

It is crucial to emphasise that both *Isolux* and *RREEFF* do not invoke *Ley 54/1997*, *Real Decreto 661/2007*, or the Spanish Judgments as authorities to solve the controversy. It is clear that liability is assessed in these cases against international law. [122] National legal provisions and jurisprudence are taken as facts for assessing expectations. [123] Certainly, in light of the circumstances and investor's representations, the tribunals show that the reasonable return is the main feature of the framework that attracts investments and thus the investor can objectively expect that the reasonable return to be preserved throughout the life span of the project.

Isolux and *RREEF* also agree with the general connection between predictable changes, expectations and due diligence. In brief, both rulings stated that the investor could have anticipated at the time the investment was made that the RESS would be subject to significant modifications.[124] Again, this *predictability* stems from the fact that the regime had been altered before (2010 decrees) the investment was made (2011–2012), and that there were rulings by the Spanish tribunals as well as political statements in this regard. [125] Both decisions reasoned that the investors (claimants)—experts in the industry—should have performed a proper legal risk assessment that would have revealed what was evident from the study of the jurisprudence, the law and the host state's *representations*, i.e. that the special regime would probably be subject to modifications. [126] Arbitrators also clarify that it is not that they are imposing an obligation of due diligence on the investor but rather confirming that expectations and reasonability/proportionality must be determined objectively, in the sense that the standard of diligence of an expert in the field provides objective grounds to measure whether the investment was predictable. [127] Although the two awards reach different conclusions concerning the liability of the claimants for breach of the FET, they do not contradict each other. In fact, while in *Isolux* the Tribunal found that the host state measures were reasonable/proportional, in *RREEF,* the tribunal found that they did not. However, this difference was not the result of inconsistency between the reasoning of the two cases but in the distinctiveness of the facts; particularly, the comparison between the projected returns of the claimant in each case and the reasonable return by reference to the capital markets. In *Isolux,* the claimant was actually obtaining higher returns after the modification than those that it projected before the modification of FIT. [128] By contrast, in *RREEF* the investor was earning less returns under the new regime than those that were

[122] *Isolux Netherlands, BV v. Kingdom of Spain (2016),* para. 793, RREEF Infrastructure (G.P.) Limited and RREEF Pan-European Infrastructure Two Lux S.à r.l. v. Kingdom of Spain (2018) paras. 259, 395–396, 399.

[123] *Ibid.*

[124] RREEF Infrastructure (G.P.) Limited and RREEF Pan-European Infrastructure Two Lux S.à r.l. v. Kingdom of Spain (2018), para. 398.

[125] *Ibid.* para. 394–399.

[126] *Isolux Netherlands, BV v. Kingdom of Spain (2016),* para. 780-781, *RREEF* (2018), para. 398.

[127] *Ibid.*

[128] *Isolux Netherlands, BV v. Kingdom of Spain (2016),* para. 808–811.

projected before the modification of the FIT and, besides, these returns were below the reasonable return determined by the tribunal.[129]

Some conclusions can be drawn from the analysis of case law on RESSs concerning the requirements of predictability and non-essentiality of the modification of the regulatory measures. First, these concepts cannot be interpreted without reference to the creation of legitimate expectations (i). Certainly, from the determination of the expectations that were created follows the identification of the essential characteristics of the framework whose alteration constitutes a disproportional/unreasonable measure (ii). Likewise, if legitimate expectations were created that the essentials of the framework will be maintained throughout project operation, it could be said that changes to these essentials are unpredictable (iii). Of course, the legitimate expectations that determine the essentials of the legal framework, must be buttressed not only by the subjectivity of the investor, but also by objective grounds (iv). Therefore, the regulation needs to be analysed in conjunction with the representations of the host state, the circumstances of the investment and the expertise of the investor to define which expectations correspond to essential characteristics and which do not (v). The circumstances of RESS disputes, in conjunction with the language of the legislation, the long-term character of PV investments, the representations of the host state through several official energy development programmes and presentations showed that the essential feature of the regulation that attracted investors was the 'reasonable return' (vi). Accordingly, the preservation of the reasonable return was reasonably and legitimately expected to be maintained throughout the projects' lifespans (vii), and its alteration would entail an unpredictable change to the essential characteristics of the regulation, i.e. its alteration would be disproportional/unreasonable (viii). This reasoning explains the results of each case; while in *Eiser, Novenergia, Antin* and *RREEF,* the host state was found liable because the measures impaired the reasonable return, in *Isolux* and *Charanne* the investor's claim was rejected because it was still obtaining a reasonable return (ix). In this vein, as a generalisation (systematisation?), one could say that changes to RESSs are expected (predictable) under a climate change context and given the investor's expertise in energy projects as well as the oscillatory dynamic of the energy regulation. However, when the regulation establishes a scheme that guarantees a reasonable return as a key (strategy?) to attract foreign investment (essential), any significant changes to the scheme that result in the cancelation of the reasonable return would be regarded as disproportional/unreasonable.

[129] RREEF Infrastructure (G.P.) Limited and RREEF Pan-European Infrastructure Two Lux S.à r.l. v. Kingdom of Spain (2018), para. 589.

7.1.3 *Ambiguity of Specific Commitments and the Need for a Formal and Material Assessment of Essentiality: Concluding Remarks Concerning the Determination of Reasonableness/Proportionality in RESS Disputes*

Finding a pattern in RESS awards to assess reasonableness/proportionality is devoid of the intention to ignore the discrepancies between the awards. Two ambiguities call for attention: the equivocal concept of specific commitments—which complicates the identification of representations capable of creating legitimate expectations—(Sect. 7.1.3.1) and the different notions of the material and formal conditions that define a relevant alteration of the essential elements of the legal framework (Sect. 7.1.3.2). Even if these intricacies do not affect the general conclusions that I have presented in the sections immediately above, [130] they require some comments to clarify the coherence of the tribunals' language.

7.1.3.1 Ambiguity of Specific Commitments: Restating the Consensus on the Rule That Informal *Representations* in Conjunction with the Context Could Create Legitimate Expectations

The tribunals in the RESS cases refer to the specific commitments in two different ways. A first approach conceives specific commitments as particular instruments, such as stability contracts or stability clauses, which unambiguously contain a declaration of the host state committing to preserve—totally or in part—the regulation existing at the time of the investment, in contrast to other types of binding assurances and to other representations that do not compel the investor to maintain conditions (strict concept). A second conception consists of grasping specific commitments as comprehensive of a broader universe that embraces any representation of the host state capable of creating legitimate expectations in contrast to those representations that do not (open concept).

The identification of the open or strict notion of specific commitments that the tribunal uses when one is reading the RESS awards is vital to disentangling which are the representations capable of creating legitimate expectations. This disentanglement becomes important in terms of understanding the rule that is embedded in several REES awards which establishes that 'in the absence of a *specific commitment* an investor cannot have the *legitimate expectation* that the regulation in place is going to remain unchanged'.[131] This is so because if one understands this rule by

[130] *See* 10.1.2.

[131] Charanne and Construction Investments v. Spain (2016), para. 499; 'absent explicit undertakings directly extended to investors and guaranteeing that States will not change their laws or regulations, investment treaties do not eliminate States' right to modify their regulatory regimes to meet evolving circumstances and public needs' [Eiser Infrastructure Limited and Energía Solar Luxembourg S.à r.l. v. Kingdom of Spain (2017), para. 362].

reference to the strict definition, one might infer that only stability clauses, stability contracts or analogous representations would entail the obligation to preserve all (or certain essential) conditions to the benefit of the investor. Conversely, if one grasps specific commitments openly, one may conclude that other representations of the investor—in addition to specific commitments, stability clauses or similar assurances—are capable of engaging its responsibility. In brief, while a strict conception of specific commitments in this rule would involve that changes to the essential features of the framework are proportional/reasonable, unless they are against stability contracts or stability clauses, etc.; an open conception would make it possible to conceive the alteration of the essential characteristics of the framework as disproportional/unreasonable even if not supported by an unambiguous representation of the host state instrumentalised in a formal declaration.

My reading of the decisions concludes that the statement—'in the absence of a *specific commitment* an investor cannot have the *legitimate expectation* that the regulation in place is going to remain unchanged'[132]—should be understood through the lens of the open concept of specific commitments. In fact, in the cases where the hosts state's conduct was found to be disproportional/unreasonable, it was recognised that the host state's obligation not to alter the reasonable return was grounded on the text of the legislation in line with the long-term character of PV investments, and the representations of the host state through several official energy development programmes and presentations, rather than in stability clauses, stability contracts, etc.

By contrast, a second common statement including the term specific commitments needs to be interpreted in line with the strict definition, to be coherent with the general reasoning of the rulings: 'the expectations of the investor need to originate from some affirmative action of the State, either in the form of *specific commitments* made by the host State to the investor (...) or in the form of *representations* made by the host State'.[133] In this passage, there is a distinction between specific commitments and other representations as two different ways of engaging the host state's obligation to maintain certain conditions of the investment. An open understanding of the term specific commitments in this statement would simply not make sense, since the general thesis of the awards is that the expectation of maintaining a reasonable return did not originate in a specific commitment, but rather in the general legislation accompanied by other representations and the circumstances of the case.

My intention in mentioning the ambiguous way in which the term specific commitments is deployed in the tribunals' decisions is not to criticise the inconsistency or to take sides for the strict or open definition of the concept, but to show how this ambiguity requires a coherent interpretation with the general reasoning of the tribunals, which is—in general terms—that the creation of the legitimate expectation of a reasonable return does not require stability clauses, stability contracts or similar

[132] *Ibid.*

[133] Antin Infrastructure Services Luxembourg S.à.r.l. and Antin Energia Termosolar B.V. v. Kingdom of Spain (2018), para. 537.

assurances (i). Moreover, the ambiguity of the term helps to conclude that the changes to the essential features of the framework of the RESSs are disproportional/unreasonable as long as the 'affirmative action of the State' [134] endorses the commitment to preserve these essential characteristics (ii). This means that, even if it is possible to find in the rulings a trend to acknowledge that the disproportionality/unreasonableness could be verified when the essentials of the legal framework are altered as an expression of the general obligation of stability embedded in the FET, there is still the need to rely on the representations of the host state to determine which are the essentials and whether the host state assumed the obligation to preserve these essentials. Nevertheless, these representations need to be interpreted in accordance with the circumstances of the investment and the context (in general) to generate the legitimate expectation of the maintenance of the essentials (iii). Overall, the tribunals' reasoning could be summarised in simple terms by positing that the proportionality/reasonability of the investor's changes to the RESS regulation requires the host state to honour the obligation of stability embedded in the FET by preserving the reasonable return that characterises the RESS in line with the 2007/2008 decrees—and the prior legislation—and other representations of the host state through official energy development programmes, official presentations, as well as the long-term character of PV investments.

7.1.3.2 Material and Formal Modification to the Essential Elements: Assessing (In)Essentiality by Reference to the Reasonable Return in the Capitals Market, the Legal and Business Environment and Identifiable Criteria

Concerning the material impact of the changes to RESSs as a criterion to assess proportionality/reasonableness, the theoretical disagreement is explicit in the decisions. While *Eiser* seems to emphasise that the total loss of the value characterises the radical alteration of the legal framework, *RREEF* and *Novenergia* posit that the loss does not necessarily need to destroy the value to be regarded as an essential change. *Antin* takes a different approach by linking the essential changes with the alteration of identifiable criteria to provide stability.

In *Eiser,* the tribunal found that the changes to the RESS were 'deeply unfair and unequal', [135] since they emptied the investment from the entire worth. [136] Accordingly, the host state was considered to be in breach of the FET because it 'drastically and abruptly revised the regime, on which their investment depended, in a way that destroyed its value'. [137] The identification of radical changes by reference to the

[134] *Ibid.* para. 538.

[135] Eiser Infrastructure Limited and Energía Solar Luxembourg S.à r.l. v. Kingdom of Spain (2017), para. 365.

[136] *Ibid.*

[137] *Ibid.* para. 387.

'devastator' [138] or 'dramatic effects' [139] of the measures was rejected in *RREEF* and *Novenergia,* but based on different arguments in each of these two disputes.

The tribunal in *RREEF*, on the one hand, contended that *Eiser* (and Novenergia) was contradictory by, simultaneously, affirming that 'Claimants were not immune from *reasonable* changes in the regime applicable to its investment'[140] and deciding 'that the Respondent ought to be integrally compensated for the losses caused by the new regime'. [141] On the other hand, the Tribunal holds that 'it seems very exaggerated' [142] to contend that the effects of the changes to the regulation deprived the investor of the investment. [143] Even if one of the assertions against *Eiser* is of a systemic nature and the other is about a fact, the basis for both is grounded on the special reasoning of the *RREEF* for assessing damages. In the perspective of the latter, the essentiality of the changes to the legal framework would be verified if there is a difference between the return of the new regime and the reasonable return that was legitimately expected by reference to the costs of money in the capital markets, i.e. if the new return is lower than the legitimately expected return the changes are essential. [144] Therefore, substantial damage is not required to determine disproportionality/unreasonableness. In the same vein, the tribunal contends that if the difference between the original and the modified return is the measure of loss, then *Eiser's* finding that the new regime destroyed the whole value of the investment to establish essentiality is an exaggeration. [145]

Novenergia also criticises requiring the destruction of the investment by establishing that the changes to the legal framework are essential. By contrast to *RREEF*, this ruling contends that, although the magnitude of the economic impact of the measures is a factor that influences the determination of the essentiality of the modification to the RESSs, it is just one of many aspects that shape its assessment. [146] Even if this ruling leaves an open space for a loss different from the destruction of the value to satisfy the examination of essentiality of the changes to the legal framework, in this case, the 'significant damaging economic effect' [147] or the 'substantial deprivation of the Claimant's investment' [148] played a definite role in

[138] *Ibid.* para. 409.

[139] *Ibid.* para. 368.

[140] RREEF Infrastructure (G.P.) Limited and RREEF Pan-European Infrastructure Two Lux S.à r.l. v. Kingdom of Spain (2018), para. 515.

[141] *Ibid.*

[142] *Ibid.* para. 517.

[143] *Ibid.*

[144] *Ibid.* para. 515–517

[145] *Ibid.* para. 517.

[146] *Novenergia II - Energy & Environment (SCA) (Grand Duchy of Luxembourg), SICAR v. The Kingdom of Spain* (2018), para. 694–695.

[147] *Ibid.* para. 695.

[148] *Ibid.*

the determination of the changes as radical and unexpected, a decrease in the revenues of '24 and 32% between 2013 and 2016'. [149]

Antin takes a more formal approach than *Novenergia, Eiser* or *RREEF*, in the sense that it relies on the concept of identifiable criteria to define weather the changes to the legal framework are essential. Certainly, the tribunal opts for refocusing the question about the essential changes to the framework by positing 'the issue at hand is not whether the New Regime provides a *reasonable* return, but rather how such "*reasonable* return" is determined'.[150] The reason for refocusing the question is to examine essentiality by reference to the rationale of stability and predictability entrenched in the FET. Accordingly, what is expected from the new regulation of the tariff, in *Antin*'s perspective, is the preservation of identifiable criteria that ensures stability and predictability of the reasonable return. [151] In this regard, the tribunal found that the new definition of reasonable return did not contain identifiable criteria because it 'seemingly depends on governmental discretion (...) in plain contrast with the relative precision of the Original Regime (...)—which provided for objective and identifiable criteria for determining the remuneration due to CSP plants'. [152]

In my view, rather than contradictory, it is possible to read all these rulings as bricks of the same bridge that constitute a twofold approach to the essentiality of the legal framework in RESSs. The examination of whether a change to the regulation of the renewable energy tariffs model altered the essential elements that incentivised the investor to establish operations in the host state focuses on the material and formal impacts (i).

The examination of materiality requires the interpreter to verify whether the tariff's modification impacts the investor's expected reasonable return by reference to the costs of money in the capital markets (ii). This means that the loss experienced by the investor does not need to destroy the investment to establish that the host state's measures altered the essential elements of the framework (iii). It could be that, given the magnitude of the damage in the case of *Eiser*, it was taken for granted that material changes to the framework could exist without the need to elaborate on whether non-substantial damages could be regarded as essential. In fact, it is doubtful that the objective of *Eiser's* reasoning in this regard was intended to deny that damages of a lesser magnitude than the destruction of the investment could constitute a change of the material essentiality.[153] Therefore, *Novenergia* and *Antin* seem to complement, rather than contradict, *Eiser* in relation to the latter's

[149] *Ibid.*

[150] Antin Infrastructure Services Luxembourg S.à.r.l. and Antin Energia Termosolar B.V. v. Kingdom of Spain (2018), para. 562.

[151] *Ibid.*

[152] *Ibid.* para. 568.

[153] 'Contrary to what the Respondent suggests, the actions of the Respondent need not have obliterated the Claimant's investment entirely in order for the Tribunal to consider that the Respondent has breached the FET. The Tribunal disagrees with the approach adopted by the arbitral tribunal in *Eiser*, if it is indeed to be interpreted in the manner suggested by the Respondent'.

reasoning. Furthermore, the requirement of materiality of the essential changes to the framework is evinced in the fact that, whereas in the awards where the tariff return did not decrease, the modifications were considered proportional/reasonable—namely *Isolux* and *Charanne*—, in the awards where the tariff return decreased, the changes were considered disproportional/unreasonable—namely, *Eiser, Antin, Novenergia, RREEF.*

The evaluation of the form requires the interpreter to verify whether the tariff's modification impacts the investor's expected reasonable return by reference to the basic characteristics of the structure of the tariff that incentivised the investor and the preservation of identifiable criteria (iv). This bifurcation of the examination of the form entails, on the one hand, to inquire whether changes in the formal structure scheme 'entirely transform[ed] and alter[ed] the legal and business environment under which the investment was decided and made' [154] and, on the other, to question whether the new regulation preserves the conditions of stability and predictability.[155]

Consequently, on radical transformations of the legal and business environment, if the structure of the tariff changes so radically as to make the factors for the assessment of the return fundamentally different from those existing at the time of investment, the host state's behaviour would be regarded as disproportional/unreasonable (v). This is clear in RESS disputes, where the basis for the calculation of the retribution that was established in the new regulation, by reference to model costs rather than to real costs, was certainly a change in the formal conditions that induced the investment under the 2007–2008 Decrees.

In addition, concerning the preservation of the conditions of stability and predictability, if the new reasonable tariff does not contain identifiable criteria that allow the investor to forecast the viability, payout and retribution of the project realistically, then the changes to the legal framework are considered disproportional/unreasonable (vi). As explained a couple of paragraphs above, this is the reasoning of *Antin*, in which the discretional power of the host state to change the retribution to the investor was considered a criterium that promotes the instability of the regulation.

There seems to be no reason to neglect the examination of the transformation of the basics of the legal structure for determining the formal essentiality of the changes, with the argument of refocusing the establishment of essentiality to the verification of identifiable criteria, as suggested by *Antin*. Both examinations—transformation of the basics and identifiable criteria—are compatible and, indeed, necessarily connected. Certainly, the lack of identifiable criteria of a new regulation in comparison to the original regulation constitutes—per se—a change in the

[*Novenergia II - Energy & Environment (SCA) (Grand Duchy of Luxembourg), SICAR v. The Kingdom of Spain* (2018), para. 694].

[154] *Novenergia II - Energy & Environment (SCA) (Grand Duchy of Luxembourg), SICAR v. The Kingdom of Spain* (2018), para. 695.

[155] *Ibid.*

essential structure of the legal framework. Indeed, this was recognised in *Antin* by including in its reasoning the argument that the new regulation of the tariff embeds a 'plain contrast with the relative precision of the Original Regime'. Moreover, a change to a regime containing identifiable criteria but whose factors for calculating the tariff are grounded on completely different premises to those of the original regulation could still be radical, abrupt and unpredictable, i.e. it could still be essential. In brief, the presence of either or both the transformation of the basic premises or the lack of identifiable criteria constitutes a change in the essential characteristics and, accordingly, an unreasonable/disproportional conduct from the host state capable of breaching the FET.

Having clarified the concepts of specific commitments and essentiality stemming from the differences in the details of the awards, I would like to end this section by outlining my understanding of the operation of the FET standard concerning RESSs:

RESS awards show that the determination of the breach of FET involves the operation of a hotchpotch of elements that could be compressed in the interaction between the obligation of stability, representations, legitimate expectations, due diligence and proportionality/reasonability (i). These elements could themselves be further unfolded in sub-elements that might end up overlapping and blurring the lines between them (ii). However, more than a contradiction this overlapping obeys the fact that the assessment of each element focusses on the same phenomena from a different angle (iii). For example, one could say either that the obligation of stability involves the host state honouring legitimate expectations, and, inversely, the investor legitimately expecting the stability of the legal framework. Moreover, it is possible to assert that the investor can legitimately expect that changes to the regulation will be proportional/reasonable, and claim that proportionality/reasonability involves the host state's changes to the regulation not running against the investor's legitimate expectations. Likewise, one could affirm that the obligation of due diligence means that, if objective circumstances show that the essential features of the legal framework might change, legitimate expectations do not arise; or one could contend that changes to the legal framework are proportional/reasonable if they could be anticipated through a proper due diligence, and, accordingly, they do not run counter to existing legitimate expectations. All of these apparently contradictive (simple) deductions (among others) are present in RESS rulings, but they do not preclude the finding of a pattern that characterises the reasoning of the tribunals (iv).

I found that the tribunals position proportionality/reasonableness as the central concept to assess whether the host state has breached the FET standard by changing the legal framework (v). Although reasonableness and proportionality are different concepts, for the purposes of defining the breach, they are characterised with the same elements; namely non-capriciousness, predictability and preservation of the essentials (vi). While the first—non-capriciousness—requires the hosts state measures to be necessary—in the sense that they serve a public interest—and suitable to the policy objective to be proportional/reasonable (vii), the other two elements—predictability and essentiality—require that host state's action does not abruptly change the characteristics of the legal framework that motivated the investor to invest (viii). These characteristics cannot be identified without the examination of the legitimate expectations of the investor (ix), which arise from the representations of the host state—interpreted—in conjunction with the particular circumstances and within the context of the investment (x). Even if it could be posited that, apart from the eventual existence of specific commitments (stability clauses, stability contracts or similar assurances) the legitimate expectation of preserving the essential elements is buttressed by the obligation of stability that is embedded in the FET, an affirmative action from the host state is needed for this legitimate expectation to emerge and to be determined (xi). Also, the essentiality of the elements of the legal framework and the unpredictability of the changes should be based on an objective examination and conform to reality (xii), in the sense that the investor cannot expect the preservation of certain elements of the framework if the circumstances anticipate that these elements might be altered.

RESS disputes coincide in detecting the reasonable return as the legitimate expectation and, accordingly, its modification as disproportional/unreasonable (xiii). A dual exam—material and formal—determines whether the essential elements were altered (xiv). The material exam does not require the destruction of the whole investment to conclude that the conduct of the host state was disproportional/unreasonable; all that needs to be done is verify a loss by reference to the capital markets (xv). The formal exam is unfolded into two, while on the one hand, it examines whether the basic structural conditions of the legal framework were preserved (xvi), on the other, it tests whether the legal framework preserves identifiable criteria to determine the tariff (xvii). Although in practice these exams analyse common indicators that reveal whether changes are essential, the verification of either material changes to the essential features of the regulation or formal changes to the regulation to the essential features of the regulation, shows that the regulatory measure is disproportional/unreasonable (xviii).

Despite the importance of these findings for my work, the pattern that I have revealed represents a coherence that is liaised to an intimate understanding of the decisions and the investment treaty law system—in general. Thus, I am not implying (at all) that the revelation of a different pattern under a different approach or stemming from a different perspective is not appropriate to understand the reasoning behind RESS disputes. The hotchpotch of elements that intervene in these decisions as well as the polysemy of the terms that are deployed (without any doubt) accept many interpretations. Nevertheless, I defend that the pattern that I have discovered in RESS disputes is also applicable to the investment treaty-climate change disputes that address the opposite situation: measures imposed by the host state that promote the climate interest by affecting investors in industries other than the green industries.

7.2 Generalisation of the Application of the Pattern of Proportionality/Reasonableness to all Investment Treaty-Climate Change Disputes

It is possible to generalise the application of the pattern of RESS disputes concerning the breach of the FET standard to all investment treaty-climate change disputes, i.e. the same methodology deployed in RESS disputes to determine the proportionality/reasonableness could be extended to controversies where host state's measures promote the goal of levelling global warming. I defend the extension of the application of the pattern (and the reasoning behind it) because, on the one hand, it conforms to the practice of the tribunals and, on the other, it represents a balanced answer to the problem of simultaneously providing stability to investors in the green industry and encouraging host states to take measures to prevent, mitigate and adapt to global warming. Both, a reasonable balance between the parties to the dispute in light of a climate change context and systematicity within the practice of investment treaty tribunals, can be achieved by examining non-capriciousness (Sect. 7.2.1), essentiality and unpredictability (Sect. 7.2.2).

7.2.1 The Determination of (Non)Capriciousness in FET Climate Change-Related Disputes: Underpinning the Standard of Arbitrariness as Threshold to Guarantee Regulatory Flexibility

As concluded above from the examination of RESS disputes, a non-capricious measure is one that is necessary and suitable.[156] In other words, if the exam of (non)capriciousness reveals that measures are unnecessary or unsuitable, the host state's activity is labelled as capricious and, therefore, disproportional/unreasonable. In these particular cases, it was found that changes to the regulation were necessary to tackle the tariff deficit and suitable to this purpose,[157] i.e. that the measures were justified by legitimate public interest considerations and that they were equipped to address the policy objective. Following the caveat adopted in RESS awards that the definition of necessity and suitability is a task that, in principle, belongs to the national authorities rather than to the investment tribunals, I argue that the high threshold of (non)capriciousness by reference to the standard of arbitrariness conforms to case law and represents a balanced reading of the FET in line with its purpose (Sect. 7.2.1.1) as well as in context with climate change (Sect. 7.2.1.2).

7.2.1.1 Consistency of the High Threshold of (Non)Capriciousness by Reference to the Standard of Arbitrariness with Case Law and the Purpose of the FET

The definition of the public or national interest (necessity) and the technical design of regulation to achieve the results it pursues, although crucial for the definition of

---

[156] *See supra* 7.1.2.1.

[157] *Ibid.*

the content of the FET obligation, is 'no[t] part of [the] Tribunal's function [...]',[158] but rather 'the respondent's sovereign right, as regulator'.[159] Accordingly, the determination of the FET standard on the interpretation of what constitutes 'the protection of public health, safety, morals or welfare, as well as other functions related to taxation and police powers of states' [160] depends—almost exclusively—on the discretion of the national authorities.[161]

Consequently, the determination of a measure as unnecessary and/or unsuitable—ergo disproportional/unreasonable—is conditioned by a very high threshold, in the sense that it should be utterly obvious that the measure is absent of any legitimate public interest or absent of any possibilities to achieve its own goal. This follows the approach of *Glamis Gold*, whose methodology to assess (non)-capriciousness consists of formulating the question of whether the behaviour of the host state represents a '(...) manifest arbitrariness, blatant unfairness (...) or a manifest lack of reasons (...)'. [162] The tribunal grounds the establishment of this high threshold on the basic reason that '[i]t is not the role of this Tribunal (...) to supplant its own judgment of underlying factual material and support for that of a qualified domestic agency'.[163]

In this vein, the necessity or suitability is not measured by reference to alternative measures, better measures or correct measures but to the standard of arbitrary measures. Undoubtedly, the study of necessity and suitability by reference to the threshold of arbitrariness certainly matches the ordinary language that defines capriciousness and disproportionality/unreasonableness—since it is evident that an arbitrary conduct is unnecessary, unsuitable, capricious and disproportional/unreasonable. Yet, defining *(non)capriciousness* through necessity and suitability, and then defining necessity and suitability as opposed to arbitrariness might appear, on the surface, as tautological. Nevertheless, if one conceives arbitrariness in the determination of (non)capriciousness as a reiteration of the caveat that the tribunal is not a co-authority that is entitled to question the accuracy of the measures,[164] the

[158] Copper Mesa Mining Corporation v. Republic of Ecuador (2016), para. 6.64.

[159] *Ibid.*

[160] *Marion Unglaube v. Republic of Costa Rica (2012)*, para. 246.

[161] In the same sense, *Saluka Investments B.V. v. The Czech Republic (2006)*, para 305–306; *Rusoro Mining Ltd. v. Bolivarian Republic of Venezuela (2016)*, para. 385.

[162] *Glamis Gold, Ltd. v. The United States of America (2009)*, para. 779.

[163] *Ibid.*

[164] 'Some difficulty might arise as to what should be the method of review in cases where either the language of the text is too vague and imprecise, as is the case with most **FET** provisions, and thus no textual indication of where the balance between investment protection and host state's regulatory power should lie, or in cases where the text does not provide any guidance (i.e. no mention of words like "necessary" or any other nexus requirement linking measures with regulatory objective-this will happen while interpreting non precluded measures provisions in BITs). In such cases also, necessity review should be used by arbitral tribunals because it will involve less discretion and thus no second-guessing about the significance of the public interest that a host state wishes to pursue': Ranjan (2014), *pp.* 853, 881.

perception is not that of running around in circles anymore. Moreover, the adoption of the standard of arbitrariness to assess necessity and suitability is of a pragmatic importance as it defines the limits to the deference admitted to host states.

In *Crystallex v. Venezuela* the concept of arbitrary measure was developed in more precise terms as 'not based on legal standards but on excess of discretion, prejudice or personal preference, and taken for reasons that are different from those put forward by the decision maker'. [165] In this dispute, the tribunal found arbitrary the declaration of a unilateral resolution of a mining contract by the government of Venezuela grounded on the suspension of the contract for one year, when it was evident that the real reason behind this was the intention to nationalise *Las Cristinas*—one of the biggest gold pits in the world—for political reasons.[166] A similar reasoning was deployed in *Copper v. Ecuador,* where the tribunal reached the conclusion that the cancelation of a copper mining contract that was formally buttressed by the inexecution of consultations to the community was an arbitrary measure, which was hiding the real intention of nationalising the investment. [167]

The interpretation of FET that establishes a high threshold of deference concerning the determination of (un)suitability or (un)necessity entails a balanced reading of its text in line with its purpose. On the one hand, it provides the host state with considerable flexibility to define its own public policy and to autonomously determine the best way to address the policy goals. The purpose of the FET standard is not to transfer sovereignty over administrative decisions to tribunals. Accordingly, the task of the tribunal—as it has been reiterated—is not to judge the appropriateness or convenience of the host state's measures and/or influence the host state's policy-making, but rather to assess whether a measure is against the proportional/reasonable treatment that the investor can legitimately expect. Certainly, it could be argued that determining necessity and suitability by reference to arbitrariness is still, to a certain extent, adjudging on the accuracy of the measure, but it is also true that a high level of deference expands the host state's power to determine its own policy and administration as much as possible, and reduces the space for the tribunal to rule on the technical and political appropriateness of regulatory measures as much as possible.

On the other hand, the high level of deference does not endorse a blank check to host states to undermine investors' interests since it would not be possible for the former to justify arbitrary measures against the latter with the argument that the determination of necessity and suitability is its exclusive prerogative. In brief, the operation of (non)capriciousness in light of a high threshold represents a balanced reading of the FET standard that ensures a considerable regulatory space to the host state and, simultaneously, preserves the investor's stability from the risk of arbitrary measures.

[165] *Crystallex International Corporation v. Bolivarian Republic of Venezuela (2016)*, para. 578.

[166] *Ibid.*

[167] *Copper Mesa Mining Corporation v. Republic of Ecuador (2016),* paras. 6.14, 6.66, 6.63.

7.2.1.2 Consistency of the High Threshold of (Non)Capriciousness by Reference to the Standard of Arbitrariness with the Objectives of the Climate Change International Legal Regime

Adopting the approach of assessing non-capriciousness in connection to a high threshold of deference concerning the determination of (un)suitability or (un)-necessity also recognises the gravitational power of the climate change international legal framework. Certainly, almost no one would question that climate change measures affecting investors in other industries—such as those phasing out conventional sources of energy—pursue a legitimate public interest and, thus, are necessary.[168] Likewise, the necessity of adjusting the implementation of RESS would be rarely controverted given the fact that the development of green energies involves a considerable level of uncertainty on the efficiency and affordability of new innovative technologies. However, someone could argue that given the particular circumstances, climate change measures are not necessary or that there are other priorities in the global and national agenda that should be prioritised over climate change decisions. Nevertheless, these assertions neglect the gravitational force of the climate change international legal framework which forms part of the context that shapes the interpretation of the FET standard. On the contrary, an expansive interpretation of (non) capriciousness—as elaborated in RESS disputes—that fixes the limits between necessary and unnecessary via the benchmark of arbitrariness, recognises the flexibility that the host state requires to incentivise the development of the objectives of the climate change legal framework.

An analogous conclusion on the expansive interpretation of necessity in the analysis of (non)capriciousness concerning investment climate change-related disputes is applicable to the determination of suitability. Determining suitability, in light of a context where the climate change international legal framework is an integral part, compels the interpreter to look for the maximisation of the flexibility of

[168] '(...) the analysis does not require much elaboration. Environmental and public health reasons are normally recognised as a legitimate public purpose and thus any measure adopted for climate change mitigation or adaptation will most likely pass through to the next stage without much discussion. When it comes to frustrating expectations of low-carbon investors, naturally a legitimate purpose is also needed. Investor-State tribunals should look for guidance in the jurisprudence of other institutions applying *proportionality* and other relevant rules of international law': Isotalo (2016), p. 29. 'Likewise, the strong scientific consensus surrounding climate change will aid host countries in defending challenged climate regulations. In applying the FET standard, tribunals assess whether the host country relied on legitimate scientific evidence as the basis for the measure. In addition, tribunals have linked *legitimate expectations* with the issue of whether the measure was enacted for a proper purpose. The reports of the IPCC—the officially constituted international body with the responsibility to gather and assess scientific evidence on this issue—certainly provide an ample basis. The most recent IPCC report states that evidence of the warming climate is "unequivocal" and that limiting climate change will require "substantial and sustained reductions of greenhouse gas emissions". Host countries can point to the dangers of climate change as evidence of a measure's proper purpose': Wilensky (2015), pp. 10683–10691.

the host state to model the climate change policy.[169] Besides, it is a fact that the host state is technically better equipped to define the best way to implement climate change policies than tribunals. Therefore, limiting the risk of incurring in capriciousness by associating suitability to the standard of arbitrariness is in line with the climate change international legal framework.

Baetens envisaged—almost 10 years ago—that 'the interpretation of the fair and equitable treatment standard could be reframed so as to protect against arbitrary or bad faith conduct but not against legitimate regulation countering climate change'.[170] Although this citation reaffirms the view that a sensible interpretation in light of climate change involves an expansive approximation towards deference that incentivises the host state to take measures to prevent, mitigate and redress global warming; it might result excessively inclined in favour of the host state if the understanding of a legitimate measure is only limited by the threshold of arbitrary measures. Indisputably, the standard of arbitrariness protects the green-energy investor's stability from regulation that evidently lacks reason or purpose, but if the concept of legitimate measure is not complemented by the analysis of proportionality/unreasonableness from the perspective of predictability and essentiality, green-energy investors may find themselves in a disbalanced position—which is the focus of the next subtitle.

7.2.2 (Un)Predictability and (In)Essentiality in FET Climate Change-Related Disputes: The Creation of Legitimate Expectations in a Climate Change Context Stemming from Non-Formal Representations

As it was concluded above from the study of RESS disputes, a non-essential and predictable change in the regulation is the one that preserves the elements of the legal framework that incentivised the investor to invest at the time of the investment. In other words, if the exam of (un)predictability and (non)essentiality reveals that the main features of the legal framework that the investor (objectively and reasonably) expected to be maintained, suddenly change, the alteration is labelled as essential and unpredictable, and thus the host state's behaviour is regarded as disproportional/unreasonable. In RESS disputes, the changes to the reasonable return that the investor expected, in general, were found to be disproportional/unreasonable.

---

[169] '(...) It has been suggested that under normal *circumstances* tribunals should defer to the State's choice of scientific evidence, as well as the evaluation of what measures may be taken in response. With regard to climate change measures, this approach is further supported by the precautionary approach': Isotalo (2016), pp. 29–30.

[170] Baetens (2010).

Despite the lack of consensus in the general case law, on the representations capable of creating legitimate expectations, I argue that—in line with RESS disputes—adopting the trend of de-formalising the creation of legitimate expectations represents a balanced reading of the *FET* in line with its purpose (Sect. 7.2.2.1), and harmonises with the *context* of climate change (Sect. 7.2.2.2).

7.2.2.1 Consistency of the De-Formalisation of the Representations with the Purpose of the FET Standard: Balancing the Host State and the Investor's Interests by Mirroring the Operation of (Un)Predictability and (In)Essentiality in RESS Disputes

RESS awards coincide in the adoption of the principle that the 'State cannot induce an investor to make an investment, hereby generating legitimate expectations, to later ignore the commitments that had generated such expectations'.[171] Although this principle is generally accepted in case law, decisions in investment treaty arbitration are inconsistent in the identification of the representations capable of creating legitimate expectations, or at least, they are not as consistent as the decisions on RESSs.

The controversies on the changes in the regulation of the gas tariff during the Argentinian economic crisis of 2000 are a good example of the different theses on the representations capable of creating expectations. In *Enron v. Argentina* [172] and *CMS v. Argentina,* [173] the tribunal established that legitimate expectations were created because the gas regulation addressed a particular type of investor. Therefore, the measures that compelled investors to recalculate the tariff in Argentinian pesos instead of US dollars as well as the measures cancelling the annual readjustment of the tariff were found to run counter to legitimate expectations.[174] Inversely, in *Metalpar v. Argentina*—which was a controversy originated from the same facts—it was held that legitimate expectations can only arise from a license, permit or contract. [175] The same reasoning was adopted in *Total* v. *Argentina*:

> In the absence of some "promise" by the host State or a specific provision in the bilateral investment treaty itself, the legal regime in force in the host country at the time of making the investment is not automatically subject to a "guarantee" of stability merely because the host country entered into a bilateral investment treaty with the country of the foreign investor. **The expectation of the investor is undoubtedly "legitimate",** and hence subject to protection under the fair and equitable treatment clause, **if the host State has explicitly assumed a specific legal obligation for the future, such as by contracts, concessions or**

[171] Charanne and Construction Investments v. Spain (2016), para. 486.

[172] *Enron Corporation and Ponderosa Assets, L.P. v. Argentine Republic (2007).*

[173] *CMS Gas Transmission Company v. The Republic of Argentina (2005).*

[174] *Ibid., Enron Corporation and Ponderosa Assets, L.P. v. Argentine Republic (2007).*

[175] *Metalpar S.A. and Buen Aire S.A. v. The Argentine Republic (2008).*

stabilisation clauses on which the investor is therefore entitled to rely as a matter of law[176] **[emphasis added]**

El Paso—another award concerning the modification of gas tariffs during the Argentinian economic crisis—adopts a broader approach than *Metalpar* and *Total* but different from *Enron* and *CMS*, on representations that can engage the host state. This decision rejected the potential of the regulation to generate specific commitments by itself—in contrast to *Enron* and *CMS*—although—in contrast to *Metalpar* and *Total*—it expressed that the creation of a specific commitment is not conditioned to any formality. [177] The tribunal's grounds for adopting this posture was that 'no general definition of what constitutes a specific commitment can be given, as all depends on the circumstances'. [178] **[emphasis added]**

Although *El Paso* avoids defining specific commitments in general terms, the tribunal suggests that representations generate binding commitments when they address specific recipients or when recipients could be identified by interpreting the object and the purpose.[179] As examples of the first group—representations with a particular recipient—the tribunal mentions statements that are made in 'a contract or in a letter of intent, or even through a specific promise in a person-to-person business meeting'.[180] On the object and purpose, the tribunal states that even if, usually, general declarations of the host state do not create binding commitments, the reiteration of a declaration suggesting stability could be interpreted according to the circumstances as generating legitimate expectations.[181]

In addition to the different theories about specific commitments deployed in the awards concerning the changes to the gas tariff during the Argentinian economic crisis, *Murphy v. Ecuador* proposes an alternative where the creation of legitimate expectations does not require specific commitments:

> An investor's *legitimate expectations* are based upon an **objective understanding of the legal framework** within which the investor has made its investment (...) Specific *representations* or undertakings made by the State to an investor also play an important role in creating *legitimate expectations* on the part of the investor but they are not necessary for *legitimate expectations* to exist. An investor may hold *legitimate expectations* based on **an objective assessment of the legal framework absent specific *representations* or promises made by the State to the investor.** [182] **[emphasis added]**

---

[176] *Total S.A. v. The Argentine Republic*, Decision on Liability (2010), para. 117.

[177] *El Paso Energy International Company v. The Argentine Republic (2011)*, para. 375.; In a similar way: *Mamidoil Jetoil Greek Petroleum Products Societe S.A. v. Republic of Albania (2015)*, paras. 643, 644.

[178] *El Paso Energy International Company v. The Argentine Republic (2011)*, para. 375.

[179] *Ibid.*

[180] *Ibid.* para. 376.

[181] *Ibid.* para. 377.

[182] *Murphy Exploration & Production Company International v. Republic of Ecuador (2016)*, para. 248.

In this case, the parties novated a service contract for hydrocarbon exploration and exploitation to a production shared agreement.[183] Under the new model, the investor assumed the risks of exploitation and exploration and obtained a percentage of the production as retribution. Accordingly, the investor assumed the risk of the changes in the market price—in the sense that if the price increases the profit increases and vice-versa, if the price decreases the profit decreases.[184] Given that the market price increased considerably, the host state imposed a 99% tax on the additional income for high prices.[185] The tribunal found that the tax went against legitimate expectations because, even in the absence of specific commitments, the legislative and regulatory reforms—which included constitutional reforms and the Investment Promotion and Guarantee Law—were enough to convince the investor that the conditions for the retribution in the event of high prices would be maintained through the whole term of the contract.[186]

The approach adopted in the RESS decisions pendulates between the reasoning of *El Paso* and the reasoning of *Murphy*. As described above, RESS awards are ambiguous in their use of the term specific commitments; sometimes deploying the term in a strict sense that restricts its meaning to unambiguous formal declarations addressed to the investor—such as stability contracts, letters of intent, stability clauses, etc.—and other times in a broader sense, which embraces any representation capable of convincing the investor that the conditions of the investment will be preserved during the project lifespan. However, regardless of the scope they assign to specific commitments, it was clear in all the RESS awards that any representations of the investor—including those comprehended in the legislation and the regulation—interpreted in conjunction with the circumstances of the investment and the context in which they were perfected, are capable of creating legitimate expectations. In the end, the notion of legitimate expectations devoid of formalities trumps the approach that identifies the legitimate expectations with unambiguous formal declarations. Nevertheless, an affirmative action is still required of the host state to create legitimate expectations.

Devoid of formalities, the concept of legitimate expectations that shapes the operation of essentiality and unpredictability in RESSs could be extended, in general, to all investment treaty-climate change disputes as it represents a balanced reading of the FET's text in line with its purpose. Otherwise, the power to regulate would be expanded, so as to render the FET standard almost useless in terms of protecting the investor's interests. Indeed, an investment treaty law system that restricts the determination of disproportionality/unreasonableness to either the event where the host state's conduct reaches the threshold of arbitrariness—non-capriciousness[187]—or the host state violates unambiguous formal representations—

---

[183] *Ibid.* para. 258.

[184] *Ibid.* para. 273

[185] *Ibid.* para. 288.

[186] *Ibid.* paras. 248, 249, 281.

[187] *See supra* 10.2.1.1.

stability contracts, stability clauses, etc.—would be clearly disbalanced in favour of the host state.

(Non)Essentiality and (non)Predictability, understood as the examination of the host state's compliance with the legitimate expectations that arise from representations that, given the circumstances and the context, acquire a binding nature, provide a considerable orbit of protection to the investor that counterbalance the power of the host state to regulate. This balance is reflected in the fact that, if, on the one hand, the host state is entitled to choose its policy and the mechanisms to implement the policy (deference), on the other, the investor is entitled to believe that the state will honour the stability of the conditions that objectively motivated the investment. Additionally, with this solution the host state will still keep for itself the general competence and freedom to modify the regulation as long as it does not change the essential features of the legal framework as they were defined through the legitimate expectations.

Additionally, having a non-formal interpretation of the representations that could constitute legitimate expectations and whose violation, accordingly, is essential and unpredictable—ergo disproportional/unreasonable—is also in line with the principle that the host state cannot induce investors to invest for later dishonouring its commitments. Certainly, it is possible to restrict the scope of this principle to formal unambiguous declarations—with the understanding that they are the only valid channel to create legitimate expectations—but one would need these restrictions to emanate expressly from the text. Tying up essentiality and predictability (and legitimate expectations) to formal unambiguous representations would be, merely, interpreting the principle against the main purpose of the FET, which is encouraging the stability of the investment.

Also—paraphrasing *El Paso*—I agree with the fact that 'no general definition of what constitutes a specific commitment [or binding commitment] can be given, as all depends on the circumstances'.[188] Certainly, what endorses the representation with a binding character is not the existence of the representation as such, but the representation understood in a given context and taking into account the particularities of its occurring. Therefore, defining the essential features of the legal framework only by reference to formal unambiguous declarations—such as stability contracts, stability clauses, etc.—would be ignoring that the context and the circumstances are what induce the investor to either rely on the host state's representations or not. Again, the purpose of the FET is to protect the investor's stability; and this includes protecting it from the alteration of the essential features, which it believed would remain untouched during the project lifespan, based on objective grounds. Consequently, restrictions to the informality of representations capable of engaging the host state should not be the general rule but the exception and, accordingly, they would need to be expressed in the BIT to be valid.

[188] *El Paso Energy International Company v. The Argentine Republic (2011)*, para. 375.

7.2.2.2 Consistency of the De-Formalisation of the Representations with the Climate Change Context: Harmonising the Climate Change Law with the Investment Treaty Law by Mirroring the Operation of (Un)Predictability and (In)Essentiality in RESS Disputes

Devoid of formalities, the concept of legitimate expectations that shapes the operation of essentiality and unpredictability in RESSs could be extended, in general, to all investment treaty-climate change disputes as it also recognises the gravitational power of the climate change international legal framework. First, the informality of the representations capable of generating binding commitments incentivise investments in the green energy industry since the investor would be confident that the essential conditions that have been offered will be maintained during the whole project. Second, the host state might extend its power to regulate non-green industries since the climate change context would impede, in many cases, the creation of the legitimate expectation that the regulation of the non-green industries will remain unchanged.

The informal approach to representations capable of creating binding commitments acknowledges the influence of the climate change context and the climate change international legal framework since it bestows a central role to the particular context in the determination of which elements of the legal framework are essential. In contrast, the opposite approach, by requiring the existence of unambiguous formal representations of the host state to create legitimate expectations, would undermine the climate change context in which representations are made. In other words, as this formal approach defines the binding character of representations through the type of instrument and the addressee—stability contract, stability clause, letter of intention, etc.—it does not provide any weight to the climate change context—or at least not as it would be given under an informal approach.

Certainly, investors today cannot ignore the existence of climate change and the natural effort of host states to address the problems associated with it. Likewise, the permanent development of the international climate change legal framework that calls and compels states to take climate change action is an evident reality that shapes the context of the investment. Therefore, it would be very difficult for investors in non-green industries to successfully argue that changes to the regulation were essential and unpredictable when these changes are related to climate change concerns and could be expected to occur in advance. Accordingly, it would be unlikely for tribunals to label changes in the regulation addressing climate change as disproportional/unreasonable.

As discussed above, legitimate expectations are determined objectively. This means that the determination of the elements of the legal framework which are expected to be unchanged during the project is dependent on the standard of due diligence. Indeed, one cannot assume the investor to be a futurologist who is bound to predict all changes, but in the case of climate change, the global concern is so evident that it would be naïve for an expert in the field to imagine that the regulation will not be modified to tackle global warming. If the climate change context does not allow the investor to conceive the absence of climate change actions against its interest as an essential element of the framework, then the regulatory changes in this regard are proportional/reasonable and do not lead to the breach of the FET standard.

Acknowledging the power of the context to impede the creation of legitimate expectations is not new in case law. In *White v. India*, for example, the tribunal did not accept that the investor had the legitimate expectation to obtain the execution of an ICC ruling in its favour from local tribunals, because 'at the time the Contract was negotiated, the Indian courts were regularly entertaining set-aside applications in respect of so called foreign awards'. [189] In the same vein, in *Mamidoil v. Albania,* the tribunal established that the legal endorsement of the legal framework's stability is conditioned to a diligent evaluation of the information and the interpretation of the local law in context. [190] Consequently, it was ruled that, in the context of the Albanian political and financial instability—that was the result of the transition from communism to liberal capitalism—the investor could not legitimately expect the same 'stability as in Great Britain, USA or Japan'.[191]

As part of the orthodox approach to the influence of the climate change international legal framework in investment treaty-climate change disputes, the scholarship in general accepts the difficulties for legitimate expectations to arise when the investor in non-green industries can see all the signs through which to predict that the legal framework will probably shift towards adopting greener policies. Vadi, for example, affirms that 'it may be difficult to argue that climate change mitigation measures are not foreseeable, as climate change has made headlines'.[192] Similarly, Wilensky posits that

> host countries may argue that investors do not have *legitimate expectations* that climate change regulations would not be implemented. Host cоuntries may point to existing international agreements under the United Nations Framework Convention on Climate Change (UNFCCC) and numerous domestic laws and policies as indicators that such measures were imminent. The intensive publicity surrounding climate regulation over the past several years also means that the adoption of such regulation should hardly come as a surprise.[193]

Moreover, the scholarship even suggests that acknowledging the climate change context as a barrier to create legitimate expectations for the investor in non-green industries is foreseeable in future tribunal decisions, given the tendency of tribunals to weight international global interests in the process of treaty interpretation. Firger and Gerrard, in this respect, contend that as 'climate change and issues such as human rights are increasingly prioritized on the international stage, a legal realist would expect arbitral tribunals to bend doctrine to changing norms and

[189] *White Industries Australia Limited v. The Republic of India (2012).*

[190] *Mamidoil Jetoil Greek Petroleum Products Societe S.A. v. Republic of Albania (2015),* para. 634.

[191] *Ibid.* paras. 625 y 626.

[192] Vadi (2015), pp. 1285, 1331.

[193] Wilensky (2015), ELR 10683, 10690–10691.

expectations'.[194] The authors add that '[t]his appears, in large measure, to be precisely what most arbitrators have done'.[195]

The absence of the legitimate expectation of the legal framework's stability would lead to the absence of essential elements; and, accordingly, to the impossibility of referring to disproportional/unreasonable conducts based on the unpredictable alteration of the essential elements that the investor considered at the time of investing. Of course, this affirmation cannot be taken in absolute terms since each case has its own particularities. It may happen that the tribunal infers that some specific representations of the host state were so strong and overwhelming that even in a climate change context they would give rise to legitimate expectations and essential elements. Also, according to Baetens and Miles, given the long-term nature of large energy projects—for example—it could happen that the investment was made before the climate change context was so compelling, so the existence of expectations would be regarded as reasonable and based on objective grounds.[196] Nevertheless, generalising the pattern of RESS disputes can lead one to conclude that it is unlikely that a host state's changes to the regulation in a climate change context would be disproportional/unreasonable, when these changes coincide with the objectives of the climate change international legal framework. This conclusion, evidently, incentivises host states to take climate change action and reduce regulatory chill concerning climate change regulation.

Against this background, it has been unveiled that the informal approach to representations that was deployed in the RESS disputes adjusts to the cases where the host state restricts investments in non-green industries to encourage actions to address climate change and to the cases where the investor in the green-industry requires the stability of the legal framework (i). This interpretation is convenient because it adapts to the investment treaty law system, preserves the balance between the parties and encourages climate change action. It also adapts to the system since it unfolds the application of the principle that the host state cannot induce investors to invest to later dishonour its commitments (ii) and it acknowledges the fact that it is not possible to establish the binding character of a commitment without reading the representations under the circumstances and the context (iii). It preserves the balance between the parties because, while the host state retains the power to modify the

[194] Firger, Gerrard (2011).

[195] *Ibid.*

[196] 'An investor could argue that this violates the *legitimate expectations* which it considered when making the investment, to such an extent that the host State can be said to have acted in breach of the fair and equitable treatment standard. More in general, for investments which were established a considerable time before the entry into force (or even the negotiations) of the Kyoto Protocol, investors could claim that the changes in regulation caused by the Protocol were impossible to foresee and yet significant enough to warrant a finding of a breach of fair and equitable treatment': Baetens (2010), p. 14; While it may be difficult to argue that climate change mitigation measures were not foreseeable for investments entered into more recently, the 'basic expectations' of investors when entering into much earlier carbon-intensive operations would certainly be affected by regulation seeking to shift economies to a low carbon approach: Miles (2008), p. 31.

legal framework, the investor is protected against changes to the conditions of the regulation that motivated it to invest (iv), as long as one can infer that these essential features are grounded on legitimate expectations. It encourages climate change action as, on the one hand, it recognises (in general terms) that investors in RESSs can legitimately expect the essential features of the legal framework to be maintained during the project (v), and on the other, it impedes (in principle) investors in non-green industries from legitimately expecting climate measures affecting the investment not to be taken by the host state against their interest (vi).

The analysis of proportionality/reasonableness in FET climate-change related disputes that I propose by mirroring RESS disputes could be explained as a process of interpretation that conforms to the orthodox approach, in the sense that it focuses on the understanding of the text in the climate change context and relies on confronting the host state's conduct with the elements that compose the definition of proportionality/reasonableness. In the end, it is a matter of verifying whether the facts correspond to the definition of proportionality/reasonableness or not. However, this is not the only perspective in which proportionality/reasonableness has been analysed by the scholarship. As a result of the influence of comparative public law in investment arbitration, some authors conceive that proportionality/reasonableness should be determined through a proportionality test weighting the different interests at stake. In the next subsection—apart from presenting some concluding remarks on the gravitation of climate change in the interpretation of the FET—I present arguments to reject this conception.

7.3 The Rejection of the Public Law-Modeled (Interest-Weighing) Proportionality Test and the Limits of the *Orthodox Approach*: Concluding Remarks on the *Orthodox Approach* in FET Climate Change-Related Disputes

As described above, the literature on climate change issues and the breach of the FET standard refers, in general, to the concepts of proportionality and/or reasonableness as the solution to balance the interests of the parties in FET climate change-related disputes. However, it must be noted that, if the interpretation of the FET standard is the subject of discussion because of its openness, the quest for the meaning of proportionality and reasonableness is not less problematic. It suffices to read the references that were made in the introduction of this chapter to detect this inconvenience: while Schill understands proportionality as a reasonable balance between the purpose, the benefit and the burden to the investor, and associates its determination to the reasonable foreseeability of the changes to the regulation, Isotalo refers to the legitimate purpose, suitability, necessity and proportionality *strictu sensu* of the climate change measure, excluding reasonableness.[197] Vadi and VanDuzer, in contrast, seem to grasp reasonableness as a concept that explains itself and, instead of providing a definition, they illustrate with cases how the concept may operate in FET climate change-related disputes.[198]

However, I have insisted that there is a pattern that can be traced in the tribunals' decisions and, particularly, in RESS decisions, which shows a logical interplay between the elements that define proportionality/reasonableness. The starting point was finding out that *(dis)proportionality and (un)reasonableness*—albeit theoretically distinguishable—share the same constitutive elements, i.e. they are both determined through the concepts of (non)capriciousness, (in)essentiality and (un)-predictability (i). Subsequently, these concepts were studied independently, grounded on the discovery that the decisions established that measures are disproportional/unreasonable if they involve either capricious changes to the legal framework or unpredictable changes to the essential features of the legal framework (ii).

Concerning the operation of the first concept—(non)capriciousness—it was found that a capricious measure is the one that is either not justified by the public interest (unnecessary) or unfit to achieve the purpose for which it was created (unsuitable) by reference to the threshold of arbitrariness (iii). This involves granting a high level of deference to the host state (iv).

On the two other concepts—(in)essentiality and (un)predictability—it was observed that their operation is inexorably linked to the creation of legitimate expectations (v). This is so because the determination of the elements of the legal framework that are essential and which the investor expects will be preserved throughout the extension of the project, depends on the objective capability of the representations of the host state in conjunction with the circumstances and the context to create those expectations. Additionally, it was detected that the decisions on RESSs—regardless of the content they assign to the concept of specific commitments—conceive that the host state's representations could potentially create legitimate expectations devoid of any formalities, i.e. legitimate expectations are not restricted to stability contracts, stability clauses, letters of intention or any other similar unambiguous formal declaration addressed to a particular investor (vi).

The (methodological) interpretation of (dis)proportionality/reasonableness through the concepts of (non)capriciousness, (in)essentiality and (un)predictability, that was deployed in the RESSs decisions, could be extended to all FET climate change-related disputes (vii). It does not only conform to the system (practice) of the investment treaty law (viii) but also provides an interpretation that balances the relationship between the investor and the host state (ix), and acknowledges the gravitational power of the climate change international legal framework (x).

On the one hand, an interpretation of (non) capriciousness—as elaborated in RESS disputes—that sets the limits of necessity and suitability through the benchmark of arbitrariness, captures the flexibility that the host state requires to incentivise

[197] Schill (2007), p. 469, 476; Isotalo (2016), pp. 29–32.

[198] VanDuzer (2016), p. 442; Vadi (2015–2016), p. 1136.

it to pursue the development of the goals of the climate change legal framework (xi) while protects the RESSs investors against policies that lack a public interest reason or that do not serve to meet the policy purpose (xii). On the other hand, an interpretation of (non)essentiality and (un)predictability that recognises the creation of legitimate expectations as a product of (informal) representations, in conjunction with the context and the circumstances, encourages climate change action by determining that investors in RESSs can legitimately expect that the essential features of the legal framework which motivated its investment in green industries will be maintained during the project (xiii) and by impeding (in principle) that investors in non-green industries legitimately expect that climate measures affecting the investment will not be taken by the host state against their interest (xiv).

Although my posture refines previous works of the scholarship by providing a systematic, balanced and methodic interpretation to the concept of proportionality/reasonableness in the FET standard that captures the gravitational power of the international climate change legal framework, it is worth considering the following two arguments that lead me to reiterate the place that my approach to FET occupies in the whole body of this book: (first) the (public law-modelled) interest-weighing proportionality test cannot provide an acceptable interpretation of the FET standard in the context of climate change (Sect. 10.3.1) and (second) my orthodox approach does not consist in the direct application of principles of the international climate change framework and it is still unsatisfactory to promote climate interest in all cases (Sect. 10.3.2).

7.3.1 *The Drawbacks of a Public Law-Modeled (Interest-Weighing) Proportionality Test in the Interpretation of the FET Standard in the Context of Climate Change: Emphasising the Importance of a High Level of Deference*

The proposal of adopting a proportionality test—by mirroring public law's interest-weighing method—which consists of a 4-step process that involves the assessment of legitimacy, necessity, suitability and proportionality *strictu sensu*, has been gaining some eco as a method to solve the tension between the host state's interests and the investor's interest in the FET. Prominent authors such as Schill[199] or Henckels[200] could be considered among the pioneers of this approach, but it is Isotalo's brilliant 2017 paper, which directly suggests the application of this type of analysis to FET climate change-related disputes.[201] To a certain extent, the analysis

[199] Schill (2010), pp. 131–133.

[200] Henckels (2015).

[201] Isotalo (2016).

of proportionality/reasonableness that I have traced in RESS awards—and extended to FET climate change-related disputes—coincides with this approach.

However, replicating public law (interest-weighing) proportionality tests to assess the breach of the FET cannot be done without depleting the high level of deference granted to the host state in investment treaty disputes and without exceeding the pliability of the FET standard. This is so obvious that Isotalo—and Henckels in a broader scope—puts a considerable effort into harmonising the public law's proportionality test with the deference in FET climate change-related disputes.[202] The success of such an effort is dubious, since the erosion of deference appears to be inherent to the public law (interest-weighing) proportionality test. A comparison between the approach that attempts to transfer, from public law, the proportionality test—represented by Isotalo's work concerning climate change—and the methodology to assess proportionality/reasonableness that was traced in RESSs decisions—my approach—serves to explain why I posit that the former approach is inadequate to interpret the FET standard in the context of climate change as well as to emphasise that the FET standard should be interpreted in accordance with a high level of deference.

First—while Isotalo establishes the legitimate purpose as an independent element of necessity,[203] I found that the legitimate purpose is what characterises the element of necessity. From my perspective, if necessity is nothing but the identification of the policy objective of the measure as (i) legitimate, then there is no point in showing a difference between the two categories: legitimate purpose and necessity. By contrast, Isotalo suggests that the evaluation of the legitimate purpose is different from the evaluation of necessity because the latter's function is to determine whether there was an alternative measure that would be less damaging to the interests of the investor. Accordingly, if it is found that there was such an alternative measure, in her perspective, the measure would be unnecessary, meaning that the host state's conduct should be labeled as disproportional.[204]

---

[202] 'There are important reasons for investment treaty tribunals to show deference to the decisions of national authorities. Some of these include the higher legitimacy of national decision-makers, their better accountability and the principle of separation of powers. National authorities are also most likely in a better position to act according to the will of the people they represent, to make assessments of relevant facts including scientific evidence, and to choose measures suitable for domestic circumstances. As noted, intrusive interpretation of treaty obligations is also seen as potentially undermining municipal climate change laws. Tribunals should therefore tread carefully and make sure that, while upholding the object and purpose of the investment treaties, they produce awards of good quality and sound reasoning and refrain from going beyond their institutional capacity. In many situations this means showing an appropriate level of deference to national decisionmakers': Isotalo (2016), pp. 20–21. 'As this book has argued, the problems inherent to proportionality analysis can, arguably, be attenuated through an institutionally sensitive approach to the standard of review and by avoiding the value-laden proportionality *strictu sensu* stage'. Henckels (2015), p. 172.

[203] Isotalo (2016), p. 29.

[204] *Ibid.*

The different approach on the categorisation (and definition) of legitimate purpose and necessity, more than representing a formal discrepancy in the language, denotes a clashing view on the operation of the FET standard concerning the level of deference. While my view—inspired by RESS disputes—intends to preserve the autonomy of the host state as much as possible, not only in choosing its own policies (necessity/legitimate purpose) but also in designing the measures to achieve these policy objectives (suitability), Isotalo's view erodes deference by attributing the role of deciding on the appropriate policy measure to the tribunal. This erosion is unavoidable in her approach since the definition of necessity that she suggests is attached to the evaluation of whether there was a better alternative measure.

Second, while Isotalo's definition of necessity overlaps the definition of suitability, my approach makes a clear distinction between necessity and suitability, and insists on preserving a high level of deference. She explicitly recognises that '[m]any of the issues relating to this step [necessity] in investment arbitration are fairly similar to the suitability inquiry'.[205] Indeed, it is very difficult to distinguish between necessity—understood as the evaluation of whether there were better measures to achieve the policy—and suitability—understood as the evaluation of whether the measures had the potential to meet the policy objective. Even if there is a distinction in the intensity of the inquiry in Isotalo's assessment of suitability and necessity, the object is still the same for both: the technical accuracy of the measure. It is evident that a measure lacking the potential to achieve the policy objective (unsuitable) is not the best possible measure (unnecessary). In contrast, in my scheme, the object of inquiry of necessity and suitability is completely different: necessity refers to the legitimate purpose of the measure and suitability, to the measure's potential to achieve the result.

Third—again—the disagreement in the conception of suitability and necessity between Isotalo and I is impregnated with a different conception about the level of deference. This has an impact on the evaluation of FET climate change-related disputes. In my perspective, climate change measures—either restricting non-green industries or promoting low-carbon industries—are in general necessary and suitable in the sense that they are only limited by the threshold of arbitrariness. This maximises the autonomy of the host state to define whether climate change policies are part of the local agenda and to determine the way in which climate change policies should be achieved through regulation. As a consequence, the host state's need to restrict RESSs or non-green industries would rarely be questioned or be regarded as capricious. Only in exceptional cases where the policy objective clearly pursues non-legitimate purposes or where the measures do not clearly tackle the policy objective that they are meant to pursue, changes to the regulation would be considered as unnecessary or unsuitable—respectively. In contrast, Isotalo's perspective attributes the tribunal a very wide discretion to define whether the climate change measures were necessary and suitable. This is true to the level that one may

[205] *Ibid*. p. 30.

think that the tribunal is exceeding its role of adjudging authority and invading the host state's space for policy-making.

Certainly, Isotalo attempts to moderate the power of the tribunal by establishing that 'under normal circumstances tribunals should defer to the state's choice of scientific evidence, as well as the evaluation of what measures may be taken in response'.[206] In her opinion, only exceptional circumstances justify 'decreased'[207] deference in the evaluation of the necessity and the suitability of the measures affecting low-carbon investors as well as in the evaluation of climate change-related measures affecting non-green investors.[208] According to her, even if the level of deference should be determined on a case-by-case basis, the following factors might indicate that it is appropriate to exclude the application of decreased deference:

(i) (...) when evaluation of the measures **requires complex factual assessments** related to matters such as science, macro-economic policy, etc., which tend to go beyond the institutional capacity of judicial bodies
(ii) Decisions based on a robust and transparent procedure warrant greater deference
(iii) Expectations that have a higher level of legitimacy warrant greater scrutiny of the interfering measures
(iv) Wide-**ranging international recognition of a given public purpose** warrants greater deference
(v) **Interference that is particularly** great warrants greater scrutiny[209] **[emphasis added]**

Although Isotalo's attempt to allow decreased deference only in exceptional circumstances has the virtue of theoretically recognising the autonomy of the host state to direct its own policy-making, it is very difficult to conceive how the power of the tribunal to define whether there were alternative measures that could have been better suited to serve the policy purpose—that she proposes—could be exercised without undermining deference as general rule.

Fourth, while Isotalo's methodology to assess proportionality comprehends the analysis of proportionality *strictu sensu*—which further weakens deference to the host state—my methodology omits this step. The *proportionality strictu sensu* that she formulates is intended to determine 'whether the importance of satisfying the legitimate purpose justifies the level of interference'.[210] In this vein, a measure could be labelled as disproportional, for example, 'where the withdrawal of the support scheme substantially harms the investment but only amounts to a relatively minor benefit to the host state'[211] or, inversely, 'In the case of an investor challenging e.g.,

[206] *Ibid.* p. 29–30.
[207] *Ibid.* p. 30.
[208] *Ibid.* p. 29–30.
[209] *Ibid.* p. 27.
[210] *Ibid.* p. 32.
[211] *Ibid.*

the newly-introduced emissions trading scheme, it is likely that (...) even a significant interference with profits would not lead to a finding of disproportionality'.[212]

In sum, in Isotalo's scheme the role of the tribunal is not only to define whether the measure was appropriate to address the policy objective (suitability) and whether it was the least restrictive measure (necessity), but also whether the public interest profits enough from the regulatory measure to justify the negative impact on the investor (proportionality *strictu sensu*). Once more, it is worth reiterating that a proportionality assessment in these terms has no other possibility than undermining deference. Even Henckels—who is a prominent supporter of the adoption of a public law-modelled (interest-weighing) proportionality test—disagrees with the adoption of the proportionality *strictu sensu* and comments that

> investment tribunals are not well placed to engage in the *proportionality strictu sensu* stage of analysis and should not do so in order to reduce the scope for subjectivity and value judgements in their-decision making—and the negative impact that this approach can have on both legal certainty and regulatory autonomy[213]

The decreased level of deference that results from the application of a public law-modelled (interest-weighing) proportionality test works against the purpose of the FET standard. It is important not to forget that the balance between the investor's interest and the host state's interest should do justice to the fact that, on the one hand, BITs are treaties that are agreed between states to develop the local economy through the granting of stability to transitional investments and, on the other hand, they still aim to preserve the host state's autonomy to decide its own policies. In this regard, it would be contradictory to read the FET standard as if its purpose were to grant the tribunal the competence to decide which climate change (or other) policy to choose, the most effective appropriate design of this policy, as well as the level of interference to the investor's interest that is justified for the public interest.

An orthodox approach to the interpretation of the FET standard concerning Climate Change measures is limited by the pliability of the standard itself. Certainly, the role of the Tribunal under the FET is to prevent the state from abusing its position, as an instrument to incentivise the flow of transnational investment. However, this role consists in determining whether the host state measures provided a fair treatment rather than defining whether the policy of the host state was the best possible policy. The decreased level of deference that is inherent to the public law-modelled (interest-weighing) proportionality test exceeds the pliability of the FET standard since its operation ends up attempting to define which the best public policy was and, consequently, reveals that the transposition of this model into investment treaty law is inadequate. In contrast, my model for assessing (dis)-proportionality in the FET does not fall into the trap of promoting a pragmatic approach that provides weight to climate change without making systemic considerations, as I explain in the next subsection.

---

[212] *Ibid.*

[213] Henckels (2015), p. 193.

7.3.2 *The Limits of the Orthodox Approach in FET Climate Change Disputes: Impossibility of Directly Applying Principles of the Climate Change International Legal Framework*

Throughout the chapter, I have proposed that the FET standard should be interpreted under an orthodox approach in a climate change context. Accordingly, the harmonisation of the FET's text with the climate change international legal framework is restricted to the task of unearthing the balance between the investor and the host state's interests as embedded in the BIT and guided by the VCLT principles of interpretation (the private dimension). As such, an analysis in these terms avoids the direct application of provisions belonging to climate change international legal framework. Yet, the gravitational power of the climate change legal framework is recognised as an element that casts light on the understanding of the standard. Two particularly important findings in the system's operation were decisive in perfecting an orthodox approach that conforms to the protection of the climate interest in FET disputes: the de-formalisation of legitimate expectations and the recognition of a high level of deference to the host state.

The de-formalisation of the creation of legitimate expectations leads to the conclusion that their determination—and consequently the concept of (dis)-proportionality/(un)reasonableness—in FET climate change disputes cannot be assessed without referring to the climate change phenomenon. Therefore, in general, one can hold that investments in the green industry will be incentivised by recognising that investors can legitimately expect that the essential RESS elements will be maintained during the project lifespan. Moreover, under the same reasoning, one can assert that investors in non-green industries cannot legitimately expect the regulation not to be modified, resulting in the host state's motivation to take climate action.

The recognition of a high level of deference guarantees the host state considerable flexibility in taking climate change measures without fearing the risk of breaching the FET standard. As a consequence, the host state will be motivated to take climate action. Although this interpretation may appear on the surface to favour the host state and disincentivise investments in green industries, that would not be true if one considers that the deference is limited by the standard of arbitrariness and that the investor will be protected against essential-unpredictable changes as defined by the legitimate expectations.

Although the orthodox approach could provide a harmonised-satisfactory solution in most of the FET climate change disputes, there might still be occasions where content-interpretation of the FET standard conflicts with the objectives of the climate change international legal framework.

In fact, the host state may need to dishonour the investor's legitimate expectations and to alter the essential elements of the legal framework to protect the climate interest—even against unambiguous-instrumentalised representations of the host state—and it will still be required to compensate. A particular case could show, for example, that the imposition of restrictions to contractual and property rights to exploit natural resources in areas protected for the preservation of the tropical forest was not legitimately expected. Likewise, the opposite situation is also feasible; the host state could act against the climate change interest but in a proportional/reasonable manner in relation to the investors' interests without breaching the FET. This could be, for example, the case of a well-organised general plan to disarticulate RESSs.

In these cases, the balance between the investor and the host state that is embedded in the FET standard and interpreted under the authority of Article 31 VCLT is not enough to promote and protect the climate interest. In the end, just as in physics the gravitational power shapes an object without changing its internal structure, the gravitational power of the international climate change framework shapes the investment treaty law system without modifying the balance between the parties that is entrenched in the text. As a result, if the interpreter wants to go beyond these limits to protect the climate interest, he is compelled to find a harmonised answer through the direct application of other sources of international law. In other words, the interpreter needs to shift perspectives from an orthodox approach to a heterodox one that promotes the protection of the climate change interest through the application of principles that are different from the *pacta sunt servanda* (public dimension), this being the topic for Part III.

References

Baetens F (2010) Foreign Investment Law and Climate Change: Legal Conflicts Arising from Implementing the Kyoto Protocol Through Private Investment. The Center for International Sustainable Development Law & The International Development Law Organization, Working Paper Series https://papers.ssrn.com/sol3/papers.cfm?abstract_id=2117950. Accessed 31 July 2022

Comisión Nacional de Energía (2008) Informe 30/2008 de la cne en Relacion con la Propuesta de Real Decreto de Retribución de la Actividad de Producción de Energía Eléctrica Mediante Tecnología Solar Fotovoltaica para Instalaciones Posteriores a la Fecha Límite de Mantenimiento de la Retribución del Real Decreto 661/2007, de 25 de mayo, para dicha tecnología. Pp. 10-12 https://energia.cnmc.es/cne/doc/publicaciones/cne108_08.pdf. Accessed 31 July 2022

Dugan C et al. (2011) Investor-State arbitration. OUP, p. 504

Firger D, Gerrard M (2011) Harmonizing climate change policy and international investment law: threats, challenges and opportunities. In: Sauvant K (ed) Yearbook on international investment law & policy 2010-2011. OUP. https://ssrn.com/abstract=1733985. Accessed 31 July 2022

Henckels C (2015) Proportionality and deference in investor-state arbitration: balancing investment protection and regulatory autonomy, CUP

Isotalo L (2016) Climate Compatible Investment Treaty Law: The Role of Legitimate Expectations. SCIL Working Paper Series http://www.scil.ed.ac.uk/__data/assets/pdf_file/0004/172039/Laura_Isotalo_SCIL_Working_Paper_6_FINAL_for_publication.pdf. Accessed 31 July 2022

Johnson L (2009) International investment agreements and climate change: the potential for investor-state conflicts and possible strategies for minimizing it. Eur Law Rev 39:11147, 11153

Miles K (2008) International Investment Law and Climate Change: Issues in the Transition to a Low Carbon World. Society of International Economic Law (SIEL) Inaugural Conference,

p. 31. https://ssrn.com/abstract=1154588 or https://doi.org/10.2139/ssrn.1154588. Accessed 31 July 2022

Ranjan P (2014) Using the public law concept of proportionality to balance investment protection with regulation in international investment law: a critical appraisal. Cambridge J Int Comp Law 3:853, 881

Schill S (2007) Do investment treaties chill unilateral state regulation to mitigate climate change? J Int Arbitr 24:469, 476

Schill S (2010) International investment law and comparative public law—an introduction. In: Schill S (ed) International investment law and comparative public law. OUP, pp. 131–133

Vadi V (2015) Beyond known worlds: climate change governance by arbitral tribunals? Vanderbilt J Transnatl Law 48(5):1286, 1292–1293

Vadi V (2015–2016) Energy Security v. Public Health? Nuclear energy in international investment law and arbitration. Geo J Int Law 47:1069, 1136

VanDuzer J (2016) The complex relationship between international investment law and climate change initiatives: exploring the tension. In: Delimatsis P (ed) Research handbook on climate change and trade law. EE, p 442

Wilensky M (2015) Reconciling international investment law and climate change policy: potential liability for climate measures under the trans-pacific partnership. Eur Law Rev 45:10683, 10691

Decisions

ICSID (2005) CMS Gas Transmission Company v. The Republic of Argentina, Case No. ARB/01/8 https://www.italaw.com/cases/288. Accessed 31 July 2022

ICSID (2007) Enron Corporation and Ponderosa Assets, L.P. v. Argentine Republic, Case No. ARB/01/3 http://www.italaw.com/cases/401#sthash.dNAY9sBJ.dpuf. Accessed 31 July 2022

ICSID (2008) Metalpar S.A. and Buen Aire S.A. v. The Argentine Republic, Case No. ARB/03/5 https://www.italaw.com/cases/documents/383. Accessed 31 July 2022

ICSID (2010) Total S.A. v. The Argentine Republic, Decision on Liability, Case No. ARB/04/01 http://www.italaw.com/cases/1105#sthash.wTdX5hQc.dpuf. Accessed 31 July 2022

ICSID (2011) El Paso Energy International Company v. The Argentine Republic, Case No. ARB/03/15 https://www.italaw.com/cases/documents/383. Accessed 31 July 2022

ICSID (2012) Marion Unglaube v. Republic of Costa Rica, Case No. ARB/08/1 https://www.italaw.com/sites/default/files/case-documents/ita1052.pdf. Accessed 31 July 2022

ICSID (2015) Mamidoil Jetoil Greek Petroleum Products Societe S.A. v. Republic of Albania, Case No. ARB/11/24 https://www.italaw.com/cases/3003. Accessed 31 July 2022

ICSID (2016a) Crystallex International Corporation v. Bolivarian Republic of Venezuela, Case No. ARB(AF)/11/2 https://www.italaw.com/cases/1530. Accessed 31 July 2022

ICSID (2016b) Rusoro Mining Ltd. v. Bolivarian Republic of Venezuela, Case No. ARB(AF)/12/5 https://www.italaw.com/cases/2048#sthash.VlFYPxlL.dpuf. Accessed 31 July 2022

ICSID (2017) Eiser Infrastructure Limited and Energía Solar Luxembourg S.à r.l. v. Kingdom of Spain, Case No. ARB/13/36 https://www.italaw.com/cases/5721. Accessed 31 July 2022

ICSID (2018a) Antin Infrastructure Services Luxembourg S.à.r.l. and Antin Energia Termosolar B.V. v. Kingdom of Spain (2018), Case No. ARB/13/31

ICSID (2018b) RREEF Infrastructure (G.P.) Limited and RREEF Pan-European Infrastructure Two Lux S.à r.l. v. Kingdom of Spain, Case No. ARB/13/30 https://www.italaw.com/cases/2317. Accessed 31 July 2022

SCC (2016a) Charanne (The Netherlands) And Construction Investments (Luxembourg) v. Kingdom of Spain. https://www.italaw.com/cases/2082. Accessed 31 July 2022

SCC (2016b) Isolux Netherlands, BV v. Kingdom of Spain, Case V2013/153 https://www.italaw.com/cases/5893. Accessed 31 July 2022

SCC (2018) Novenergia II - Energy & Environment (SCA) (Grand Duchy of Luxembourg), SICAR v. The Kingdom of Spain, Case No. 2015/063 https://www.italaw.com/cases/6613. Accessed 31 July 2022

UNCITRAL (2006) Saluka Investments B.V. v. The Czech Republic, ad hoc, partial award https://www.italaw.com/cases/961. Accessed 31 July 2022

UNCITRAL (2009) Glamis Gold, Ltd. v. The United States of America, ad hoc http://www.italaw.com/cases/487#sthash.xCSTGU3d.dpuf. Accessed 31 July 2022

UNCITRAL (2011) White Industries Australia Limited v. The Republic of India, ad hoc https://www.italaw.com/cases/documents/1170. Accessed 31 July 2022

UNCITRAL (2016a) Copper Mesa Mining Corporation v. Republic of Ecuador, Case No. 2012-2 https://www.italaw.com/cases/4206#sthash.tqKhkan6.dpuf. Accessed 31 July 2022

UNCITRAL (2016b) Murphy Exploration & Production Company International v. Republic of Ecuador, Case No. 2012-16 https://www.italaw.com/cases/documents/4251. Accessed 31 July 2022

Part III
Osmosis of the Global Climate Change International Legal Framework into Investment Treaty Law Through Law Ascertainment: Heterodox Approach to Harmonisation

Chapter 8
Heterodox Approach and the Sources of Law in Investment Treaty-Climate Change Disputes

Multi(syncretic)lateralization—as it was revealed in Part I[1]—has the virtue of unearthing the existence of conflicts of laws between the BIT clauses and the different instruments within the fragmented multiverse of international law. It also has the advantage of embracing tensions between the opposing global interests that are represented in these different instruments. If the process of *multilateralization* is not shaped through *syncretism*, then, any clash between provisions of different legal systems is not conceivable. In other words, it would be impossible for provisions belonging to BITs or other international law rules to conflict with the sources embedding global public interests, if the latter set of rules does not hold the pedigree of formal law in investment treaty law. *Syncretism*, while recognising the relevance of global interests and the authority of instruments belonging to the fragmented multiverse of international law in the investment treaty law system, captures the intra-systemic conflict of rules and tensions of interests within its operation.

Although global public interests and international law instruments are clearly entrenched in the operation of investment treaty tribunals, scholars rarely recognise that provisions of the climate change international legal framework apply directly to investment treaty-climate change disputes. Against this abulia, one could argue that the circumstances of the particular case might suggest that the appropriate solution would not be found in the BIT's content-interpretation but rather in the ascertainment of a provision of international climate change law. This approach corresponds to, what I have labelled in Part I, as the public dimension of the tension of the investment treaty law[2] since the focus is to reveal the prevalence of one provision representing a global public interest over another.

This layer of law ascertainment succeeds the pre(logical)interpretative phase that evaluates whether the climate change international legal framework is materially and formally relevant—which was described (proposed) in Part I. Also—for practical

[1] *Supra* Chap. 3, 3.3.
[2] *Ibid.*

T. Restrepo Rodríguez, *Investment Treaty Law and Climate Change*,
https://doi.org/10.1007/978-3-031-18655-4_8

reasons—the adoption of a heterodox approach is only necessary if content-interpretation—as illustrated in Part II—has failed to provide a harmonised answer between the investment treaty law and the international climate change law. Besides, an orthodox approach enjoys a better reception in the community and gives the impression of being more systematic than the heterodox approach.

In contrast to content-interpretation of the BIT proposed in Part II, which achieves harmonisation with the climate change law by finding the meaning of the standards in a climate change context and without relying on the application of external elements (gravitational power of the climate change international legal framework); the law ascertainment of climate change provisions may modify the structure of the investment treaty law through the penetration of external rules that might be perceived as counter rational to the system. That is why the title for Part III makes an analogy to the biological process of osmosis,[3]—or even with the colloquial meaning of osmosis.[4]

In this section, I support the direct application of the precautionary principle as well as the application of the common but differentiated responsibilities and respective capabilities (CBDRRC) principle to investment treaty-climate change disputes. However, according to the methodology of this book—reductionist legal positivism—the application of these principles requires a conforming to the system's operation.

Investment treaty law—as illustrated above[5]—is not self-contained and the practice of the tribunals shows that rules appertaining to international law may apply directly to investment treaty disputes; that is the case, for example, for the principles of due process, [6] good faith [7] or unclean hands. [8] In fact, one could say that while the rules that tribunals choose to establish the liability of the host state—with very few exceptions—belong to international law (Draft Articles on State Responsibility—DASR),[9] the host states' obligations are sourced exclusively in

[3] '[M]ovement of a solvent (such as water) through a semipermeable membrane (as of a living cell) into a solution of higher solute concentration that tends to equalize the concentrations of solute on the two sides of the membrane': Merriam-Webster, Osmosis https://www.merriam-webster.com/dictionary/osmosis.

[4] '[A] process of absorption or diffusion suggestive of the flow of osmotic action especially: a usually effortless often unconscious assimilation'. *Ibid.*

[5] *Supra* Chap. 3, 3.1.2.

[6] *Corona Materials LLC v. Dominican Republic (2016)*, para. 251.

[7] *Churchill Mining PLC and Planet Mining Pty Ltd v. Republic of Indonesia (2017)*, paras. 488, 528.

[8] *Hulley Enterprises Limited (Cyprus) v. The Russian Federation (2014)*, para. 1358.

[9] For example, [On full compensation] *Rusoro Mining Ltd. v. Bolivarian Republic of Venezuela (2016)*, para. 640; *Crystallex International Corporation v. Bolivarian Republic of Venezuela (2016)*, paras. 846–849. [On certainty and burden of proof] *Rusoro Mining Ltd. v. Bolivarian Republic of Venezuela (2016)*, para. 641; *Khan Resources Inc., Khan Resources B.V., and Cauc Holding Company Ltd. v. The Government of Mongolia (2015)*, para. 388.; (Specific performance) *Mohammad Ammar Al-Bahloul v. The Republic of Tajikistan (2010)*, para. 47. [On Causation] *Crystallex International Corporation v. Bolivarian Republic of Venezuela (2016)*, para 860. [On proportionality] *Merrill and Ring Forestry L.P. v. Canada (2010)*, para. 427. [On Damages]

the BIT. However, this openness to international law might be limited by the BIT's choice of law and the special provisions of the BIT, *lex specialis*.

The direct application of international climate change rules is binding to the tribunal, when the case at hand shows that the global climate interest should prevail over the global interest in transnational investment or any other global interest connected to the latter.[10] Under Article 38(1) ICJ—as the foundational provision of law ascertainment [11]—conventions, international customary law and general principles of law, are binding sources, while judicial decisions and doctrine appear to be interpretative instruments.[12] Although the authority of Article 38(1) has been contested in some decisions,[13] one could affirm that in practice tribunals adopt this traditional division—with a certain degree of customisation—to identify the international law that applies to disputes and, incidentally, to distinguish between binding and non-binding international instruments.

There is no question as to the binding character of the treaties or the conventions that are simultaneously ratified by the investor's state and the host state. The overwhelming reception of the VCLT in investment treaty-based disputes when adopted by the parties to a BIT evinces the accuracy of this statement. For instance, if the Paris Agreement and the Kyoto Protocol are adopted by BIT's parties, it could be concluded that these conventions are *law* to the tribunal and, accordingly, that the climate change precautionary principle and the CDRRC principle must determine the outcome of the dispute. The wide-ranging reception of the climate change international legal framework among states,[14] facilitates the penetration of the

---

Murphy Exploration & Production Company International v. Republic of Ecuador (2017), para. 424. [On the loss of a chance] *Chevron Corporation (USA) and Texaco Petroleum Company (USA) v. The Republic of Ecuador (2009)*.

[10] 'In addition to any specific treaty (e.g. the terms of a BIT under which a claim is brought), international arbitral tribunals may have to apply other general rules of international law, as expressed, for example, in the UN Assembly resolutions or other texts. Note, however, that there is no international legislature, and that questions of international law are not determined for individual states by non-binding resolutions or general declarations. These may well be relevant and authoritative, but there is always a process of assessment and evaluations'. Bishop et al. (2005), p. 717.

[11] *Merrill and Ring Forestry L.P. v. Canada (2010)*, para. 184: 'The meaning of international law can only be understood today with reference to Article 38(1) of the Statute of the International Court of Justice (...)'.

[12] 'The Court, whose function is to decide in accordance with international law such disputes as are submitted to it, shall apply: a. international conventions, whether general or particular, establishing rules expressly recognized by the contesting states; b. international custom, as evidence of a general practice accepted as law; the general principles of law recognized by civilized nations; subject to the provisions of Article 59, judicial decisions and the teachings of the most highly qualified publicists of the various nations, as subsidiary means for the determination of rules of law': Art. 38, United Nations, *Statute of the International Court of Justice* (1946).

[13] Liman Caspian Oil BV and NCL Dutch Investment BV v. Republic of Kazakhstan (2010), para. 172.

[14] 'To this date, 187 Parties have ratified of 197 Parties to the Convention' United Nations Climate Change, 'Paris Agreement - Status of Ratification'.

climate change precautionary principle and the CBDRRC principle into the investment treaty law system.

Conversely, treaties or conventions not ratified by the parties to the BIT agreement and soft law instruments would not be binding unless they are characterised by the tribunal's practice with the pedigree of formal-binding law.[15] The DASR, for instance, despite it lacking the status of treaty and pertaining to state-to-state liability instead of state-to-investor liability, is consistently applied by tribunals. [16] Regardless of how these instruments are framed—e.g., customary international law, [17] general international law, [18] principles of international law, [19] 'the most authoritative statement' [20]—the dynamic that is important to capture is that tribunals have conferred authority to the DASR.

Indeed, in Chap. 9, I argue that the international defence of necessity defined in Article 25 of the DASR may determine the application of the precautionary principle because tribunals often rely on the authority of this defence. Furthermore, the labelling of Article 25 as customary law, principle or simply international law is irrelevant for the purpose of its integration into the system, since the justification of its application lies in the authority that tribunals bestow to this article, rather than on the elements indicating that the norm appertains to one or the other category among the sources. My argument in Chap. 10 deserves the same considerations; the proposal of applying the principle of equity and Article 27 of the DASR, as rules of compensation that simplify the translation of the CDRRC into investment treaty law, is not dependent on the classification of these norms as principles or customary law, but on the binding authority that these norms represent for tribunals.

The direct application of the climate change international legal framework must be in line with the parties' choice of law. This restriction stems from the consent of the parties to the investment arbitration.[21] Although BITs are not explicit in expressing that the FCCC is within the applicable law, they directly or indirectly establish—either by reference to (model or customised) procedure rules or rules of

[15] *See supra* Introduction, iii. Shielding the Harmonisation from Arbitrariness: Reductionist Legal Positivism (Methodological Package).

[16] *Mr. Kristian Almås and Mr. Geir Almås v. The Republic of Poland (2016)*, para. 206.; *Mohammad Ammar Al-Bahloul v. Tajikistan* (2010), paras. 42, 165*; Quiborax S.A., Non Metallic Minerals S.A. and Allan Fosk Kaplún v. Plurinational State of Bolivia (2015)* paras. 326, 327.; Gold Reserve Inc. v. Bolivarian Republic of Venezuela (2014), paras. 678–679; Vestey Group Ltd v. Bolivarian Republic of Venezuela (2016), para. 326.

[17] *Chevron Corporation v. Ecuador* (2009), para. 159; Mohammad Ammar Al-Bahloul v. Tajikistan (2010) para. 42.

[18] *Liman Caspian Oil BV and NCL Dutch Investment BV v. Kazakhstan* (2010), para. 326; *Merrill v. Canada* (2010), para. 183; *Vestey Group Ltd. v. Venezuela* (2016) para. 115.

[19] *Quiborax v. Bolivia* (2015), para. 327; *Gold Reserve Inc. v. Venezuela* (2014), para. 679.

[20] *Mohammad Ammar Al-Bahloul v. Tajikistan* (2010), para. 165.

[21] *See* Introduction to Part I, Text to *Notes* 8-13.

an arbitration centre[22]—that the international law is the governing law. As the international law embraces the FCCC, there should be no problem in assuming that the choice of law is not restricting the legitimacy of the latter in investment treaty disputes. Furthermore, the preambular language of some BITs, in addition to the text of BIT provisions, where environmental, sustainable development and climate change considerations are relevant, simplifies the Tribunal's decision to apply rules from the climate change international legal framework.[23] Tribunals appear to have broad-ranging discretion for ascertaining provisions from the multiverse of international law,[24] including those belonging to the climate change international legal framework.[25]

In the simplest scenario—through the analysis of tribunals' practice that corresponds to my positivist approach—I conclude that the climate change legal framework is applicable law because—directly or indirectly—the BIT provides so. In a more complicated scenario, where the parties are silent on the applicable law, I assume that the climate change international legal framework is applicable law because the process of *multi(syncretic)lateralization* reveals that—according to the practice of tribunals—the balancing of global public interests is a fundamental

[22]For example, Article 42(1) ICSID 'The Tribunal shall decide a dispute in accordance with such rules of law as may be agreed by the parties. In the absence of such agreement, the Tribunal shall apply the law of the Contracting State party to the dispute (including its rules on the conflict of laws) and such rules of international law as may be applicable' [Convention on the Settlement of Investment Disputes Between States and Nationals of Other States, International Centre for Settlement of Investment Disputes (1966), p. 159]; Article 54 Arbitration Additional Facility Rules '1) The Tribunal shall apply the rules of law designated by the parties as applicable to the substance of the dispute. Failing such designation by the parties, the Tribunal shall apply (a) the law determined the conflict of laws rules which it considers applicable and (b) such rules of international law as the Tribunal considers applicable' [ICSID Additional Facility Rules, Schedule C Arbitration (Additional Facility) Rules International Centre for Settlement of Investment Disputes (2006)]; Article 33 1. UNCITRAL 'The arbitral tribunal shall apply the law designated by the parties as applicable to the substance of the dispute. Failing such designation by the parties, the arbitral tribunal shall apply the law determined by the conflict of laws rules which it considers applicable' [Arbitration Rules, United Nations Commission on International Trade Law (2001) UN Doc A/31/98, 31st Session Supp No. 17]; Article 27 SCC; 'Applicable law(1) The Arbitral Tribunal shall decide the merits of the dispute on the basis of the law(s) or rules of law agreed upon by the parties. In the absence of such agreement, the Arbitral Tribunal shall apply the law or rules of law that it considers most appropriate' [Arbitration Rules of The Arbitration Institute of the Stockholm Chamber of Commerce (2017)].

[23]Cfr. Levashova (2011).

[24]Even in the cases where there is not direct or indirect reference to the applicable law, the application of international law seems unavoidable. According to Spierman 'Irrespective of whether there is a specific treaty provision, it is necessary to apply international law to treaty claims. To the extent that the host State has undertaken treaty obligations relevant to the dispute, it may be presumed that they will be applied directly or by an arbitral tribunal, in addition to the implications derived to the principle *pacta sunt servanda.* When claimants bring such treaty claims, they obviously rely on international law, and there is no way for a competent arbitral tribunal but to apply international law'. Spierman (2015), p. 1385.

[25]*See infra* Chap. 9, 9.2.2.; Chap. 10, 10.2.1.2.

feature of the system of investment treaty arbitration. Following the same logic, in a third scenario, where the parties to the treaty exclude the application of the climate change international legal framework, the lack of consent leads me to conclude its inapplicability. Fourth, in the scenario where the treaty provides that the domestic law (or any other (sub)system different from international law) is exclusively the applicable law, the analysis of the application of the climate change international legal framework would be undertaken under national law and outside the investment treaty law system. Similarly, in a final scenario, where the parties exclude the applicability of international law in general, the analysis of the application of the climate change international legal framework falls outside the investment treaty law system, since a dispute as such would lack of one of the basic features of the investment treaty law: the balancing of global common interests.

The interaction between the BIT and the climate change international legal framework could create conflicts of rules and tensions between opposing interests. The autonomy of the BIT's parties to regulate their own interests legitimises the application of the BIT as *lex specialis*,[26] thus excluding the application of the climate change international law.

The application of the *lex specialis* is common in investment treaty disputes. [27] In *Liman Caspian v. Kazakhstan*, for example, the Tribunal concluded that since the ECT is *lex specialis* to International law, the appropriate rule that determines the standard of treatment was Article 10(1) ECT, which purports a 'specific standard of fairness and equitableness above the minimum standard' [28] of public international law. [29] Similarly, in *Hamadi v. Oman,* the Tribunal chose not to apply the principle of attribution as contained in the DASR, because Article 10.1.2. US-Oman FTA—which was *lex specialis*—established a narrower test of attribution. [30]

The principle of *lex specialis* is relevant in Chaps. 9 and 10 in terms of limiting the direct application of the climate change international legal framework in certain cases. In Chap. 9, I claim that general exceptions may exclude the application of the international defence of necessity and the direct application of the climate change precautionary principle, if their text expresses that they have the effect of precluding the wrongfulness of the breach, rather than impeding the breach itself. Otherwise, it is understood that there is no *lex generalis-lex specialis* relationship between general

[26] 'Virtually all applicable law rules honor the principle of party autonomy. In an investment treaty arbitration the treaty is the first port insofar as the law chosen by the parties is concerned. The classic formulation is that the state offers to arbitrate according to the terms contained in the treaty and the investor accepts that offer by commencing arbitration': Bjorklund (2014), p. 273; *Chevron v. Ecuador* (2009), para. 242–243.

[27] *Khan Resources Inc., Khan Resources B.V., and Cauc Holding Company Ltd. v. The Government of Mongolia (2015)*, para. 388.; Chevron Corporation *Chevron v. Ecuador* (2009). para. 242; *El Paso Energy International Company* (2011), para. 552–554.

[28] *Liman Caspian Oil BV and NCL Dutch Investment BV v. Kazakhstan* (2010) para. 263.

[29] *Liman Caspian Oil BV and NCL Dutch Investment BV v. Kazakhstan* (2010); *Vestey Group Ltd v. Venezuela* (2016), para. 327.

[30] *Adel A Hamadi Al Tamimi v. Sultanate of Oman (2015)*, para. 317–321.

exceptions and the international defence of necessity, since the two rules do not share the same substance; while general exceptions prevent the perfection of the breach, the international defence precludes the wrongfulness of the breach. In Chap. 10, I contend that the differentiation model of the CBDRRC does not apply to the compensation for lawful expropriations, given that the ordinary BIT expropriation clause establishes—as *lex specialis*—that compensation in this event is assessed by reference to the market value of the investment.

However, the truth of the matter is that the conflicts of rules between the climate change legal framework and the BIT are rare, since the provisions of these systems rarely regulate the same substance. A different problem is that the interests behind the BIT and the international climate change framework may enter into tension without the existence of a conflict of rules. Here, I am specifically referring to the tension between the global climate interest and the promotion of transnational investment, which—in the end—is the problem that I am dealing with: the public dimension.

In the case of tension between global public interests, the solution does not involve ruling out, on a general basis, one of the conflicting provisions, such as when the specific rule on the compensation for lawful expropriation excludes, on a general basis, the application of a differential approach to compensation. Unlike the *lex specialis-lex generalis* approach, the tension of interests involves weighting one against the other and applying the rule that represents the prevailing interest. In this regard, the rule that does not apply to the case at hand is not excluded for every single case of tension between the interests that lie behind.

Against this backdrop, in Chap. 9, I illustrate the tension between the climate interest, represented in the precautionary climate change principle on one hand, and the interest of transnational investment, represented in the *pacta sunt servanda,* on the other. The international defence of necessity is deployed in this chapter as a rule of law ascertainment that may result in the application of the precautionary principle instead of the *pacta sunt servanda* if the climate interest prevails. By contrast, in Chap. 10 I illustrate the tension between the climate interest, represented in the CBDRRC on the one hand, and the interest of transnational investment, represented in the full compensation principle, on the other. If the tribunal opts to privilege the climate change interest, then the compensation would be proportional to the particular conditions of the parties to the dispute before the climate change law.

The heterodox approach that follows is the last attempt to harmonise investment treaty law with the climate change international legal framework.

References

Bishop R et al. (2005) Foreign investment disputes. Kluwer Law, p. 717

Bjorklund A (2014) Applicable law in international investment disputes. In: Giorgetti C (ed) Litigating in international investment disputes. Brill Nijhoff, p 273

Levashova Y (2011) Role of sustainable development in bilateral investment treaties: recent trends and developments. J Sustain Finance Investment (1):3–4, pp. 222–229 https://www.researchgate.net/publication/254317188_Role_of_sustainable_development_in_Bilateral_Investment_Treaties_recent_trends_and_developments. Accessed 1 August 2022

Spierman O (2015) Investment arbitration: applicable law. In: Marc Bugenberg M et al. (eds) International investment law. C.H. Bech/Hart/Nomos, p. 1385

Decisions

ICSID (2010a) Liman Caspian Oil BV and NCL Dutch Investment BV v. Republic of Kazakhstan, Case No. ARB/07/14 https://www.italaw.com/cases/1977#sthash.1BmlOYMa.dpuf. Accessed 1 August 2022

ICSID (2010b) Merrill and Ring Forestry L.P. v. Canada, ICSID Case No. UNCT/07/1 https://www.italaw.com/cases/669#sthash.blsKTHTM.dpuf. Accessed 1 August 2022

ICSID (2014) Gold Reserve Inc. v. Bolivarian Republic of Venezuela, ICSID Case No. ARB(AF)/09/1 https://www.italaw.com/cases/2727#sthash.VgHhhAmn.dpuf. Accessed 1 August 2022

ICSID (2015a) *Adel A Hamadi Al Tamimi v. Sultanate of Oman*, Case No. ARB/11/33 https://www.italaw.com/cases/1070#sthash.TlMdDISS.dpuf. Accessed 1 August 2022

ICSID (2015b) Quiborax S.A., Non Metallic Minerals S.A. and Allan Fosk Kaplún v. Plurinational State of Bolivia, Case No. ARB/06/2 https://www.italaw.com/cases/885. Accessed 1 August 2022

ICSID (2016a) Corona Materials LLC v. Dominican Republic, Case No. ARB(AF)/14/3 https://www.italaw.com/cases/2690#sthash.rsUjjcwi.dpuf. Accessed 1 August 2022

ICSID (2016b) *Crystallex International Corporation v. Bolivarian Republic of Venezuela*, Case No. ARB(AF)/11/2 https://www.italaw.com/cases/1530#sthash.FnlLIerj.dpuf. Accessed 1 Aug 2022

ICSID (2016c) *Rusoro Mining Ltd. v. Bolivarian Republic of Venezuela*, ICSID Case No. ARB (AF)/12/5 https://www.italaw.com/cases/2048#sthash.VlFYPxlL.dpuf. Accessed 1 August 2022

ICSID (2016d) Vestey Group Ltd v. Bolivarian Republic of Venezuela, Case No. ARB/06/4 https://www.italaw.com/cases/3616. Accessed 1 August 2022

ICSID (2017) Churchill Mining PLC and Planet Mining Pty Ltd v. Republic of Indonesia, Case No. ARB/12/14 and 12/40 https://www.italaw.com/cases/1479#sthash.Psc1DedU.dpuf. Accessed 1 August 2022

SCC (2010) *Mohammad Ammar Al-Bahloul v. The Republic of Tajikistan*, Case No. v. (064/2008) https://www.italaw.com/cases/65#sthash.J9gKN91T.dpuf. Accessed 1 August 2022

UNCITRAL (2009) *Chevron Corporation (USA) and Texaco Petroleum Company (USA) v. The Republic of Ecuador*, PCA Case No. 34877 https://www.italaw.com/cases/251#sthash.J8vcHWpC.dpuf. Accessed 1 August 2022

UNCITRAL (2014) *Hulley Enterprises Limited (Cyprus) v. The Russian Federation*, PCA Case No. AA 226 https://www.italaw.com/cases/544. Accessed 1 August 2022

UNCITRAL (2015) Khan Resources Inc., Khan Resources B.V., and Cauc Holding Company Ltd. v. The Government of Mongolia https://www.italaw.com/cases/604#s.thash.rD3oSpJd.dpuf. Accessed 1 August 2022

UNCITRAL (2016) *Mr. Kristian Almås and Mr. Geir Almås v. The Republic of Poland*, PCA Case No. 2015-13 https://www.italaw.com/cases/4288#sthash.h9gU32IF.dpuf. Accessed 1 August 2022

UNCITRAL (2017) *Murphy Exploration & Production Company International v. Republic of Ecuador*, PCA Case No. 2012-16 (formerly AA 434) https://www.italaw.com/cases/1198#sthash.7LfAYnFV.dpuf. Accessed 1 August 2022

International Legal Instruments

ICSID (2006) Additional Facility Rules, Rules International Centre for Settlement of Investment Disputes ICSID/11 https://icsid.worldbank.org/sites/default/files/AFR_2006%20English-final.pdf. Accessed 1 August 2022

International Centre for Settlement of Investment Disputes (1966) Convention on the Settlement of Investment Disputes between States and Nationals of Other States 575 UNTS 159 https://jusmundi.com/en/document/treaty/en-convention-on-the-settlement-of-investment-disputes-between-states-and-nationals-of-other-states-icsid-convention-1965-thursday-18th-march-1965?gclid=EAIaIQobChMIvfyntYW69QIVA7KGCh2mmACdEAMYASAAEgKfAPD_BwE. Accessed 1 August2022

The Arbitration Institute of the Stockholm Chamber of Commerce (2017) Arbitration Rules https://sccinstitute.com/media/1407444/arbitrationrules_eng_2020.pdf. Accessed 1 August 2022

United Nations (1946) Statute of the International Court of Justice https://www.refworld.org/docid/3deb4b9c0.html. Accessed 1 August 2022

United Nations Climate Change. Paris Agreement - Status of Ratification https://unfccc.int/process/the-paris-agreement/status-of-ratification. Accessed 1 August 2022

United Nations Commission on International Trade Law (2010) Arbitration rules. UN Doc A/31/98, 31st Session Supp No. 17. https://uncitral.un.org/sites/uncitral.un.org/files/media-documents/uncitral/en/uncitral-arbitration-rules-2013-e.pdf. Accessed 1 August 2022

Chapter 9
Direct Application of the Precautionary Principle in Investment Treaty-Climate Change Disputes: Deploying the International Defence of Necessity as a Rule of Law Ascertainment

The precautionary principle epitomises the spirit of the Climate Change International Legal Framework. Thus, Article 3 of the Framework Convention on Climate Change (FCCC)[1] compels states to 'anticipate, prevent or minimize the causes of climate change and mitigate its adverse effects [. . .]'[2] even in cases where there is a 'lack of full scientific certainty [. . .]'.[3] However, the application of this principle might result in the breach of some of the BIT obligations, mainly expropriation and FET standards. Indeed, host states' emergency measures against climate change that affect foreign investors only occasionally conform to the conditions established as general exceptions in BITs.

Consequently, a heterodox approach is required to harmonise the investment treaty law and the Climate Change International Legal Framework in cases where a BIT's content-interpretation is unable to provide a satisfactory answer to defences grounded on the urgency of the host state's climate change measures. Choosing the application of the precautionary principle over a BIT's application—law ascertainment—becomes the only option at hand. Yet, the methodological approach of this research—reductionist legal positivism—prevents me from directly applying the precautionary principle and discarding the application of the BIT without a proper filter to the system. In this scenario, Article 25 of the DASR emerges in the investment treaty law system as an ascertainment rule that is appropriate for balancing the interests that are behind the precautionary principle and the *pacta sunt servanda*, thus determining the preeminence of one or the other in the particular controversy.

The application of necessity to investment treaty environmental-related disputes—and even climate change-related disputes—has been proposed by other

[1] United Nations Framework Convention on Climate Change (1992).

[2] *Ibid.*

[3] *Ibid.*

T. Restrepo Rodríguez, *Investment Treaty Law and Climate Change*,
https://doi.org/10.1007/978-3-031-18655-4_9

authors.[4] According to Viñuales, although there are doubts in the definition of the elements that characterise the operation of the defence,[5] 'the fact remains that a threat to an environmental interest may potentially trigger the customary defense of necessity or a public emergency clause'.[6] Likewise, Di Benedetto states that necessity 'represent[s] a useful tool in excluding the wrongfulness of emergency environmental measures, especially when the applicable IIA does not provide any reference to non-commercial issues'.[7] Furthermore, this author concludes that it is plausible to adopt a normative model of general exceptions that embraces environmental issues, 'which may even crystalize into a customary international rule'.[8] In the same vein, Monebhurrun[9] infers from investment arbitration decisions the existence of an 'environmental defence' intended to preclude liability or lower compensation 'by arguing that what the investor presents as a measure infringing the applicable investment agreement is simply a national environmental regulation which can also have been adopted in order to enforce an international environmental agreement'.[10]

On investment treaty-climate change disputes, Johnson posits that the disconformity between host states' climate change measures and BIT obligations could be justified under the realm of the national security interest exception, 'even if the exception is absent from the text of the IIA'[11]—because 'customary international law similarly allows States to take measures *necessary* to protect their national security interests'.[12] Marshall appears to adopt a very similar view, when she contends that the application of necessity—even if 'not expressly included in the IIA'[13]—could stem from host states' invocation of the defence of Article 25 of the DASR because

[4]De Brabandere holds a similar approach but on human rights: 'Regulations and laws adopted by host states in the exercise of their public authority in order to ensure respect for these human rights can nevertheless enter into conflict with certain provisions of investment treaties or investment contracts which States have negotiated and signed with foreign investors. The question then is whether the human rights obligations of the state can be invoked as a defense to justify the breach of an international investment agreement. In essence, the state can assert that the protection of human rights can function as a form of 'necessity', which precludes the wrongfulness of the breach of the investment treaty, either as matter of treaty law ('non-precluded measures clauses') or through the application of the general rules on circumstances precluding wrongfulness in the 2001 Draft Articles on State Responsibility': De Brabandere (2018).

[5]Viñuales (2015), pp. 384–389.

[6]*Ibid.*, p. 389.

[7]Di Benedetto (2013), p. 175.

[8]*Ibid.*, p. 213.

[9]Monebhurrun (2013), p. 195.

[10]*Ibid.*, p. 200.

[11]Johnson (2009), pp. 11147–11150.

[12]*Ibid.*

[13]Marshal (2010a).

'climate change jeopardizes both their essential security interests and their public order'.[14]

As we can see, these general approximations to necessity—on environmental and climate change measures—appear to assimilate the international defence of necessity that is established in Article 25 of the DASR and the general exceptions clauses included in some BITs. Although this assimilation is acceptable, since both defences operate in exceptional circumstances to inhibit the liability of the host state and they could—eventually—conflate, it must be clear that they have a different legal nature and consequences. While the international defence of necessity is a secondary rule—in the language of the Law Commission[15]—which operates as a mechanism of law ascertainment by precluding the wrongfulness of the breach to an international obligation, the general exceptions are primary rules—entrenched in BITs—that prevent the perfection of the breach.

Considering the difference between the international defence of necessity and general exceptions, this chapter argues that *necessity*—as established in Article 25 of the DASR—might endorse the application of the climate change precautionary principle over the BIT, thus precluding the wrongfulness of the host state climate change measures (Sect. 9.1). However, the activation of the international defence of necessity would not be required, if it is found that the precautionary measure did not breach any obligations because of falling into the general exceptions of the agreement (Sect. 9.2). The chapter concludes that, while the deployment of general exceptions achieves harmonisation through a typical operation of content-interpretation, the exceptional application of the international defence of necessity involves a systematised ponderation of global interests that leads to the direct application of the climate change precautionary principle (Sect. 9.3).

---

[14] *Ibid.*

[15] 'These articles seek to formulate, by way of codification and progressive development, the basic rules of international law concerning the responsibility of States for their internationally wrongful acts. The emphasis is on the secondary rules of State responsibility: that is to say, the general conditions under international law for the State to be considered responsible for wrongful actions or omissions, and the legal consequences which flow there-from. The articles do not attempt to define the content of the international obligations, the breach of which gives rise to responsibility. This is the function of the primary rules, whose codification would involve restating most of substantive customary and conventional international law': International Law Commission (2001) para. 1–2.

9.1 The International Defence of Necessity in Investment Treaty-Climate Change Disputes: Article 25 of the DASR as a Rule to Determine the Direct Application of the Precautionary Principle

The operation of *necessity* in investment treaty law is uncontestable. Both 'the customary basis of the international defense of necessity, as defined in Article 25 of the Draft Articles on State Responsibility [DASR], is nowadays widely acknowledged'[16] and it has been considered in many investment treaty tribunals. The evolution of necessity in the investment treaty case law shows that the concept can embrace *global public interests* within its realm that are different from the protection to the existence of the host state, self-defence or self-preservation, making it possible to conceive the direct application of the precautionary principle in investment treaty-climate change disputes (Sect. 9.1.1). Yet, the successful application the precautionary principle through necessity requires a flexible interpretation of the conditions of Article 25 of the DASR that encompasses its function of balancing global public interests (Sect. 9.1.2).

9.1.1 The Recognition of Non-Economic Interests Within the Scope of Necessity in Case Law: Embracing the Necessity of Precautionary Climate Change Measures Within the Operation of the Investment Treaty Law System

Leaving aside for now the discussion on the interaction between Article 25 DASR and general exceptions—addressed below[17]— it could be said that the international defence of necessity was acknowledged for the first time in the cases concerning the Argentinian economic crisis of the early 2000s. Although some decisions applied the defence and other decisions did not, all of them conceived the international defence of necessity as part of the legal framework applicable to investment treaty disputes. Generally, the basis for its application was the protection of the national security interest. But environment, health and human rights were also considered in some of these cases, as interests that could potentially trigger Article 25 DASR. Subsequent decisions have also contemplated the application of the *necessity* in investment treaty law and have extended the catalogue of non-economic interests that it may comprehend, including environmental interests. The reception of non-economic interests through the concept of necessity in case law paves the way to the

[16] Viñuales (2015), p. 381.

[17] *Infra* Sect. 9.2.1.

application of the precautionary principle in climate change-related disputes. It is worth grouping the cases—to illustrate this evolution[18]—in the decisions on the Argentinian crisis (Sect. 9.1.1.1) and other decisions that contemplate the notion of the international defence of necessity (Sect. 9.1.1.2).

9.1.1.1 Evolution of the Scope of *Necessity* in the Disputes Concerning the Argentinian Economic Crisis: Paving the Way for the Reception of the Climate Change Precautionary Principle by Considering the Protection of Non-Economic International Interests

At the beginning of the century Argentina experienced a deep economic crisis leading to the collapse of the economy, with appalling indicators for unemployment, poverty, social service spending per capita, debt, risk of default and inflation. [19] This was translated, in turn, into social distress with a failed healthcare system, unaffordable food prices, violent demonstrations and suspension of public transport services, among others.[20] Thus, the Argentinian government was compelled to make drastic legislative and regulatory reforms to try to get the local economy back on track. Among these measures, the 2002-Emergency law provided a change in the methodology for calculating the retribution tariff to gas distributors by cancelling the determination in dollars of the tariff—before converting to Argentinian pesos—and freezing the annual increase of the tariff that was attached to the PPI (US producer price index). [21] Additionally, the Argentinian government eliminated the regulatory exchange ratio of 1:1 between the US dollar and the Argentinian peso, leading to the devaluation of the peso.[22] These measures harmed the interest of foreign investors in gas distribution companies and, therefore, generated a considerable number of investment treaty disputes. Likewise, analogous modifications in the regulation of water service providers also led to some investment treaty controversies. [23] In the centre of these disputes, *necessity* was invoked as defence by Argentina and considered as applicable law by tribunals—yet, in some of the cases, the defence was not successful as the BIT application was privileged.

[18] For good analysis of the state of necessity in these cases, *see* Paddeu (2018), pp. 404–409.

[19] *LG&E Energy Corp., LG&E Capital Corp., and LG&E International, Inc. v. Argentine Republic (2007)*, para. 234–237.

[20] *Ibid.*

[21] *LG&E Energy Corp., LG&E Capital Corp., and LG&E International, Inc. v. Argentine Republic (2007)*, para. 23.

[22] *Ibid.*

[23] *Impregilo S.p.A. v. Argentine Republic (2011)*; *Suez, Sociedad General de Aguas de Barcelona, S.A. and Vivendi Universal, S.A. v. Argentine Republic (2010); Suez, Sociedad General de Aguas de Barcelona S.A., and InterAguas Servicios Integrales del Agua S.A. v. The Argentine Republic* (2010); *Urbaser S.A. and Consorcio de Aguas Bilbao Bizkaia, Bilbao Biskaia Ur Partzuergoa v. The Argentine Republic (2016).*

In *LG&E*—the first award where plea of necessity succeeded—it was highlighted the Tribunal's obligation to 'determine whether the measures implemented by Argentina were necessary to maintain public order or to protect its essential security interests, albeit in violation of the Treaty'. [24] This citation incorporates two of the messages that I have proposed to deliver in this chapter: first, that the defence of necessity is applicable law to investment treaty disputes and, second, that the success of the necessity may lead to the suspension of the application of the BIT, as the breach of the obligation becomes irrelevant by losing its wrongfulness.[25] These messages are confirmed when appropriate in all the awards concerning the disputes related to the Argentinian economic crisis.

Although the tribunal's decision in *LG&E* to exclude liability seems to be grounded on the conflation of the general exceptions of the Argentina-USA BIT with the international defence of necessity—as described below[26]—it contends that the conditions of Article 25 DASR were met, since the circumstances of the Argentinian crisis represented a 'serious threat to its existence, its political and economic survival, to the possibility of maintaining its essential services in operation, and to the preservation of its internal peace'. [27] Furthermore, in the *obiter dicta,* the award acknowledges that the success of the defence of necessity is not restricted to the cases where the existence of the state is at risk, but embraces a broader universe that comprehends 'any danger seriously compromising its internal or external situation',[28] including threats to 'ecology'.[29]

The conclusion of *LG&E* on the Argentina's absence of liability was contrary to *CMS,* a prior decision concerning the same changes to the regulations of gas tariffs. [30] Yet, the differences in the outcome of these two cases lay more in the appreciation of the facts[31] rather than in legal reasoning. Indeed, *CMS* confirms that the international defence of necessity is applicable law to investment treaty disputes [32] and that Article 25 of the DASR could be triggered on a general basis when the host state

[24] *LG&E Energy Corp. (2007)*, para. 205.

[25] '(. . .)Thus an act otherwise in breach of an international obligation ("not in conformity" with it) is not considered wrongful, and does not therefore entail the secondary obligations attached to an illicit act, thanks to the "exceptional" presence of one of the conditions that under international law preclude wrongfulness, here necessity (. . .)': *Ibid.*, para. 166.

[26] *Infra* (n. 180).

[27] *LG&E Energy Corp,* Decision on Liability (2007), para. 257.

[28] *Ibid.*, para. 251.

[29] *Ibid.*

[30] *CMS Gas Transmission Company v. The Republic of Argentina (2005).*

[31] In *CMS,* it was considered that the contribution of the host state was substantial to the state of necessity (paras 328–329) and that the host state measure was not the only measure available (para. 324). Consequently, the plea of necessity failed: *CMS Gas Transmission Company* (2005). A similar conclusion was adopted in: *Enron Corporation and Ponderosa Assets, L.P. v. Argentine Republic* (2007) para. 308–313.; *Sempra Energy International v. The Argentine Republic* (2007), paras. 350–351, 353–354.

[32] *CMS Gas Transmission Company* (2005), para. 315–331.

measure is taken to 'prevent major breakdowns' [33] rather than only restricted to the occasions when the measure is aimed at safeguarding the existence of the state.

The openness of the catalogue of interests that trigger the international defence of necessity seems to be a pattern in these cases. Indeed, the reasoning of these awards point to the direction that a severe economic crisis legitimises the application of Article 25, even if some decisions found that the situation did not fulfill all the requirements of the provision; e.g., non-contributing to the situation of necessity or being the only available measure. Diverse interests were labelled as essential in connection to the Argentinian economic crisis, including the political, social and economic stability, [34] the security and affordability of the electricity and gas supply, [35] the provision to 'the population with water and sewage facilities' [36] and the 'protecting of Argentina's indigent population'. [37]

Moreover, even if it was not always material to the outcome of the particular case, some of the awards on the Argentinian economic crisis—just like *LG&E* linked *necessity* to the threat to ecology[38]—conceived the applicability of Article 25 to situations where measures are taken for the promotion of interests that are closely related or could be assimilated to the climate interest. This is the case of *Continental* and *Impregilo*—disputes on the impact of the Argentinian government's measures on an insurance company and a water service supplier, respectively—where the tribunals acknowledged that necessity could be envisaged for 'safeguarding the environment' [39] or promoting 'environmental stability'. [40] Likewise, the twin-rulings of *Suez*—awards that also addressed the pesification of the retribution-tariff for water supply and the freezing of its adjustments—recognised that the host state measures pursued an essential interest since they were 'vital to the health and well-being of nearly ten million people'; [41] and they implicitly embraced the suggestion that the host state, by taking the regulatory measures, was complying with 'Human Rights Obligations'. [42] Yet, the tribunal barred the application of necessity, as it

[33] *Ibid.*, para. 319.

[34] *Impregilo S.p.A. v. Argentine Republic (2011)*, paras. 336, 346, 350.

[35] *Total S.A. v. The Argentine Republic (2010)*, paras. 345, 484.

[36] *Impregilo S.p.A. v. Argentine Republic (2011)*, para. 346.

[37] *EDF International S.A., SAUR International S.A. and León Participaciones Argentinas S.A. v. Argentine Republic (2012)*, para. 1172.

[38] Above, n. 29.

[39] *Continental Casualty Company v. The Argentine Republic (2008)*, para. 166. The notion of essential interest in the interpretation of general exceptions in *Continental* also embraces the environment (para. 176).

[40] *Impregilo S.p.A. v. Argentine Republic (2011)*, para. 346.

[41] *Suez, Sociedad General de Aguas de Barcelona, S.A. and Vivendi Universal, S.A.* (2010), para. 260; *Suez, Sociedad General de Aguas de Barcelona S.A., and InterAguas Servicios Integrales del Agua S.A.* (2010) para. 238–240.

[42] *Suez, Sociedad General de Aguas de Barcelona, S.A. and Vivendi Universal, S.A.* (2010) par. 262; *Suez, Sociedad General de Aguas de Barcelona S.A., and InterAguas Servicios Integrales del Agua S.A.* (2010), para. 240.

considered that, in the case at hand, the obligations to promote health and protect human rights were not 'inconsistent, contradictory or mutually exclusive' [43] with the obligations of the host state under the BIT. The same essential character of health and human rights was recognised in *Urbaser,* [44] a case that also concerned the impact of the measures during the crisis on water suppliers. However, contrary to the twin-rulings—on the claimant's argument against the application of necessity that was based on the duty of the host state to simultaneously comply with human rights and the BIT's obligations—the tribunal highlighted that

> [...] Claimants' argument is too short. It does not resolve the conflict between the obligation to guarantee the Concessionaire's right under the Concession and the access of the poor and vulnerable population to water when this cannot be ensured otherwise than by failing to comply with the host State's obligations toward the Concessionaire.[45]

Urbaser represents the conclusion of a long process of reflection of investment arbitration tribunals on the application of *necessity* concerning the Argentinian economic crisis, that shows interesting results about the effectivity of non-economic interests in investment treaty law. Since the first award—*LG&E*—the international defence of necessity was conceived as an expression of customary law with the potentiality of precluding the wrongfulness of non-compliance with BIT obligations. Yet, none of the awards proclaimed the success of the international defence in preventing liability by itself. In fact, in *LG&E* and *Continental,* Article 25 of the DASR was deployed as an instrument to interpret the BIT's general exception rather than as a direct source to support the rejection of Argentina's liability. In other cases, it was considered that the Argentinian regulatory measures that were taken during the crisis did not fulfill the requirements of Article 25 DASR.

Nevertheless, since the *LG&E* award, this group of cases acknowledged that the concept of essential interest that triggers *necessity* embraces a broader universe than the preservation of the state's existence, such as the threat to national security interests or to the political and economic stability, which were evidently at risk in the context of the crisis. Furthermore, tribunals moved from considering *in abstracto* non-economic international interests such as ecology, environment, health and human rights as possible bases for invoking *necessity*, to actually integrate them as part of their reasoning in the Argentinian crisis disputes. *Impregilo,* the twin-rulings of *Suez* and *Urbaser,* not only considered that the action of the Argentinian government was intended to comply with the host state's obligations to protect human rights and health, and guarantee access to water, but also envisaged the potential conflict between these obligations and the BIT. Whichever the decision of each of these awards was, confronting these non-economic international obligations with the

[43] *Suez, Sociedad General de Aguas de Barcelona, S.A. and Vivendi Universal, S.A.* (2010) para. 262; *Suez, Sociedad General de Aguas de Barcelona S.A., and InterAguas Servicios Integrales del Agua S.A.* (2010), para. 240.

[44] *Urbaser S.A. and Consorcio de Aguas Bilbao Bizkaia, Bilbao Biskaia Ur Partzuergoa v. The Argentine Republic (2016).*

[45] *Ibid.*, para. 720.

obligations under the BIT through the concept of necessity—or, in any case, verifying whether there was a conflict calling for the application of Article 25 DASR—indicates that the international defence of necessity could serve as a path for the application of the climate change precautionary principle in investment treaty-climate change disputes.

Following the awards on the Argentinian economic crisis, annulment decisions and dissenting opinions in this regard, as well as awards related to disputes that arise from other themes, confirm the applicability of the defence of necessity in investment treaty law disputes, and its potential to promote community interests in investment treaty law, such as the direct application of the precautionary principle.

9.1.1.2 Expansion of the Scope of Necessity in Investment Treaty Disputes Beyond the Argentinian Crisis Awards: Paving the Way for the Reception of the Climate Change Precautionary Principle by Recognising International Obligations that Promote Global Public Interests

None of the annulment decisions that followed the considerable number of investment arbitration decisions on the Argentinean economic crisis denied the relevance or rejected the applicability of *necessity* to investment treaty disputes. The annulment decision of *CMS*—in accordance with the ILC Commission—confirmed that Article 25 DASR is a 'secondary rule of international law' [46] that was applicable to the dispute. In *Enron*, 'the Committee finds that the Tribunal did not err' [47] in conceiving the application of the international defence of necessity. On the Annulment Committee of *Sempra*, although it was found that the tribunal failed to apply the general exception of Article XI, the Committee did not controvert the potentiality of Article 25 DASR to preclude wrongfulness in investment treaty disputes. [48]

Similarly, decisions on topics other than the Argentinian economic crisis have considered the likeliness of *necessity* to prevent the liability of the host state. In *Fenosa v. Egypt*, the tribunal analysed whether the measures restricting the supply of gas to the investor fulfilled the conditions of Article 25 DASR. [49] While rejecting the claimant's assertion that the imbalance of the gas supply was caused (and justified) by the Egyptian revolution—in the context of the so-called *Arab Spring*[50]— the tribunal openly accepted that:

[46] *CMS Gas Transmission Company v. The Republic of Argentina (2007),* para. 134.

[47] *Enron Corporation and Ponderosa Assets, L.P. v. Argentine Republic (2010),* para. 356.

[48] *Sempra Energy International v. The Argentine Republic (2010),* paras. 186–210.

[49] *Unión Fenosa Gas, S.A. v. Arab Republic of Egypt (2018),* para. 8.1.–8.62.

[50] *Ibid.*

> Under customary international law, necessity is one of the circumstances that the state can invoke to preclude the wrongfulness of an act in breach of an international obligation of that state, including a breach of its obligations under a bilateral treaty.[51]

Devas v. India and *Deutsche Telecom v. India*—two cases that dealt with the host state's measures restricting the use of S-band to investors in telecommunications for national security reasons—also contemplated the application of the international defence of necessity in investment treaty law, but in the end, opted to apply the general exceptions clauses of the respective BITs rather than Article 25 DASR. [52]

Furthermore—as shown in *Impregilo,* the twin-rulings of *Suez* and *Urbaser* concerning the Argentinian crisis—recent cases have demonstrated that the operation of *necessity* in investment treaty law embraces the tension between host states' international obligations—that represent global interests—and the application of the BIT standards. Thus, two specific cases are worth the attention:

Bernhard von Pezold v. Republic of Zimbabwe

In this case, the von Pezold family and the other claimants were expropriated under local laws as a result of the host state's long-term policy that aimed to legitimise the re-location of indigenous inhabitants in an area that formerly belonged to the colony of Southern Rhodesia, which is today the State of Zimbabwe. During the period of colonialism (late 1923 to 1980) the land was divided between 'white commercial farms'—property of European Colonisers—and 'Tribal Trust Lands' where the natives lived. This division was the representation of the oppression suffered by the locals as a result of colonialism. In the agreement of liberation that gave birth to the current State of Zimbabwe (the Lancaster House Agreement), the new independent government committed not to alter property tittles for ten years.

In 1992, the efforts for resettlement began with the expedition of the Land Acquisition Act 1992, which established the possibility for the host state to expropriate the land against the payment of fair compensation. However, until the year 2000, the land reform was proven to be unsuccessful, which led to the frustration of the Zimbabwean that fought for liberation. Given that the failure of the reform was in part owed to the lack of funds for compensation, the Constitution was reformed in 2000, limiting compensation to improvements to the land. Parallel to this reform, the locals commenced to illegally 'invade' the white farmers' lands with the complicity of the host state's authorities and encouraged by an anti-European discourse. An additional reform in 2005, facilitated the government expropriation by only

[51] *Ibid.*, para. 8.2.

[52] *CC/Devas (Mauritius) Ltd., Devas Employees Mauritius Private Limited and Telecom Devas Mauritius Limited v. India (2013),* para. 219–256.; *Deutsche Telekom v. India (2017),* para. 225–239.

requiring to its perfection a notice of acquisition for resettlement from the authorities, which the proprietors were not entitled to challenge. [53]

Against this background, the von Pezold family and the other claimants argued that the properties lost all value and, accordingly, submitted a claim under the BIT. One of the defences that Zimbabwe invoked was *necessity,* to redistribute the land and repair the Zimbabwean indigenous population. Although the tribunal found that the respondent did not prove that the requirements of Article 25 were met—and attributed the loss of the investor to mere political motives rather than *necessity*—some considerations on the duty of the state to adopt affirmative action requires attention:

> [...]it is accepted by the international community that situations will arise where racial discrimination is justified and will remain so for as long as is necessary. Policies that discriminate in favour of the aboriginal inhabitants of a particular State (affirmative action) may, generally speaking, fall within this category as justifiable. These policies aid in promoting, inter alia, the health and well-being, general knowledge, social integration, skills, and commercial know-how of, and added employment and business opportunities for, indigenous persons and their broader communities.[54]

According to this passage, the need to promote the equality of the indigenous population—in light of the historical discrimination and oppressive context of colonialism—could justify affirmative-discriminatory racial measures. Moreover, these measures could be regarded as an 'international community'[55] interest that suits the conditions of Article 25 DASR, since '[t]he discrimination in favour of indigenous Zimbabweans was necessary in order to remedy the unconscionable anti-aboriginal policies implemented throughout the Rhodesian era'.[56]

In fact, the tribunal does not question the need for discriminatory redistribution policies between the adoption of the Lancaster House Agreement (1979) and 2000.[57] However, it considered that the discriminatory character of the measures taken afterwards was only based on racist and partisan-political opportunistic reasons rather than on legitimate redistribution and affirmative action concerns. As a result, the award establishes that the last package of measures was not the only available alternative but also that it impaired the international community's interest—and its 'obligation *erga omnnes*'[58]—of non-discrimination.[59]

---

[53] *Bernhard von Pezold and Others v. Republic of Zimbabwe (2015),* paras. 91, 163.

[54] *Ibid.*, para. 652.

[55] *Ibid.*

[56] *Ibid.*, para. 653.

[57] *Ibid.*

[58] *Ibid.*, para. 657.

[59] *Ibid.*

Bear Creek Mining v. Peru (Dissenting Opinion)

This dispute concerns the operation of a silver mining that was granted in concession to the investor in an area that was populated by local communities. The relationship between the investor and the communities was tense from the beginning. From time to time, there were agreements with some communities, but demonstrations, riots and even violent manifestations were usual in the area of the project. Because of this situation, the government instructed the investor to implement a social participation plan. However, the plan was considered to be unsatisfactory for the communities because of the project's environmental impact. Accordingly—after the dramatic deterioration of the security situation in the area and the failure to conciliate with the communities—the government, first, suspended the approval of the Environmental Social Impact Assessment for 12 months, and, two months later, issued a decree declaring its engagement 'in a new round of consultations with communities, in accordance with the ILO Convention'.[60] This was followed by a legal battle between the parties at the Peruvian courts on the validity of the mining operations and the host state's measures, which ended with the investor withdrawing the claims.[61]

Although the tribunal dismissed the application of Article 25 of the DASR because it concluded that the list of the BIT's general exceptions were *lex specialis* and exhaustive,[62] the dissenting opinion of Professor Philippe Sand conceived the plausibility of deploying *necessity* as a means to ponder the international obligation of the host state to protect local communities against the application of the BIT. [63] The Professor maintained that it was among the hosts state's functions to ensure the implementation of Article 15 of ILO Convention 169,[64] which comprehends the rights of the local community members to

> 'participate in the use, management and conservation' of resources 'pertaining to their lands'; (2) to be consulted 'with a view to ascertaining whether and to what degree their interests would be prejudiced, before undertaking or permitting any programs for the exploration or exploitation of such resources pertaining to their lands'; and (3) 'wherever possible [to] participate in the benefits of such activities, and [to] receive fair compensation for any damages which they may sustain as a result of such activities'.[65]

On the articulation of these guarantees, the Professor made it clear that the responsibility of the host state is not 'to hold an investor's hand and deliver a "social license"'.[66] Furthermore, his opinion clarifies that the BIT is not an 'insurance policy

[60] *Bear Creek Mining Corporation v. Republic of Peru* (2017), para. 203.

[61] *Ibid.*, para. 119–216.

[62] *Ibid.*, para. 473.

[63] *Bear Creek Mining Corporation v. Republic of Peru (2017).*

[64] *Ibid.*, para. 37.

[65] *Ibid.*, para. 13.

[66] *Ibid.*, para. 37.

against the failure of an inadequately prepared investor to obtain such a license'.[67] This position does not suggest that the investor is expected to assume the functions that belong to the host state, nor that the international law is imposing direct obligations to the investor. However—inspired by *Urbaser*[68]—it acknowledges that the investor has some duties and burdens that necessarily complement the obligation of the host state to comply with ILO Convention 169.[69]

In addition to the recognition of the obligation of the host state to implement ILO Convention 169, the dissenting opinion was vehement in asserting that the indigenous and tribes that conformed the local communities had rights under international law that were as important as the investor's interests,[70] placing the global public interests at the same hierarchical level as the protection to the rights under the BIT. Moreover, in this respect, the Arbitrator highlights that the case at hand requires a solution that harmonises the application of the BIT with international law, and specifically, with the rights recognised in ILO Convention 169.[71]

The dissenting opinion concludes that, even if in the concrete case the international defence of necessity would not have been successful since there were other measures that were available for the government of Peru, the BIT did not prevent the possibility of applying Article 25—'which continues to function as a secondary rule of international law'[72]—to preclude wrongfulness. Yet, apart from Professor Sands' conclusion, what is more important in terms of his approach, is the recognition that international obligations that promote global public interests—such as the protection of indigenous communities—could be effective through *necessity*.

It has been shown in this section that a considerable variety of international obligations that represent global interests—on some occasions more determined than on others—has been recognised within the realm of *necessity* in investment treaty disputes: promoting anti-discriminatory action and preventing risks to health; protecting the autonomy and equality of indigenous communities; ensuring the provision of water and energy services to the citizens; or preventing violation to human rights as well as damages to the ecology and the environment. Certainly, within this constellation, the climate change precautionary principle could find a place as an obligation of the host state that has the potentiality to preclude the wrongfulness of the BIT and prevent the effectiveness of the BIT standards.[73] Yet, it is evident from the prior analysis of case law that the success of the defence of necessity is exceptional. This exceptionality is grounded on a strict interpretation of

[67] *Ibid.*

[68] 'In Urbaser v. Argentina, the Tribunal noted that human rights relating to dignity and adequate housing and living conditions "are complemented by an obligation on all parts, public and private parties, not to engage in activity aimed at destroying such rights"': Ibid. para. 37.

[69] *Ibid.*, para. 10–11.

[70] *Ibid.*, para. 36.

[71] *Ibid.*, para. 10–11.

[72] *Ibid.*, para. 41.

[73] De Brabandere (2018), pp. 19–21.

its conditions. In this respect, the next section explores the adaptability of the climate change precautionary principles to the requirements of Article 25 DASR in investment treaty-climate change disputes.

9.1.2 *The Adaptability of the Climate Change Precautionary Principle to the Conditions of Article 25 DASR: Adopting a Flexible Interpretation of the Requirements of the International Defence of Necessity*

The exceptionality of necessity is unchallenged. This is not only clear from the commission's special report,[74] but is also incisively repeated in the investment treaty awards. [75] The reason behind this exceptionality is to prevent the states from abusing *necessity* and disregarding international commitments. [76] To achieve this purpose, on the one hand, the international commission drafted Article 25 DASR as a negative proposition rather than a right[77] and, on the other, established very restrictive conditions to admit the success of the defence:

> Necessity **may not be invoked** by a State as a ground for precluding the wrongfulness of an act not in conformity with an international obligation of that State unless the act:
> (a) is the **only way** for the State to safeguard an **essential interest** against a **grave and imminent peril**; and
> (b) does **not seriously impair an essential interest** of the State or States towards which the obligation exists, or of the international community as a whole.
> 2. In any case, necessity may not be invoked by a State as a ground for precluding wrongfulness if:

[74] 'The term "necessity" (état de nécessité) is used to denote those exceptional cases where the only way a State can safeguard an essential interest threatened by a grave and imminent peril is, for the time being, not to perform some other international obligation of lesser weight or urgency. Under conditions narrowly defined in Article 25, such a plea is recognized as a circumstance precluding wrongfulness': International Law Commission (2001), para. 1–2.

[75] *LG&E,* Decision on Liability (2007), para. 228; *Impregilo* (2011), para. 344.

[76] *LG&E,* Decision on Liability (2007), para. 248; *Impregilo* (2011), para. 344. ILC, International Law Commission (2001), para. 2.

[77] 'In an attempt to anchor necessity in positive law, Anzilotti argued that a defense based on necessity was inherent in all rules of law, because no state would consent to be subjected an impeding rules in such situations. However all these views ultimately lead to the suggestion that sovereignty is a subjective right. (rejected in the case of *SS Wimbledon*, among others). Consequently the effect of the above reasoning is to undermine the binding character of international law: to confuse necessity in the sense of a political, moral, or other imperative with what the ILC claims is merely an exception of necessity that, far from being a subjective right, simply permits, under certain circumstances, the temporary non-execution of an international obligation': Heathcote (2010), pp. 491–492.

> (a) the international **obligation in question excludes the possibility of invoking necessity**; or (b) the State has **contributed to the situation of necessity**. (Art. 25 DASR)[78] [emphasis added]

There is convincing evidence to confirm that the operation of the climate change precautionary principle matches the concept of necessity and adapts to the requirements of Article 25 DASR. This is not a surprise since—in the words of *LG&E*—'[t] he concept of state of necessity and the requirements for its admissibility lead to the idea of prevention'. [79] An action from the host state that aims to 'anticipate, prevent or minimize the causes of climate change and mitigate its adverse effects [...]'[80] appears to safeguard an essential interest of the international community (Sect. 9.1.2.1) against an imminent and grave peril (Sect. 9.1.2.2). However, it is important to be aware that the effectiveness of the climate change precautionary principle through *necessity* would be at odds if some of the requirements from Article 25 DASR are interpreted in a very restrictive-literal manner, namely that the host state's measure should be the 'only way' to protect the interest (Sect. 9.1.2.3) and that the host state should have not 'contributed to the situation' (Sect. 9.1.2.4).

9.1.2.1 The Climate as an Essential Interest: Encompassing Article 25 DASR with Global Common Interests

As revealed above,[81] investment arbitration tribunals have envisaged health, environment and ecology within the scope of the interests that trigger Article 25 DASR. The similar characteristics of these interests with the climate interest lead us to confirm that the promotion of the latter through *necessity* could conform to the system's operation. Furthermore, the recognition of *necessity* as an instrument to preclude the application of the BIT when the host state's measures are needed for complying with international obligations that represent global common interests—such as those embedded in ILO Convention 169 or the Universal Declaration of Human Rights[82]—allows me to reaffirm that the host state's international obligation to take precautionary measures, as long as they promote the global climate interest, represents an essential interest under the realm of Article 25 (1)(a).

Extending the concept of essential interest from state interest to international community interest is crucial to circumscribe the climate interest within the text of Article 25 (1)(a). Otherwise, the fulfillment of this requirement risks to fail, since the climate change is generally framed as a common threat to the international community rather than a threat to a particular state—and, in these terms, the climate interest

[78] International Law Commission (2001).

[79] *LG&E*, Decision on Liability (2007), para. 248.

[80] United Nations Framework Convention on Climate Change (1992) Art. 3.

[81] *Supra* Sect. 9.1.1.

[82] *Supra* Sect. 9.1.1.2.

is grasped more as an international common interest rather than as a typical interest of the state.

Moreover, an interpretation of essential interest that is embracive of community interest makes it possible to invoke the defence of necessity to promote the climate interest without the—often very complicated—method of arguing that there is a direct connection between the risk to the climate interest and the national security, health or environmental interests of the particular host state.[83] This does not mean that I deny that these interests may eventually overlap with the climate interest. The only intention is to highlight the independence of the global climate interest as a basis for triggering Article 25 DASR.

The (international) communitisation of the essential interest is admissible as it conforms, not only to the openness of the ordinary meaning of the text of Article 25 (1)(a) DASR, but also to the purpose. In fact, one of the modifications of this provision in comparison to its predecessor—Article 33 DASR (1980)[84] —consisted in changing the expression 'essential interest **of the state**'[85] for the more general expression of 'essential interest'.[86] Thus, the Commission was clear in its commentary that the Article 25 DASR 'extends to particular interests of the State and its people, as well as of the **international community as a whole**'.[87] This modification has been considered to be a not so minor change[88] that attempts to reflect contemporary international law.[89]

The character of global interest protected under the Climate Change International Legal Framework certainly constitutes a common concern of humanity that falls into the issues of contemporary international law that motivated the modification of the concept in Article 25 (1)(a). Therefore, there should be no doubt that the invocation of the climate change precautionary principle as basis for the defence of necessity suits the requirement of safeguarding an essential interest.

[83] *Infra* Sect. 9.2.2.

[84] '(a) the act was the only means of safeguarding an essential interest of the State against a grave and imminent peril' (emphasis added): International Law Commission (1981), p. 34.

[85] *Ibid.*

[86] '(a) is the only way for the State to safeguard an essential interest against a grave and imminent peril': ILC (2001), Art. 25 (1)(a) DASR.

[87] International Law Commission (2001) para. 15.

[88] Paddeu (2018), pp. 395–396.

[89] *Ibid.*, p. 396.

9.1.2.2 The Grave and Imminent Peril of Global Warming and Its Effects: Overcoming Scientific Uncertainty as a Barrier to the Application of Article 25 DASR

The precautionary principle requires the state to 'anticipate, prevent or minimize the causes of climate change and mitigate its adverse effects [...]'[90] and to act in the events where 'threats of serious or irreversible damage, lack of full scientific certainty [...]'.[91] In other words, the principle compels the state to take climate action when the loss is certain or uncertain. This particularity of the precautionary principle, in addition to the fact that the effects of a given activity that are prejudicial to the climate (generally) manifest themselves in the long run, leads to the question of whether the threats that lack full scientific certainty and/or that are not materialised in the short-term could be regarded as imminent in the terms of Article 25 DASR (1)(a).

Investment arbitration tribunals have not addressed the question of immediacy in depth. However, it is possible to find some references to it. In *LG&E v. Argentina,* it was established that the peril 'must be imminent in the sense it will soon occur'. [92] This approach would entail that risks that materialise in the long-term—such as those that threaten the climate—would not satisfy Article 25 DASR (1)(a). However, other rulings have associated immediacy more to the causal proximity between the action of the host state and the prevention of the damage rather than by reference to the proximity in time. That is the case of *CMS, Enron* and *Sempra* where an imminent peril was associated with the need to 'prevent the worsening of the situation' [93] that otherwise would be 'out of control' [94] or 'unmanageable'.[95] Similarly, *Impregilo* adopts the causal proximity approach, but the decision goes beyond these other three awards by expressly taking the posture that the uncertainty of the loss does not rule out the imminence of the peril. [96]

Impregilo bases its conclusion in the ILC commentary to the DASR,[97] which in turn refers to the ICJ decision in the *Gabčíkovo-Nagymaros* case:[98]

> the ILC Commentary explains that "the peril has to be objectively established and not merely apprehended as possible", and that "[i]n addition to being grave, the peril has to be imminent in the sense of proximate". According to the Commentary, the International Court of Justice explained in the *Gabčíkovo-Nagymaros Project* case that "the invoking State could not be

[90] United Nations Framework Convention on Climate Change (1992) Art. 3.

[91] *Ibid.*

[92] *LG&E,* Decision on Liability (2007), para. 253.

[93] *Enron* (2007), para. 307, *Sempra* (2007) para. 349. In *CMS* the text is slightly different: 'prevent a worsening of the situation' [*CMS* (2005) para. 322].

[94] *Enron* (2007) para. 307, *Sempra* (2007) para. 349.

[95] *Ibid.*

[96] *Impregilo* (2011), para. 347.

[97] *Ibid.*

[98] International Law Commission (2001), para. 15–16.

the sole judge of the necessity, but a measure of uncertainty about the future does not necessarily disqualify a State from invoking necessity, if the peril is clearly established on the basis of the evidence reasonably available at the time" [99]

As the citation illustrates, apart from acknowledging that a future and uncertain loss could be labelled as imminent, the commentary and the *Gabčíkovo-Nagymaros Project case* establish that the standard to determine immediacy of the peril is the *evidence reasonably available*. This posture coincides with Professor James Crawford's special report on the version of the DASR that precedes the current version. In this respect, the special rapporteur expressly addresses the concerns about the scientific uncertainty and the precautionary principle[100] as well as these concerns in connection to the compliance of principle 15 of the Rio Declaration—which mirrors the precautionary principle of Article 3(3) FCCC in environmental international law.[101] He concludes that immediacy is verified 'if the peril is established on the basis of the *evidence reasonably available* at the time (as based, for example, on a proper risk assessment procedure)'.[102]

Here—as in the analysis of essential interest—the independence of the global climate interest from neighboring interests is crucial for the success of invoking the defence of necessity in investment treaty-climate change disputes. In fact, the threat to the climate in itself is what legitimises the interpreter to label the peril as imminent rather than the threat to the national security, the health, the environment or any other interest that may be indirectly in danger as a result of climate change. Under these conditions of independence, the immediacy of the peril to the climate increases its chances to be verified, since nowadays there is *evidence reasonably available* that strongly suggests that a considerable number of industrial and non-industrial activities represent an actual risk to climate—in comparison to the certainty of the *ricochet* effects that a given activity may have on neighboring interests of generally undetermined victims.

Against this background, it seems reasonable to assert that, despite the scientific uncertainty that is natural to the application of the climate change precautionary principle, the peril could be regarded as imminent under Article 25 (1)(a) if the *evidence reasonably available* suggests so.

[99] *Impregilo* (2011), para. 347.

[100] Crawford (1999) para. 290–291.

[101] 'Principle 15. In order to protect the environment, the precautionary approach shall be widely ap plied by States according to their capabilities. Where there are threats of serious or irreversible damage, lack of full scientific certainty shall not be used as a reason for postponing cost-effective measures to prevent environmental degradation': United Nations (1992).

[102] Crawford (1999), para. 291.

9.1.2.3 The 'Only Way for the State to Safeguard' the Global Climate Interest: Adopting an Interpretation Within the Limits of Reasonability and Relaxing the Burden of Proof

Proving that the host state measure was the only way to safeguard the global climate interest could be the most difficult barrier to *necessity* succeeding in investment treaty-climate change disputes.[103] This is so because there are big chances that a tribunal may not find any other choice but to interpret the expression the only way in a *restrictive* manner. In other words, said expression the *only way* appears to be restrictive in itself since it—literally [104]—poses the burden on the host state to prove that there were no alternatives to overcome the situation but to disregard the obligation—which might be an impossible task in many cases.

The restrictive approach seems to be relatively broadly accepted in investment treaty disputes. Thus, tribunals frequently draw on the International Law Commission's Comment that '[t]he plea is excluded if there are other (otherwise lawful) means available, even if they may be more costly or less convenient'.[105] In *CMS*, for example, the Tribunal found that the measures were not the 'only means' [106] because—to solve the Argentinian economic crisis—the hosts state had a set of possible alternatives by which to modify gas tariffs such as 'the dollarization of the economy, [or]granting of direct subsidies to the affected population or industries [...]'. [107] Similarly, in *Enron* and *Sempra* it was said that 'a rather sad world comparative experience in the handling of economic crises, shows that there are always many approaches to address and correct such critical events'. [108]

Such a restrictive interpretation makes it impossible to invoke necessity in the cases of an economic crisis, since there will always be many possibilities available and a plethora of opinions on which the correct alternatives are. Furthermore, this reasoning implies that, generally, where the host state faced an emergency situation that could have been solved through different measures—as almost always—choosing one of these would conduce to the failure of the defence of necessity. Briefly,

---

[103] This view is shared by Lise Johnson: 'If, however, tribunals strictly applied a requirement that the challenged climate change-related measures be the "only" measures available to address the national security threat, such requirement could be an insurmountable barrier preventing States from successfully invoking the defense. Indeed, given the complexity of the climate change challenge and the multitude of strategies countries are contemplating adopting in order to tackle the problem, it appears unrealistic to request States to establish that there were no other alternatives to the measure they chose to adopt.125 The current strict formulation of the customary international "necessity" test thus seems ill-suited for analyzing the legitimacy of measures designed to address climate change': Johnson (2009), pp. 11158–11160.

[104] *Enron*, Decision on the Application for Annulment of the Argentine Republic (2010), para. 369.

[105] International Law Commission (2001), para. 15.; *CMS* (2005) para. 324, *Impregilo* (2011) para. 352.

[106] *CMS* (2005) para. 324.

[107] *Ibid.*, para. 323.

[108] *Enron* (2007), para. 308; *Sempra* (2007), para. 350.

taking a mere literal interpretation of the expression *only means to safeguard the interest* leaves the host state defenceless.

Specifically, in investment treaty climate change-related disputes, the restrictive interpretation would unavoidably lead to the failure of the plea of necessity, given the particularities of the climate change action: first, the host state has within its toolbox many instruments to tackle climate change; second, there is scientific uncertainty as to which would be the only way to address global warming and, third, the action of the host state alone does not amount to the protection of the climate; it requires collective action. Accordingly—under a restrictive approach to the requirement—first, it could be argued that the action was not the only way because there were other instruments available to achieve the same climate objectives (e.g., reducing emissions from household consumers rather than emissions from the investor's industry);[109] second, it could be argued that the action was not the only way since there was no scientific certainty about its success in comparison to other measures,[110] and third, it could be argued that the isolated action of the state does not safeguard the interest because it would require the action of other states to prevent the loss.

Fortunately for the protection of the global climate interest, investment tribunals have also adopted, in some cases, a less stringent reading of Article 25 (a)(1). This approach promotes the examination of the requirement of the *only way to safeguard*

[109] The reasoning of *Fenosa v. Egypt* on the reduction of gas supply could be used to illustrate how the restrictive approach would deal with the political decision of affecting one segment of the economy rather than other: 'the "act" was not "the only way" for the Respondent "to safeguard an essential interest against a grave and imminent peril', using the words in Article 25(1)(a) of the ILC articles. In contrast to the Claimant and the Damietta Plant, other commercial users of gas remained preferred by the Respondent. This is confirmed by the disproportionality in the reductions for delivered gas between the Damietta Plant and other commercial users in 2012–2013; namely: 12% reductions for Industrial users (receiving 14.8 Bcma), 14% for Power users (28.9 Bcma), 14% for Idku (5.7 Bcma) and 60% (0.3 Bcma) for the Damietta Plant (from a total production of 58.8 Bcma); and the Plant's reduction resulted in a curtailment of 96% in the Claimant's contractual volumes for export.60In 2012–2013, if the "act" had been "the only way", it would have been equally necessary for the Respondent to curtail supplies to other commercial users to a similarly minimal level as the Damietta Plant and the Claimant. That was not done': *Fenosa* (2018), para. 8.46.

[110] The problem of determining alternatives when there is no certainty of the effectiveness of the measures was highlighted by the Annulment Committee in *Enron* concerning economic measures: 'In adopting measures to safeguard an essential interest, a State may in practice not be in a position to know with certainty whether a given measure will prove to be effective, and reasonable minds may judge that some measures are likely to be more effective than others. For instance, suppose that there are two possible measures that a State might take in order to seek to safeguard an essential interest. One is 90 per cent probable to be 90 per cent effective to safeguard that essential interest, while the other is 50 per cent probable to be 60 per cent effective. Suppose that the former measure would (subject to the potential application of the principle of necessity) be inconsistent with obligations of the State under international law, while the latter measure would not. Would the State be precluded from invoking the principle of necessity if it adopted the former measure, on the basis that there was an alternative available? Or could the State claim that the measure taken was the "only way" that stood a very high chance of being very effective?': *Enron*, Decision on the Application for Annulment of the Argentine Republic (2010), para. 371.

as limited to the measures that were *reasonably available.* [111] In *Urbaser*, the tribunal highlighted that this requirement 'has to be taken in observing its reasonable proportions' [112] as it involves weighting the general interest of the community with the investor's interests.[113] Accordingly, the tribunal affirmed that the alternatives to the cancellation of the contract that were proposed by the claimant, namely, providing subsidies to low-income users or granting financial equilibrium to the investor:

> [. . .] [a]t the time when the emergency measures were taken and in the immediate after-math [of the crisis], none of such solutions was possible or even seriously to be taken into consideration.[114]

It is noteworthy that these alternatives would have been considered to inhibit the application of Article 25 under the restrictive interpretation—since they are technically possible—but the tribunal opted to ignore them since they are not *reasonably available* under the particular circumstances.

In the same vein, the Annulment Committee in *Enron* admits that the expression '*only measure*' of Article 25 is open to an interpretation that goes beyond the restrictive approach. [115] According to the Committee, a plausible interpretation is that the alternative measures that preclude *necessity* should not include measures that involve a similar breach or a graver breach than the one experienced by the investor:[116]

> Under this interpretation, if there are three possible alternative measures that a State might adopt, all of which would involve violations of the State's obligations under International law, the State will not be prevented from invoking the principle of necessity if it adopts the

[111] *Urbaser* (2016), para. 719. The same expression *reasonably available* was deployed in Continental concerning Article XI NAFTA: 'In general terms, within the economic and financial situation of Argentina towards the end of 2001, the Measures at issue (the Corralito, the Corralon, the pesification, the default and the subsequent restructuring of those debt instruments involved here) were in part inevitable, or unavoidable, in part indispensable and in any case material or decisive in order to react positively to the crisis, to prevent the complete break-down of the financial system, the implosion of the economy and the growing threat to the fabric of Argentinean society and generally to assist in overcoming the crisis. In the Tribunal's view, there was undoubtedly "a genuine relationship of end and means in this respect [. . .]". The Tribunal is also guided by the principles highlighted above in its analysis of whether Argentina had *reasonably* available alternatives, less in conflict or more compliant with its international obligations, "while providing an equivalent contribution to the achievement of the objective pursued to the Measures challenged by Continental as inconsistent with the BIT [. . .]". In evaluating whether these alternatives were in fact reasonably available and would have avoided the adoption of the challenged Measures, the Tribunal is mindful that it is not its mandate to pass judgment upon Argentina's economic policy during 2001–2002, nor to censure Argentina's sovereign choices as an independent state': *Continental* (2008), para. 197–199.

[112] *Urbaser* (2016), para. 716.

[113] *Ibid.*

[114] *Ibid.*, para. 729.

[115] *Enron*, Decision on the Application for Annulment of the Argentine Republic (2010), para. 369–370.

[116] *Ibid.*, para. 370.

measure involving the least grave violation of international law. Under this interpretation, the principle of necessity will only be precluded if there is an alternative that would not involve a breach of international law or which would involve a less grave breach of international law.

Furthermore, the Decision on the Annulment of *Enron* highlights various factors that a tribunal should determine to apply the law and provide enough reason for its decision that concern the interpretation of the *only way to safeguard the interest.* In this respect, a sufficient reasoning for the application of Article 25 (1)(a) would entail expressing whether, in the view of the tribunal, the available measures include those that would be less effective to overcome the situation according to probabilities;[117] making clear whether the tribunal's evaluation of the alternative measures considers the knowledge of the host state at the time of taking the measure;[118] unearthing whether the tribunal determines the existence of alternative measures according to the standard of a 'reasonable and appropriately qualified decision maker';[119] and revealing whether the tribunal recognises any level of deference to the host state.[120] Moreover, the Committee of Annulment determined in this case, that the tribunal failed to assess the *only way* through legal reasoning but rather adopted an economic approach with the consequence of failing to apply the law.[121] In sum, even if the Committee abstains from taking a position on the correct interpretation of Article 25 (1)(a) in the case at hand, contemplates the possibility of adopting an interpretation within the boundaries of reasonability and beyond a literal interpretation.

We must conceive the analysis of *the only way to safeguard the essential interest* through the lens of *reasonable available* measures to apply the climate change precautionary principle in investment treaty disputes. The particular context of the emergency situation that compels the host state to take precautionary climate measures as well as the limitation of the alternative possible measures to those that would be coherent from a public policy perspective should enlighten the reasoning of tribunals. This involves acknowledging that the sole existence of technical alternatives to the precautionary climate change measure that was adopted does not cancel *ipso facto* its legal character of being the only way to safeguard the interest.

On the contrary, a legal analysis of this requirement entails—as suggested in the Decision on the Annulment of *Enron*—that a *reasonable available measure* with the potentiality of precluding the application of Article 25 on climate change action is expected not to breach an international obligation or, if it does, the breach should be less grave than the one caused by the climate measure taken by the host state. Additionally, when studying whether the climate precautionary action was the

[117] *Ibid.*, para. 371.

[118] *Ibid.*, para. 372.

[119] *Ibid.*

[120] *Ibid.*

[121] *Ibid.*

only way to promote the climate interest, it would be reasonable for a tribunal only to label as available measures those that could have equaled the effectiveness of the measures taken by the host state in promoting the climate interest or that could have reached better results. Other features that would indicate whether there were reasonably available measures are the following: considering only the measures that could be envisaged at the time when the climate measure was taken according to the host state's knowledge and determining the existence of alternative measures according to the standard of a 'reasonable and appropriately qualified decision maker'.[122]

Importantly, even if the requirement of the *only* way is interpreted through the limits of *reasonable available measures*, it will be impossible for the host state to prove in absolute terms that the precautionary climate measure was the only reasonable alternative to safeguard the climate interest. This is so because the host state would need to show that an endless list of conceivable climate measures that have not been tested were not at hand or that their application was unreasonable. Additionally, the lack of scientific certainty that characterises the effect of climate change measures hinders the establishment of which other measures would have given the same or better results to surpass the situation of necessity. These circumstances give rise to the question of who has the burden of proving whether the precautionary climate measure was the only one reasonably available.

Apart from the general principle that the *onus probandi* corresponds to the party that makes the assertion, the exceptional character of Article 25 and the negative grammar structure of its formula (*necessity **may not be invoked** [...]unless the act*) suggest that the host state has the burden of proving that the precautionary measure was the only one available. [123] Therefore, several tribunals have ruled that the burden of proof belongs to the host state. [124] In *Total*, for instance, one of the reasons for rejecting the defence of necessity was that 'Argentina has not discussed nor did it provide evidence, as it was its burden, that there were no reasonable alternatives to the freezing of the tariffs'. [125] Likewise, in *von Pezold,* the tribunal made it clear that it is 'the Respondent's burden to prove that it took the "right" or "correct" decision' [126] and added that the host state did not prove that the only available measure to tackle the violence and the relocation of locals tribes was abstaining from using its police powers to intervene.[127]

Contrarily, in *Urbaser,* the tribunal seems to alleviate the burden of the host state by requiring it to merely prove on a general basis that the measure was the only one available and by inverting the *onus probandi*:

> Respondent has made more than a **prima facie** showing that the emergency measures taken were the only ones available to the Argentine Government at the time, taking into account

[122] *Ibid.*

[123] *Total* (2010) para. 221.

[124] *Total* (2010) para. 223; *von Pezold* (2015), para. 645.

[125] *Total* (2010) para. 223.

[126] *von Pezold* (2015), para. 645.

[127] *Ibid.*

> the extreme economic, institutional and social disturbances suffered by the country and its population. It would have been incumbent on Claimants to offer at least a serious indication as to the nature of other measures that had been available to the Government at that time. [128]

A third approach—different to this one but also different from the interpretation that puts the burden on the host state—is found in the Decision on the Application for Partial Annulment of *Continental.* [129] In this case, the Committee, after affirming that the ICSID convention does not provide anything concerning the burden of proof, established that tribunals are not compelled to adopt any particular position on the burden of proof. Furthermore, the Committee suggested that instead of adopting a rule or standard on who bears the *onus probandi,* a tribunal may concentrate its efforts 'simply to making findings of fact on the basis of the evidence before it'.[130] Consequently, the investor's plea for annulment was dismissed because Argentina did not necessarily have the burden of proving the situation of necessity.[131]

The posture of the Committee in *Continental* does not contradict the *Urbaser* decision to lighten the host state's burden to demonstrate that the measure was the *only way to safeguard the interest*. Indeed, the Annulment Committee, instead of adopting one rigid rule concerning the *onus probandi*, injects flexibility to the ponderation of the burden of proof. Within this dynamic, the particular characteristics of investment treaty-climate change disputes lead to conclude—in the language of *Urbaser*—that while the host state has the duty to prove *prima facie* that the measure was the only reasonable alternative, the investor has the burden to prove the contrary. In other words, once the host state has proven on a general basis that the climate action was necessary to overcome the situation of necessity, the burden of proof is transferred to the investor.

Adopting the rigid rule that the *onus probandi* belongs to the host state is illogical in climate change disputes for two reasons. First, it is impossible for the host state to prove that it has taken the only possible (reasonable) measure since this would involve the same logical difficulties of proving a negative general statement—that there were no other reasonable measures available. In fact, the best thing the host state could actually do is to explain its reasoning based on the fact that it considered its choice as the only climate measure available, given the circumstances and the limited number of alternatives pondered by the host state for protecting the climate interest. Second, the uncertainty that is inherent to the application of precautionary climate measures calls for granting a certain level of deference to the host state. On this occasion, deference is not justified on the grounds of sovereignty but on the convincement that the host state is better positioned than a tribunal to establish that it had no other choice but to take the climate measure that conflicts with the BIT's obligation.

---

[128] *Urbaser* (2016), para. 717.

[129] *Continental,* Decision on the Application for Partial Annulment (2011) para. 135–136.

[130] *Ibid.*, para. 135.

[131] *Ibid.*, para. 136.

In sum, it could be affirmed that despite the fact that some tribunals interpret the requirement of the *only way to safeguard the interest* in a literal-restrictive fashion and allocate the burden of proof on the host state, a more open interpretation that limits the alternative measures to those that are reasonably available and a more dynamic conception on the *onus probandi*—that merely requires the host state to prove on a general basis that it took the *only reasonable measure*—would allow precautionary climate change measures to fulfill this requirement.

9.1.2.4 The Non-Contribution to the Situation of Necessity: Adopting an Interpretation Within the Limits of Reasonability and Inverting the Burden of Proof

Determining and proving the *non-contribution* of the host state to the situation of necessity that motivated the precautionary climate measure face similar barriers to those that were addressed above when dealing with the requirement of being the only way to safeguard the interest. This is so because the sole-literal interpretation of Article 25 (2)(b) DASR poses on the host state the impossible task of proving that it has not contributed to the climate change.

The literal interpretation of *non-contribution* has been softened through the investment treaty law practice. This softening consists of understanding the expression contribution as *substantial* contribution. The idea of limiting the requirement of Article 25 (2)(b) DASR to *substantiality* stems from the Law Commission's commentaries to the DASR.[132] Yet, the question of what is supposed to be a 'sufficiently substantial and not merely incidental or peripheral'[133] contribution remains open and requires further interpretation.

In most of the Argentinian economic crisis investment-treaty disputes, the misadministration of the host state's economy—in the 1980s and the 1990s—was found to substantially contribute to the situation of necessity.[134] *LG&E* seems to identify the substantiality of the contribution with the idea of *fault*, in the sense that 'If there is **fault** by the State [...] the causal relationship between the State's act and the damage caused is produced'.[135] In the same line, *Sempra* and *Enron* associate the requirement of *non-contribution* with the principle that 'prevent[s] a party taking legal advantage of its own **fault**'.[136] Similarly, even if *El Paso* does not use the technical expression fault, it seems to embrace within the concept of significantly

---

[132] *CMS* (2005), paras. 328–329; *Suez and InterAguas* (2010), para. 241; *Suez and Vivendi* (2010), para. 263; *Impregilo* (2011), para. 357.

[133] International Law Commission (2001), paras. 20, 84.

[134] *CMS* (2005), paras. 328–329; *Enron* (2007), paras. 311–313; *Sempra* (2007), paras. 353–354; *El Paso Energy International Company v. The Argentine Republic (2011)*, para. 611.; *Suez and InterAguas* (2010), para. 243; *Suez and Vivendi* (2010), para. 265; *Impregilo* (2011), para. 358.

[135] LG&E, Decision on Liability (2007), para. 256. [emphasis added]

[136] *Enron* (2007), para. 311; *Sempra* (2007), para. 353. [emphasis added]

substantial contribution some elements that characterise a typical attribution based on subjective elements:

> that Treaty cannot possibly allow for the possibility that if the host State itself has caused or significantly helped to cause, **intentionally or by omission**, the situation and the consequences complained of, that State may shirk its obligations under the BIT [137]

Impregilo, in the same vein, rejects the idea that the contribution to the situation of necessity requires to 'be specifically intended or planned'. [138] According to this tribunal, it is enough that the situation of necessity stems from 'Ill-conceived policies',[139] even if they are 'well intended'.[140] However, the posture of this decision on whether the contribution demands the action of the host state to be qualified as 'reckless or negligent, or even caused by a lesser degree of fault'[141] remains ambiguous. The decision does not discard these labels nor does it associate them with the concept of ill-conceived policy.

All these rulings show that it is not enough to verify a causal link between the conduct of the host state and the situation of necessity to preclude the defence. In line with the theory of 'contributory negligence',[142] the requirement of Article 25 (2) (b) DASR also needs to be evaluated through the analysis of the diligence of the host state. There is no doubt that the intentional causation of the state of necessity is contrary to the requirement of non-contribution. Yet, how to assess the diligence of the host state when it contributes to the state of necessity remains blurry:[143] fault,

[137] *El Paso* (2011), para. 615. [emphasis added]

[138] *Impregilo* (2011), para. 356.

[139] *Ibid.*

[140] *Ibid.*

[141] *Ibid.*

[142] 'This notion of contributory negligence seems more accurate to analyze the situations of necessity. When a norm provides a situation of necessity and when a State manifestly invokes her without any basis, we can imagine the perfection of an *abus de droit,* which is defined in international law as "the State's use of a right, power or competence in a way or towards an objective that do not belong to the purpose of that power or competence (. . .)" This notion is clearly implicit in the different norms that provide a state of necessity, just like the principle of good faith. It appears that, in the end, what it is required from a State to not be able to ground the state of necessity is a strict causal relationship; a simple contribution from his part to the state of necessity cannot exclude this qualification. We can then distinguish between two different questions and two levels of argumentation: the contributive negligence of a State can constitute one of the sources of the situation of necessity and this is frequently the case. In the case of *abus de droit*, the situation of necessity does not really exist, it is in its entirety created by the State who wants to avoid its international obligations': Cassella (2011), pp. 168–169.

[143] The Annulment Committee of Enron formulates the same question but on the concept of fault: 'However, that begs the question as to what will amount to "fault" in this context. Must the conduct of the State in question be deliberate (in the sense of being deliberately intended to bring about the situation of necessity), or does it suffice that the conduct was reckless or negligent, or is some even lesser degree of fault sufficient?': *Enron*, Decision on the Application for Annulment of the Argentine Republic (2010), para. 389.

omission, ill-conceived policies, reckless, negligent, lesser degree of fault, are some of the notions that have been deployed by these tribunals.[144]

The doubts on the determination of a host state's diligent behaviour in the analysis of *non-contribution* in the investment treaty disputes concerning the Argentinian crisis can be enlightened—although not completely dissipated—with the notion of foreseeability that was deployed in the cases *Fenosa* and *von Pezold.* In *Fenosa,* the tribunal affirmed that the element of contribution requires a 'common-sense for interpretation' [145] that involves 'placing the event in chronological proximity to the situation of necessity'.[146] If this placing suggests that the situation of necessity was a foreseeable result from the host state action, it should be concluded that the host state has contributed. Accordingly, the tribunal found that the plea of necessity succeeds in complying with the condition of non-contribution, not only because 'the Respondent did not subjectively intend that social unrest should take place, leading to the Egyptian revolution and its consequences of 2011-2014'[147] but also because '[i]t did not cause or **foresee** the consequences of the Global Financial Crisis of 2007-2008' [148] that affected the supply of gas to the investor. A similar reasoning was adopted in *von Pezold*, where the tribunal held that the government contributed to the economic crisis[149] because it

> **could not have been completely unaware** that the compulsory expropriation and redistribution of land from the seasoned and skilled farm owners to the unskilled Settlers/War Veterans would have a detrimental impact on the State's agricultural exports.[150]

Foreseeability appears to be better suited than other notions to defining the subjective element of a significantly substantive contribution. Some other concepts such as fault or omission might be too general and their interpretation could be difficult to grasp for some legal traditions. Models such as recklessness or negligence may provide a less stringent threshold to comply with the requirement of non-contribution than the one that is expected from the exceptional character of the defence of necessity. Conversely, *foreseeability* is a more precise term that is easy to understand, and it is still sufficiently strict to match the exceptional nature of

---

[144] *See* text to *Notes* 135–142.

[145] *Fenosa* (2018), para. 8.60.

[146] *Ibid.*

[147] *Ibid.*

[148] *Ibid.* [emphasis added]; A similar conclusion related to economic crisis was found in *Urbaser*: 'For such a demonstration to be successful, it should be shown that the Government's acts were such that they either were directed towards a crisis resulting in the emergency situation that the country experienced in early 2002, or at least of such a nature that the Government must have known that such crisis and emergency must have been the outcome of its economic and financial policy (...) Economic policies evaluated as wrong as they allegedly were in the 1990s in Argentina were not of a kind that they could lead to a crisis and emergency of such a magnitude as it blew up in the second half of 2001'. *Urbaser* (2016) paras. 711, 713.

[149] von Pezold (2015), para. 667–668.

[150] *Ibid.*, para. 666. [emphasis added]

necessity. Admittedly, all these other concepts may overlap with foreseeability and lead to the same conclusions concerning contribution.

In the context of investment treaty-climate change disputes, the compliance of the requirement of non-contribution needs to be interpreted through a subjective approach. Otherwise, it would be an impossible-to-meet condition, since every country has contributed to climate change. In other words—paraphrasing *Fenosa*—'[t]o an extent, [the] situation of [climate change] can always be traced back, as a matter of history, to political and economic mistakes made by a State, years, if not decades, earlier'. [151] Therefore, the only option for a climate precautionary measure to conform with the condition of Article 25 (b) (1) is understanding the notion of significantly substantial contribution as a link that lies beyond the causal relationship between the action of the host state and climate change.

In investment treaty-climate change disputes the interpretation that encapsulates *contribution* within the boundaries of chronological proximity may provide subjective grounds to distinguish between the conducts that fulfill the conditions of Article 25 (2) (b) DASR and those that do not. Hence, for the contribution of a host state to global warming to be *significantly substantial*, it needs to be foreseeable. This reasoning eliminates long projections on the historical involvement of some states in climate change as a basis for rejecting conformity with this condition. However, it is not possible to say the same about their participation in the recent history of global warming. Of course, while in the times of the industrial revolution and the subsequent massification of fossil fuels, the scientific community was not aware of the impact of many activities on climate change, since the 1990s the impact of some activities on global warming has been evidently foreseeable. Consequently, if a host state did not decrease its participation in global warming in proportion to the admissible limits (global standards or goals?) during the past 30 years, the element of non-foreseeability makes it impossible for it to argue that it did not contribute to the state of necessity that motivated the climate change precautionary measures.

It is noteworthy that *non-foreseeability* by itself is not enough to verify *non-contribution* in investment treaty-climate change disputes. The analysis of Article 25 (2) (b) DASR falls short if the notion of foreseeability is not combined with an objective element of materiality in the causation. Even if the question on which conduct is *material* to the situation of necessity should be solved on a case-by-case basis, case law embraces the commission's position that the level of participation of the host state that is required is located below the level of participation that is required for an event of force majeure.[152] This means that the impact of the host state's behaviour on climate change needs not to be regarded as the exclusive cause to significantly contribute to the situation of necessity. However, on the other hand, this entails that the host state's participation needs to be more than just a condition to global warming. It places the threshold of *contribution* to global warming at the (complicated) level co-causation.

[151] *Fenosa* (2018), para. 8.60.

[152] International Law Commission (2001), para. 20.

In a globalised world where the transboundary trade reveals a diffuse interrelation between producing, transforming, distributing and consuming, it is very difficult to apportion a specific participation of each of the states to global warming and, thus, it is very difficult to determine which have significantly contributed in terms of Article 25 (b) (1) DASR. Even if it is true that industrialised countries have materially contributed to global warming since most of the activities that affect climate are developed in these countries or, in any case, have made them wealthier than developing countries; developing countries also profit from exporting natural-polluting resources and consume products that result from industrial transformation, not to enter into a debate about the role of developing states in deforestation. Consequently, it could be said that, in natural terms, both southern and northern hemispheres are concomitant causes of climate change.

However, this debate on who has materially contributed to climate change has been held since the beginning of the negotiations of the Climate Change International Legal Framework.[153] As a result, there appears to be consensus that developed countries bear a higher historical responsibility in global warming that is reflected in the international instruments, which embrace the relevance of concepts such as climate justice, equity, sustainable development or the principle of common but differentiated responsibilities and respective capabilities.[154] This differentiation between states could serve as an interpretative guidance for investment treaty tribunals to decide whether the contribution of a host state has been material to climate change. For example, one could say that, while the special circumstances of small islands or the least developed countries suggest that they did not materially contribute to climate change, superpowers such as the USA or China did. In these terms, there would be a large probability that the plea of necessity invoked by superpowers would fail because of their contribution to the state of necessity and, inversely, the plea of necessity of least developed countries and small islands is likely to succeed.

The discussion on the burden of proof concerning the element of non-contribution has been less intense than the one that was described regarding the element of *the only way to safeguard the interest*, but it could be said that the controversy lies in the same arguments. Some tribunals consider that the burden of proof is on the respondent because of the exceptional character of the plea of necessity, the negative formulation of Article 25(2) [155] ('necessity **may not be invoked** by a State as a ground for precluding wrongfulness **if**') and 'the categorical terms of paragraph 2(b) of Article 25 DASR'. [156] Other tribunals, contrarily, establish that the burden of proof is on the investor because, even if the host state is the one that invokes the defence of necessity, it is the investor that asserts that the state of necessity is self-inflicted. [157]

---

[153] *See infra* Chap. 10, Sects. 10.2.2.1 '*The Application of the Principle of Equity*' and 10.2.2.2.

[154] *Ibid.*

[155] *Fenosa* (2018), para. 8.38.

[156] *National Grid* (n 23) para. 260.

[157] *LG&E,* Decision on Liability (2007) para. 256; El Paso (2011), para. 625.

The reasoning of the Decision on the Application for Partial Annulment of *Continental* that was quoted above when addressing the condition of the *only way*,[158] can also be applied to the allocation of the burden of proof regarding the element of non-contribution. This reasoning allows tribunals to consider a dynamic approach to the *onus probandi* and suggest that they focus on the evidence at hand rather than on the technicality of who bears the burden of proof.[159] Bearing in mind this flexibility, I am of the opinion that the *onus probandi* corresponds, in principle, to the host state.

This conclusion contrasts my approach when addressing the condition of the only way[160] given two considerations. On the one hand, unlike the proof of the *only way*—where the host state is unable to expose the reasons for discarding an endless list of alternatives to the climate precautionary measures whose effectiveness is unknown—*non-contribution* could be proven by demonstrating that the efforts that the host state has made to reduce global warming conform with the standards that the Climate Change International Legal Framework has been mastering for the last 30 years. On the other hand, while the uncertainty in terms of *reasonable alternatives* justifies a deferential approach that only requires the host state to prove *prima facie* that the climate precautionary measure was the only way to safeguard the interest, when determining *non-contribution* deference is not justified because the Climate Change International Legal Framework already provides guidance on which is the acceptable participation of host states in climate change.

Admittedly, the measures, duties and goals that correspond to each particular state concerning global warming are still very general—at least those that are established in the Climate Change International Legal Framework. Consequently, even if this does not inhibit the possibility for the host state to prove that its participation in global warming conforms to the general expectations of the international community, the rule of the *onus probandi* in practice cannot be very strict. This easing of the burden of proof on *non-contribution* does not differ that much from the host state's duty to prove *prima facie* its assertion, which is the approach conceived above regarding the condition of the only way.

Briefly, it could be affirmed that there is a general consensus that the contribution of the host state to preclude the success of the plea of necessity needs to be significantly substantive. This substantiality has a subjective element and an objective element. I sided with the approach that characterises the subjective element with foreseeability, as I considered that it is more precise than the concept of fault or omission, and more stringent than recklessness or negligence. On the element of materiality, I mentioned that the phenomenon of co-causation is what better describes the materiality of the contribution, understood as a participation that ranks a level below the absolute character of the causation attributed to the *force majeure*. The concrete application of these two elements in investment treaty-climate

[158] *Continental,* Decision on the Application of partial Annulment (2011).

[159] *Ibid*., para. 135.

[160] *Supra* Sect. 9.1.2.3.

change disputes means, on the one hand, that it is impossible for host states to argue that global warming of the past 30 years was unforeseeable. On the other, that it is more likely for least developed countries than for developed countries to succeed in the plea of necessity given the fact that there is an agreement that the latter group of states bears a major portion of responsibility for global warming—as reflected in the international climate change legal framework. Finally, I conclude that the burden of proof of the condition of non-contribution belongs to the host state, which will need to prove that its behaviour conforms to the standards of acceptable contribution to the climate change established in the Climate Change International Legal Framework.

In this section (5.1), I have demonstrated that there is a possibility for the climate change precautionary principle to apply directly through the defence of necessity and to preclude the application of BIT obligations. Even if the exceptional character of Article 25 DASR and the consequent stringency of its elements continue to be recognised as paramount for its application, the evolution of the operation of necessity that can be traced in investment treaty awards, shows that global public interests that are different from the national essential security interest are within the scope of this article, and it acknowledges that the elements of Article 25 DASR can be interpreted beyond a literal-restrictive understanding of the text. As part of this interpretation, *necessity* has been conceived as a duct to superpose the application of international obligations that represent global common interests over BIT obligations; e.g., ILO Convention 169 or the Universal Declaration of Human Rights. This context suggests that the application of the climate precautionary principle—which represents the common concern of humanity of tackling global warming—might inhibit the effectiveness of the BIT.

While the *essentiality* of the climate interest does not admit much discussion since it clearly suits the text of Article 25 (1)(a) and the conscious drift that the drafting commission took to embrace international community interests within the realm of the concept of *essential interest*, the other elements require further interpretative efforts to embrace the climate change precautionary principle.

The practice of investment treaty tribunals has fertilised the ground for the adaptation of the climate change precautionary principle to the elements of the imminence of the peril, *the only way to safeguard the interest* and the *non-contribution of the state to the situation of necessity.* On the *immediacy of the peril*, the uncertainty that is inherent to the climate precautionary principle is not a problem, since this element is considered to be evaluated according to the reasonably available corroborating evidence. Likewise, the practice of determining the *only way to safeguard the interest* by comparison to the *reasonable available alternatives*, eliminates the barrier of assessing this element by reference to an endless number of alternatives that would inhibit the application of the climate change precautionary principle. Similarly, on the element of non-contribution to the situation of necessity, limiting its failure only to occasions when it is found that the host state made a significantly substantive contribution, provides the climate precautionary principle with more chances to fulfill this requirement.

Moreover, I posit the effectiveness of the direct application of the climate change precautionary principle through necessity requires some special considerations that stem from the particular characteristics of the climate interest:

First, the climate needs to be conceived as an interest that is independent from neighboring interests. This enhances the protection of the climate interest by facilitating its conformity to the elements of essentiality of the interest and imminence of the peril, since the verification would not be dependent on the link or rebound effects that the impact of the climate has on interests such as the environment, health or human rights.

Second, the reasonable alternatives that might justify rejecting the argument that the application of the climate change principle was the only way to safeguard the interest should be exclusively restricted to those measures that would entail the non-breach of international obligations or a less grave breach of international obligations, as well as to measures that are equally or more effectives than the one that was adopted. Otherwise, the climate precautionary measure would not be recognised as the *only way* to avoid the detriment to the climate interest, given the fact that the scientific limitations on the accuracy of climate measures will always bring to the table a plethora of alternatives, and different hypotheses on the success of the alternatives.

Third, the definition of the contribution of the host state to the situation of necessity as *substantively significant* should be limited to the foreseeability of the climate change and to the materiality of the participation of the host state in global warming. If on the one hand, this conclusion overcomes the problems of a literal interpretation, it still provides interpretative value to the stringent character of the element of non-contribution. Accordingly, developed states that have contributed to global warming since the 1990s without adapting to the standards that the global community has established for fighting climate change would not be able to comply with the requirement of non-contribution. Contrarily, countries that have succeeded in these efforts or that have not materially contributed to climate change because of their low level of industrialisation would fulfill the requirement of non-contribution when adopting climate change precautionary measures.

Lastly, host states would only be required to prove on a general basis that the host state adopted the *only reasonable measure available* and that it did not *materially* contribute to the situation of global warming. This alleviation of the *onus probandi*—that in principle belongs to the host state because of the negative formulation of Article 25 DASR—overcomes the impossibility to technically prove with absolute certainty that there were not reasonable alternative measures to the climate change precautionary measure that was adopted, and permeates the difficulties in demonstrating that the participation of the host state in global warming conforms to the expectations of the international community, which are reflected in the climate change international framework.

Having demonstrated that climate change precautionary measures have the potential to preclude the wrongfulness through the defence of necessity, it must be highlighted that the operation of *necessity* would not be required if the climate change precautionary measures are proven to be included in the general exceptions that prevent the breaching of BIT obligations. This is the subject of the next section.

9.2 General Exceptions in Investment Treaty-Climate Change Disputes: Exploring the Absence of Breach of BIT Obligations When Adopting Climate Change Precautionary Measures

The deployment of 'general exceptions' or the international defence of necessity (Article 25 DASR) may prevent the liability of the host state. Although investment tribunals distinguish between these two mechanisms—as the former impedes the breach of the BIT obligations, while the latter precludes the wrongfulness of the breach[161]—one cannot deny that from a pragmatic point of view the function of both is the same. In the end, general exceptions as well as necessity excuse or justify the conduct of the host state for nonconformity with particular obligations; such as expropriation, FET, MFN, etc. [...].

Accordingly, the operation of the general exceptions hinders the effectiveness of the international defence of necessity, in the sense that if these exceptions are verified, there would be no room for the operation of Article 25 DASR, i.e. if there is no breach then the concept of necessity is useless.[162] However, the relationship between these two layers could be far more intricate than that. Since general exceptions are tailor-made, there would be cases in which the drafting of the particular text of general exceptions might lead to conflict or conflation with the international defence of necessity. Other cases could be characterised by complete independence.

Therefore, apart from depending on the specific interests that the parties agreed to promote or protect through the clause (Sect. 9.2.2), the conformity of climate precautionary measures to the general exceptions depends on the (non)relationship between the exception and Article 25 DASR that stems from the particular provision (Sect. 9.2.1). It is noteworthy that this subsection turns back from law ascertainment to content-interpretation, where the focus of the inquiry is to unearth whether the climate interest has a place within the agreement according to the balance between the parties as entrenched in the BIT text.

[161] *Infra* Sects. 9.2.1 and 9.3.1.

[162] As the Tribunal asserted in *Continental*: 'the application of Art. XI in the present case (if warranted) may be such as to render superfluous a detailed examination of the defense of necessity under general international law applied to the particular facts of the present dispute': *Continental* (2008), para. 162.

9.2.1 The Conformity of Climate Precautionary Measures to the General Exceptions Stemming from the (Non)Relationship Between the Exception and Article 25 DASR: Interpretation of the Subject Complement of the Measure

Necessity and general exceptions appear to have a different formal source, structure, operation and effects. While general exceptions are embedded in the BIT, the international defence of necessity that is established in the DASR has the character of customary law; while the former accepts many formulas according to the particular text that the parties agreed on, the latter is limited to one formula (Article 25 DASR); while the former determines the obligations of the parties (primary law), the latter determines the liability of the parties (secondary law) and; while the verification of the former inhibits the breach of BIT obligations, the verification of the latter precludes the wrongfulness of the breach.

Despite these differences, when the text of the *general exception* contains a subjective complement that describes the measure as necessary or needed,[163] it might lead the interpreter to redirect its view to Article 25 DASR and to wonder to what extent the understanding of the text is linked to the international defence of necessity. An example of this type of clause is Article XI of the USA-Argentina BIT:

> This Treaty shall not preclude the application by either Party of *measures* **[subject]** *necessary for* **[subject complement]** the maintenance of public order, the fulfillment of its obligations with respect to the maintenance or restoration of international peace or security, or the protection of its own essential security interests.[164] [emphasis, bold and brackets added]

Article XI—as described below[165]—generated an ample debate on the role of the elements of the defence of necessity in interpreting the text. From this experience—and not intending to assert that this discussion is settled—one could affirm that climate change precautionary measures would have less possibilities of conforming

[163] For example, 'The provisions of this Treaty (...) shall not preclude any Contracting Party from adopting or enforcing any measure (i) necessary to protect human, animal or plant life or health': Art. 24.2. (b), Energy Charter Treaty (1994). In the same line, Art. 24 2(b) ECCOWAS Energy Protocol (2003); Art. 15 (a) Agreement Between the State of Israel and Japan for the Liberalization, Promotion and Protection of Investment (2017); Arts. 17 (a) and 17(b), Agreement between the Government of Canada and the Government of the Hong Kong Special Administrative Region of the People's Republic of China for the Promotion and Protection of Investments Canada Hong (2016); Art. 11, Agreement between the Slovak Republic and the Islamic Republic Of Iran for The Promotion And Reciprocal Protection of Investments (2016); Art. 12.24, Free Trade Agreement Between the Republic of Singapore and the Republic of Turkey (2015); Art. 9 (a) and (b), Free Trade Agreement between The Government of Australia and The Government of the People's Republic of China (2015).

[164] Article 11, Treaty Between United States of America and the Argentine Republic Concerning the Reciprocal Encouragement and Protection of Investment (1991).

[165] *See* Sect. 9.2.1.1.

to the general exception when the text redirects to *necessity*, than in those cases where the *subjective complement* qualifies the measure in a more general way—e.g., '[t]he provisions of this Agreement shall not in any way limit the right [...] **directed to** [...]',[166] '[n]othing [here]in shall be construed to prevent a Party from adopting, maintaining or enforcing any measure otherwise consistent with this Chapter **that it considers appropriate to** ensure that',[167] 'nothing in this Agreement shall be construed to prevent the adoption or enforcement by any Member of measures [...] **designed and applied** [...]',[168] etc.

Thus, the conformity of climate change precautionary measures to clauses that contain an explicit reference to *necessity* in their text (Sect. 9.2.1.1) should be analysed separately from the conformity to those clauses that use more general expressions to condition the triggering of general exceptions (Sect. 9.2.1.2).

9.2.1.1 Conformity of Climate Change Precautionary Measures to Clauses that Contain an Explicit Reference to Necessity in Their Text: Between Confluence, *Lex Specialis*, Primary-Secondary Law Application

The interpretation of the expression 'necessary to' of Article XI of the Argentina-USA BIT as well as the relationship between the clause and Article 25 DASR was at the centre of the debate in many of the investment treaty disputes related to the Argentinian crisis. The posture of the tribunals was thus far from cohesive. These different views on the relationship between Article XI Argentina-USA BIT and Article 25 DASR were grouped by Kurtz—in an insightful article—[169] in 3 categories: confluence, *lex specialis* and primary-secondary application.[170]

Confluence consists of the operation of defining the term *necessary* by reference to the content of Article 25 DASR. This is what the author conceives as 'conflating the two standards',[171] a sort of fusion between the general exception and the international defence of necessity. The adoption of this approach obeys to the need to fill the gap in Article XI concerning the conditions for applying the exception.[172]

---

[166] Art. 11 (3), Republic of Mauritius and The Government of The Republic of India for The Promotion and Protection of Investments (1998).

[167] Art. 1114 (1), North American Free Trade Agreement (1994).

[168] Art. 5(1)(a), Agreement Between The Government of the Republic of Rwanda and The Government of The Republic of Turkey Concerning the Reciprocal Promotion and Protection of Investments (2016).

[169] Kurtz (2010), pp. 325–371.

[170] *Ibid.*, pp. 329–330.

[171] *Ibid.*, p. 341.

[172] 'The problem here, however, is that the Treaty itself did not deal with the legal elements necessary for the legitimate invocation of a state of necessity. The rule governing such questions will thus be found under customary law': *Sempra* (n 31) para. 378; 'While this analysis concerning Article 25 of the Draft Articles on State Responsibility alone does not establish Argentina's defense,

Although confluence—as noticed by Kurtz—was the basis of the reasoning in *Enron*,[173] *CMS*[174] *and Sempra*;[175] subsequent awards seemed not to have followed this trend and even the Annulment Committee of *CMS* and *Sempra* disagreed with defining the relationship between Article XI and Article 25 DASR using this method.[176]

The second method that Kurtz found in case law—*lex specialis*—entails the privileging of the application of Article XI to the study of the emergency situation over Article 25 DASR. The premise for the application of this method is that the two provisions at hand regulate the same substance at the same level but, simultaneously, they reveal some inconsistency when comparing one to the other[177]—such as 'the inclusion of the public order exception in the treaty provision'.[178] According to the author, the priority of the general exception could be analysed through two divergent theses: either total or partial exclusion to the application of Article 25 DASR.[179]

While with the partial exclusion of Article 25 DASR the preference for the treaty provision would still tolerate a residual application of the international defence of necessity, the total exclusion cancels any possibility for applying Article 25 DASR.[180] Although the author—as he admits[181]—struggles to present evidence

it supports the Tribunal's analysis with regard to the meaning of Article XI's requirement that the measures implemented by Argentina had to have been necessary (…)': LG&E (2004), para. 258. 'The expert opinion of Dean Slaughter and Professor Burke-White expresses the view that the treaty regime is different and separate from customary law as it is lex specialis. This is no doubt correct in terms that a treaty regime specifically dealing with a given matter will prevail over more general rules of customary law. Had this been the case here the Tribunal would have started out its considerations on the basis of the Treaty provision and would have resorted to the Articles on State Responsibility only as a supplementary means. But the problem is that the Treaty itself did not deal with these elements. The Treaty thus becomes inseparable from the customary law standard insofar as the conditions for the operation of state of necessity are concerned. As concluded above, such requirements and conditions have not been fully met in the instant case': *Enron* (2007), para. 334.

173 *Enron* (2007), paras. 333–334.

174 CMS (2005), paras. 355–378. Even in CMS, the Tribunal does not specifically affirm that confluence was required, its reasoning consists in conflating Article XI with the elements of Article 25 DASR.

175 *Sempra* (2007), pp. 373–391.

176 *CMS* Decision of the Ad Hoc Committee on the Application for Annulment of the Argentine Republic, (2007), paras. 128–136; *Sempra* Decision of the Ad Hoc Committee on the Application for Annulment of the Argentine Republic (2010), paras. 186–210.

177 Kurtz (2010), p. 352.

178 *Ibid.*

179 *Ibid.*, pp. 353–354.

180 *Ibid.*

181 'In sum, the LG&E award is a candidate for a lex specialis approach but it is not a convincing one. The problem lies in the internal contradiction of that award. If the treaty exception is an expression of lex specialis, then an adjudicator should determine the scope of priority to be accorded to the treaty text. This would mean either applying the customary norm in a residual fashion (where not in conflict) or displacing it entirely. The LG&E Tribunal appears reluctant to choose between

of the adoption of the partial-*lex specialis* methodology in the disputes concerning the Argentinian crisis, there are some recent cases that have deployed this method.[182] That is the case of *El Paso*, where the tribunal found that 'only if Article XI is found by the Tribunal not to apply in the case would an analysis of Article 25 be performed' [183] because 'Article XI is the lex specialis, Article 25, the lex generalis'.[184] Similarly, on the total-*lex specialis* methodology, apart from *UPS*—which is the example that the author used—there are recent awards that embrace this approach. In *Bear v. Peru*, for instance, the tribunal reached the conclusion that[185]

> In substance, in view of the very detailed provisions of the FTA regarding expropriation (Article 812 and Annex 812.1) and regarding exceptions in Article 2201 expressly designated to "Chapter Eight (Investment)", the interpretation of the FTA must lead to the conclusion that no other exceptions from general international law or otherwise can be considered applicable in this case.[186]

The last of the methodologies spotted by Kurtz in investment treaty disputes—primary-secondary application—rejects the contact between the two provisions.[187] The basis for the instruments' independence from one another lies in the different scope that they embrace. While general exceptions are primary law that define the breach of the agreement, the international defence of necessity is secondary law that deals with the preclusion of the wrongfulness once the breach has been verified.[188]

---

these approaches, perhaps recognizing the strength of the former (within the internal logic of the lex specialis method) but also its implications (the likely preclusion of most claims to invocation). Its own poorly conceived test of 'legitimacy' appears driven more by pragmatism than principle and will hardly aid in guiding successive tribunals in this difficult area': Kurtz (2010), p. 356.

[182] In the case of *Impregilo*, the Tribunal found that 'the lex specialis of Article 4 does not preclude Argentina from invoking the necessity plea'—Article 4 established that if the losses of the investor are due to political-economic events (within others) the host state must provide an equal treatment to the one provided to national or to international investors from other nationalities: Impregilo (2011), para. 355. *See* also *El Paso* (2011), para. 55. cf. *Total* (2010) paras. 229–230; Cf. EDF (2012), p. 1157.

[183] *El Paso* (2011), para. 552.

[184] *Ibid.*

[185] 'Article 2201.1 of the FTA provides for the following "Exceptions": 3. For the purposes of Chapter Eight (Investment), subject to the requirement that such measures are not applied in a manner that constitute arbitrary or unjustifiable discrimination between investments or between investors, or a disguised restriction on international trade or investment, nothing in this Agreement shall be construed to prevent a Party from adopting or enforcing measures necessary: (a) to protect human, animal or plant life or health, which the Parties understand to include environmental measures necessary to protect human, animal or plant life or health; (b) to ensure compliance with laws and regulations that are not inconsistent with this Agreement; or (c) the conservation of living or non-living exhaustible natural resources': Bear (2017), para. 472.

[186] Bear (2017), para. 473. Cf. *Bear v. India*, Dissenting opinion (2017), para. 41.

[187] Kurtz (2010), pp. 354–355.

[188] 'There is a third methodology that overcomes the deficiencies in the approaches we have considered so far. This would begin by characterizing all of the treaty provisions at issue, both forms of obligation and exception, as primary legal standards that determine whether a State has committed a wrongful act at international law. It is only if breach is determined by the composite

The author illustrates this methodology with *Continental* and the decision of Annulment for *CMS.*[189] Importantly, after he wrote the paper (2010), other decisions have embraced the primary-secondary application methodology, [190] which is the most frequently applied approach today. Thus, the Committee of *Sempra*'s annulment—for example—was clear in separating Article XI Argentina-USA BIT from the international defence of necessity:

> Article 25 is concerned with the invocation by a State Party of necessity "as a ground for precluding the wrongfulness of an act not in conformity with an international obligation of that State". Article 25 presupposes that an act has been committed that is incompatible with the State's international obligations and is therefore "wrongful". Article XI, on the other hand, provides that "This Treaty shall not preclude" certain measures so that, where Article XI applies, the taking of such measures is not incompatible with the State's international obligations and is not therefore "wrongful". Article 25 and Article XI therefore deal with quite different situations. Article 25 cannot therefore be assumed to "define necessity and the conditions for its operation" for the purpose of interpreting Article XI, still less to do so as a mandatory norm of international law.

The success of the adequation of climate change precautionary measures to the clause of general exceptions depends on which, among the three methodologies—*confluence, lex specialis (total or partial)* or *primary-secondary application*—has been chosen to interpret the text.

It is clear, for instance, that the confluence methodology would involve less possibilities for precautionary climate change measures to be recognised as an exception than those that the primary-secondary application would entail. This is because while the conflation of Article 25 DASR with the general exception compels the interpreter to examine the *necessity* of the climate change precautionary measures under the stringency that is inherent to the elements of the international defence, the independence of general exceptions from Article 25 DASR—that the *primary-secondary application* presupposes—grants the interpreter more freedom to construe the conditions for triggering the exception.

Similar conclusions should derive from the comparison of *confluence* and primary-secondary application and the comparison between *confluence* and *lex specialis,* concerning the potential of climate change precautionary measures to suit the general exception clause. Contrary to *confluence*, *lex specialis* would allow a more flexible interpretation of the conditions to justify the non-conformity of the climate change precautionary measures with the BIT standards, since the content of the text is considered to be tailor-made for the parties instead of linked to the strict elements of Article 25 DASR.

---

application of these rules that an adjudicator would examine the secondary possibility of the customary defence [sic] of necessity to preclude the finding of a wrongful act': *Ibid.*

[189] Kurtz (2010), p. 357; *Continental* (2008), paras. 166–167; *CMS,* Decision of the Ad Hoc Committee on the Application for Annulment of the Argentine Republic (2018), para. 129.

[190] *El Paso* (2011) para. 553–554; *Deutsche Telekom* (2017), paras. 225–229; *Bear v. India,* Dissenting opinion (2017), para. 41.

However—from a different perspective—the probability of success of precautionary climate change measures as a basis for excusing liability under the application of *lex specialis* could be jeopardised if one adopts the total-*lex specialis* approach. This is so because, unlike the partial-*lex specialis* methodology—where the disconformity to the general exceptions still permits Article 25 DASR to exclude the wrongfulness of the breach—the total-*lex specialis* methodology completely eliminates the application of the international defence of necessity. The relevance of this restriction appears to be a serious barrier for the verification of the exception when invoking climate change precautionary measures, given the fact that—in most BITs—general exceptions clauses do not expressly include climate within the interests that justify the breaching of BIT obligations. Consequently, if this is the case, the climate change precautionary measure risks not only to fall outside the general exception but also to be out of the realm of Article 25 DASR.

If the goal is to find an answer that harmonises with the climate interest, then the drawbacks of adopting the total-*lex specialis* or the confluence methodologies, suggest considering the partial-*lex specialis* and primary-secondary application as the only plausible alternatives. Unlike confluence, these two methodologies recognise the independence of the general exceptions from Article 25 DASR and leave open the possibility to analyse the need for climate change precautionary measures under lighter conditions than those of the international defence of necessity. Furthermore, contrary to the total-*lex specialis* methodology, the partial-*lex specialis* and the *primary secondary application* methodologies make it possible to prevent liability through the operation of the Article 25 DASR, in the event that the climate change precautionary measures fail to comply with the conditions of or to fall within the interests embraced by the general exception.

Nevertheless, the harmonisation of the Climate Change International Legal Framework with investment treaty law cannot ignore systemic considerations. In this sense, one could ask how the partial-*lex specialis* and *primary secondary-application* differ, given the finding that both allow a flexible interpretation of the general exceptions and still allow the residual application of Article 25 DASR. The short answer would be that, while partial-*lex specialis* involves the recognition of a conflict of laws, primary-secondary application does not. In other words, the partial-*lex specialis* methodology assumes that the international defence of necessity and the general exception share the same substance, but there is an inconsistency between them that stems from the particular shaping that the parties give to the circumstances that justify the non-conformity with the BIT standards. Conversely, primary secondary-application methodology accepts that the general exceptions and Article 25 are different in substance and comprehend different scopes.

The key question to choose between the application of either the partial-*lex specialis* methodology or the primary secondary-application methodology would be whether the defence of necessity and the general exception clauses have the same substance or regulate the same phenomena. A general answer would be that they do not. Even those decisions that describe the relationship between the international defence and the general exceptions as partial-*lex specialis* incur in the contradiction of describing the two concepts as different.[191] These rulings accept the main difference—repeated through this section—that while one impedes the breach (primary law) the other precludes the wrongfulness once the breach has been verified (secondary law). Therefore, from a systemic perspective, the only possible methodology defining the relationship between Article 25 DASR and the general exceptions that harmonises with the Climate Change International Legal Framework principle is the primary-secondary application.

It is important to mention that general exceptions could theoretically be drafted with a formula that suggests that the clause regulates the same substance as Article 25 DASR. Such a formula would consist of expressing that the general exceptions preclude the wrongfulness of the breach, rather than impeding the breach itself. This would not only place the two provisions at the same level but also in conflict. In this hypothesis, *lex specialis* would be the method to apply—more likely the *total-lex specialis* methodology.[192] Similarly, general exceptions could theoretically be drafted with a formula that clarifies that the expression 'necessity'—or similar expressions—should be interpreted according to the conditions of Article 25 DASR. If this is the text, there is no other option but to adopt the confluence methodology and accept the supplementary role of the international defence of necessity.

By and large, the analysis of the decisions that deal with the relationship between general exceptions that contain the term 'necessary to'—or similar expressions—and the international defence of necessity shows that the *primary-secondary application* methodology, on the one hand, harmonises with the Climate Change International Legal Framework and, on the other hand, adapts to systemic considerations. In fact, this methodology both acknowledges the substantial difference between the two instruments, and conduces to consider climate change precautionary measures under more flexible conditions to suit the general exception than those of the international defence of necessity, and even preserves the residual application of Article 25 DASR. It is worth mentioning that this preference for the primary-secondary application is not an obstacle to apply the confluence or the *lex specialis* methodology if the text of the general exception is clear. Furthermore, vacating the expression 'necessary to'—or similar terms—contained in the general exception of the elements of the international defence of necessity calls for an interpretation that is more inclined to assimilate such a clause with the general exceptions that contain more general expressions—such as 'directed to', 'designed to', 'applied to', etc. [. . .].

---

[191] *See* text to *Notes* 186–189.

[192] It is doubtful that the concept of *partial lex-specialis* in itself is devoid of contradictions. If the concept of *lex-specialis* is precisely to determine the outcome of a conflict of norms that could not be solved by any other way than ruling out one of them, then there the only possible of understanding *lex specialis* is by reference to the *total-lex specialis.* In other words, considering that a clause is *lex specialis* equals to consider that the parties had the intention to separate from the general rules of customary law and regulate the situation in a different manner.

9.2.1.2 Clauses that Embrace Climate Change Precautionary Measures When General Exceptions Are Not Linked to 'Necessity': Between Self-Judging Clauses, Measures 'Necessary To' and Additional Requirements

Apart from the term 'necessary to',[193] multiple expressions are used in the different models of general exceptions to determine the *subjective complement* to the measures that justify non-compliance with BIT obligations. These expressions—which connect the legitimate objectives to the measure—include, for example, measures that 'are directed to',[194] 'in pursuance of',[195] that the '[host state] considers necessary for',[196] '[host state] considers appropriate for',[197] that 'are intended to',[198] 'relating/ed to'[199] 'in pursuing',[200] 'imposed to'[201] 'that provide for'[202] or 'applied and designed to'.[203] Although it could be concluded that some of these terms are interchangeable, the deployment of one particular expression could determine the host state's level of discretion in defining the validity of the exception in comparison to the others.

Indeed, it could be said that conditioning the precautionary measure's validity as an exception to the requirement of being 'necessary to' promote the climate interest, is different from conditioning its validity to the requirement of being 'directed to' or 'in pursuance of' the climate interest. While the expression 'necessary to' entails that the circumstances unavoidably lead to the adoption of the climate precautionary

[193] *Supra* Sect. 9.2.1.1.

[194] Mauritius-India BIT (1998) Art. 11 (3); Art. 11, Agreement between The Government of the Republic of Singapore and The Government of the Republic of Peru on the Promotion and Protection of Investments (2003).

[195] Article 3, Draft Agreement for the Promotion and Protection of Investment between the Republic of Austria and (2010).

[196] Article 15.1 (a), Agreement Between the Government of the Republic of Korea and the Government of Japan for the Liberalisation, Promotion and Protection of Investment (2002).

[197] North American Free Trade Agreement (1994) Art. 1114; Art. 11.16, Central American Free Trade Agreement (2004); Art. 9.16., Trans-Pacific Partnership.; Art. 13, Reciprocal Investment Promotion and Protection Agreement Between The Government of the Kingdom of Morocco And The Government of the Federal Republic of Nigeria (2016).

[198] Bilateral Agreement for the Promotion and Protection of Investments between The Republic of Colombia and_ Colombian Model (2007).

[199] North American Free Trade Agreement (1994) Art. 904(1); Art. 9 (f), Agreement on Investment among the Governments of the Hong Kong Special Administrative Region of the People's Republic of China and the Member States of the Association of Southeast Asian Nations (2017); Israel-Japan BIT (2017) Art. 15 (f); Rwanda-turkey BIT (2016) Art. 5(1)b, Hong Kong-Canada BIT (2016) Art. 17 (c); Australia-China FTA (2015) Art. 9.

[200] North American Free Trade Agreement (1994) Art. 904 (2).

[201] ASEAN-Hong Kong BIT (2017) Art. 9(e); Australia-China FTA (2015) Art. 9 (c).

[202] Art. 24.3. Comprehensive Economic and Trade Agreement (CETA) between Canada, of the one Part, and The European Union, of the other part (2014).

[203] Rwanda-Turkey BIT (2016) Art. 5(1)(a).

measure, the expressions 'directed to' or in 'pursuance of', simply mean that the object of the precautionary measure is to promote the climate. These last two expressions clearly entail more chances for the host state's climate change precautionary measures to excuse nonconformity to BIT standards than the expression 'necessary to'.

Similarly, comprehending within the radio of the exceptions the precautionary measures that the host state 'considers appropriate for' promoting climate, is different from defining its scope by reference to measures 'designed for or applied to' this end. While the term 'considers appropriate for' suggests that the subjective perspective of the host state on the effectiveness of the precautionary measure might be enough to define the validity of the exception, 'designed or applied to' appears to represent a property of the precautionary measure that could be judged in an objective manner by a third party. Accordingly, contrary to an exception that contains 'designed or applied', an exception that includes the terms 'considers appropriate' or 'considers necessary' would not accept any other standard to measure its validity than the assessment of the host state, i.e. it would be self-judging.

The aim of contrasting these different *subjective complements* is neither to define in absolute terms each of the models nor to establish with precision the degree of host state's discretion that each of the models entails. On the contrary, the intention is to illustrate that general exceptions accept different formulas of *subjective complements* that extend or restrict the host state's space to adopt climate change precautionary measures and its power to assess the validity of the exception. Of course, these types of expressions need to be interpreted in accordance with the whole text of the particular clause as well as in connection to the object, purpose and context of the concrete BIT, i.e. be interpreted in the terms of Article 31 VCLT.[204]

The tribunal in *Devas v. India*—after highlighting that the interpretation of general exceptions should be perfected in accordance with Article 31 VCLT [205]—placed the expression **'directed to'** [*the protection of its essential security interest*] as a middle term between 'necessary to' and 'self-judging clauses'.[206] Accordingly, it was concluded that, although the host state was not required to prove that 'the measure adopted was the only one it could resort in the circumstances',[207] it still had the burden to demonstrate that the measure is **'related to** its essential security interest',[208] i.e. the tribunal equated the expressions 'directed to' and 'related to'. Moreover, it was added that under the expression 'directed to' the host state has not only 'considerable freedom as to the action it can take'[209] and a 'wide measure of deference',[210] but also that the claimant would face a high threshold to deprive the

[204] United Nations, *Vienna Convention on the Law of Treaties* (1969), p. 331.

[205] *Devas* (2013), paras. 230–232.

[206] *Ibid.*, paras. 241–242.

[207] *Ibid.*, para. 243.

[208] *Ibid.*

[209] *Ibid.*, para. 235.

[210] *Ibid.*, para. 244.

exception from validity 'such as bad faith, absence of authority or application to measures that do not relate to essential security interests'.[211]

In general terms, it appears that the discretion for the host state to take precautionary climate change measures would be extended to its maximum expression when the BIT explicitly contains a self-judging clause.[212] These clauses, which condition the validity of the exception to the measures that the host state 'considers necessary for'[213] or 'considers appropriate for'[214] the promotion of the legitimate objective, seem to be limited only by the examination of good faith.[215] Contrarily, it is likely that the clauses that deploy only the expression 'necessary to'—devoid of any reference to the subjectivity of the host state—would shrink to its minimum expression, the discretion for the host state to take precautionary climate change measures. Nevertheless, this minimum space of discretion—as explained above[216]—does not restrict the exception's possibilities of success as much as the strict elements of the international defence of necessity would.

In *Deutsche Telekom v. India*, the tribunal dealt with a general exception clause on measures 'to the extent necessary to' protect the security interest. [217] It posited that the standard required for triggering the general exception was higher than verifying that the measure was 'related to' the security interest,[218] yet lower than 'requiring the state to prove that the measure was the "only way" to achieve the stated purpose'.[219] Accordingly, the tribunal 'undoubtedly recognize[d] a margin of deference'[220] to the host state for assessing the necessity of the measure and reserved for itself the determination of

> whether the measure was principally targeted to protect the essential security interests at stake and was objectively required in order to achieve that protection, taking into account

---

[211] *Ibid.*, para. 245.

[212] 'The Tribunal is convinced that when States intend to create for themselves a right to determine unilaterally the legitimacy of extraordinary measures importing non-compliance with obligations assumed in a treaty, they do so expressly': *CMS* (2005), para. 370. In the same line, *Sempra* (2007), para. 379; *El Paso* (2011), para. 590, *Deutsche Telekom* (2017), para. 231.

[213] Investment promotion and protection agreement between Japan, Republic of Korea and China (2012). Art. 18.1 (a).

[214] NAFTA (1994) Art. 1114; CAFTA (2004) Art. 11.16; TPP (2018) Art. 9.16., BIT Morocco-Nigeria (2016) Art. 13.

[215] United Nations Conference on Trade and Development, The Protection of National Security In IIAs (2009) UNCTAD Series on International Investment Policies for Development 91–92; *LG&E,* Decision on Liability (2007), para. 214; CMS (2005), para. 374; *El Paso* (2011) para. 595. *Enron* could be understood as suggesting that self-judging clauses are contrary to the object and purposes of BITs: Cfr. *Enron* (2007) paras. 331–332. In the same line of *Enron:* Cfr. *Sempra* (2007), paras. 373–374.

[216] *Supra* Sect. 9.2.1.1.

[217] *Deutsche Telekom* (2017), para. 23.

[218] *Ibid.*, para. 238.

[219] *Ibid.*

[220] *Ibid.*

whether the state had reasonable alternatives, less in conflict or more compliant with its international obligations.[221]

Whichever the level of discretion that the *subjective complement* determines for the host state's action, some general exceptions establish additional conditions for the effectivity of the exception that might define whether a precautionary climate change measure prevents liability. The most common condition is requiring exceptional measures not to 'constitute arbitrary or unjustifiable discrimination between investments or between investors, or disguised restriction on the international trade or investment'.[222] This formula is also applicable when the general exception clause redirects '*mutatis mutandi*' to the GATT.[223] Other language modulating the scope of the general exceptions includes restricting the validity of the exception to 'circumstances of extreme emergency in accordance with its laws normally and reasonably applied on a non-discriminatory basis',[224] to 'a genuine and sufficiently serious threat [. . .]';[225] or by forbidding to 'use such measure as a means of avoiding its obligations';[226] among others.[227]

Within this universe of tailor-made general exceptions, it is difficult to predict the embracing of precautionary climate change measures. However, the consensus that expressions different from 'necessary to' represent lesser restrictions for the host state's measures to rank as exceptions, certainly widens the probabilities of climate change precautionary measures to fall under the conditions of the text. If one adds to this the trend to understand the clauses that adopt 'necessary to' as independent from the strict elements of Article 25 DASR, the chances of climate change precautionary measures to succeed as exceptions extend even further.

An expansionist perspective on the regulatory discretion—that potentially harmonises general exceptions with the climate change international framework—is confirmed by the awards that confer a considerable deference to the host state to define the content of the *subjective complement* of the measure, such as whether it

[221] *Ibid.*, para. 239.

[222] Art. 2201.1, Canada-Colombia Free Trade Agreement (2008); Art. 2201.1, Canada-Peru Free Trade Agreement (2008); Art. 13, Agreement between Japan and the Islamic Republic of Iran on Reciprocal Promotion and Protection of Investment (2016); Israel-Japan BIT (2017) Art. 15; Iran-Slovakia BIT (2016) Art. 11.

[223] 'Article 15 General Exceptions: For the purposes of this Agreement, Article XX of the GATT 1994 and Article XIV of the GATS are incorporated into and form part of this Agreement, mutatis mutandis: Art. 15, Agreement between The Argentine Republic and Japan for the Promotion and Protection of Investment (2018); Art. 18, Agreement between Japan And The Kingdom Of Cambodia for the Liberalization, Promotion And Protection of Investment (2007); ASEAN-Hong Kong BIT (2017) Ch. 11, Art. 2.

[224] Art. 12, Agreement between The Republic of Hungary and The Republic of India for the Promotion and Protection of Investments (2003).

[225] Korea-Japan BIT (2002) Art. 16 (1).

[226] China-Japan-Korea BIT (2012) Art. 18 (2).

[227] For example, '[. . .]shall be duly motivated and shall not nullify or impair any benefit one or more other Contracting Parties may reasonably expect under this Treaty to an extent greater than is strictly necessary to the stated end': ECT (1994) Art. 24 (2).

was 'necessary to', 'directed to' or 'designed to' promote the legitimate objective. Yet, the degree of deference might be altered by the parties' choice to introduce additional modulations to the general exceptions establishing a *chapeau*—such as an arbitrariness or discrimination threshold—or adding requirements, such as the condition of a sufficiently serious threat to the interest.

Additionally, the parties very often decide to restrict the range of legitimate objectives that the exception covers to specific interests. The conformity of precautionary climate change measures to the legitimate objectives that general exceptions embrace is the topic of the next subsection.

9.2.2 The Conformity of Climate Change Precautionary Measures to the Exceptions Stemming from the Range of Legitimate Objectives Covered by the Text: Direct and Indirect Embrace of the Climate Interest

There appears to be no BITs that include as part of their *general exceptions* explicit language exempting the host state from liability for taking measures to promote or protect the climate. Consequently, only two alternatives emerge as a method to harmonise the climate interest with the *general exceptions* clause. On the one hand, precautionary climate change measures could conform to a text that includes a non-exhaustive list of exceptions, or an open formula that embraces, generally, any legitimate regulatory objectives[228] (Sect. 9.2.2.1), but yet, such a text is uncommon. On the other hand, the connection between general exceptions and precautionary climate change measures might be construed through the connexity of the climate interest with other interests that are explicitly included in the text of the BIT (Sect. 9.2.2.2). However, how close or direct the connection between the climate interest and the interests that are explicitly included in the *general exceptions* should be, remains uncertain to prevent liability.

[228] 'Nothing in this Chapter shall be construed to prevent a Party from adopting or maintaining any measure otherwise consistent with this Agreement that it considers appropriate to ensure that investment activity in its territory is undertaken in a manner sensitive to its environmental, health, or other regulatory objectives': Art. 19 (6), Pacific Agreement on Closer Economic Relations (PACER) (2017). A similar provision TPP (2018) Art. 9.16. Morocco-Nigeria BIT (2016) Art. 23 (1): 'Except where the rights of Host State are expressly stated as an exception to the obligation of this Agreement, a Host State's pursuit of its rights to regulate shall be understood as embodied within a balance of the rights and obligations of Investors and Investments and Host States, as set out in the Agreement': Art. 10, The Reciprocal Promotion and Protection of Investments between The Argentine Republic and The State Of Qatar (2016); Iran-Slovakia BIT (2016) Art. 11(b).

9.2.2.1 Open Exceptions

Open exceptions allowing host states to take regulatory measures without of breaching their obligations are exceedingly rare but they do exist. For example, Article 10 of the Qatar-Argentina BIT provides that

> For the purpose of this Agreement, the Parties recognise their right to regulate in their territories in order to achieve legitimate public policy objectives, such as national security, the protection of public health, safety, the environment, public morals, social and consumer protection, or the promotion and protection of cultural diversity.[229]

This model of clause, by using the formula 'legitimate public objectives such as' would surely legitimise a climate change precautionary measure as an exception in an investment treaty climate change-related dispute. However, what remains puzzling concerns the extent to which such a provision opens the floodgates for any regulatory measure to rank as an exception. It could be argued that such an unlimited general exception would leave the investor unprotected and, accordingly, the BIT's rationale of encouraging the flow of transnational investment might be disrupted. This is probably why other models of open exceptions condition their effectivity to additional conditions.

One of these other open exceptions models conditions the validity of the exception to consistency with the BIT. Articles 9.16 TPP and 19 (6) PACER—in an analogous fashion to Article 10 Qatar-Argentina BIT[230]—provide general exceptions through a non-exhaustive list of legitimate objectives ('environmental, health and other regulatory objectives'[231]) but—contrary to Article 10 Qatar-Argentina BIT[232]—limit the effectiveness of the exceptions to the measures that are 'consistent with this[BIT] agreement'.[233] Yet, a limitation such as this poses a difficult interpretative challenge since it could be understood as if the regulatory measure could be excused from not conforming to the standard only when it is consistent with the obligations of the agreement. In other words, the text could lead to a contradiction: regulatory measures are exceptions to the breach of the BIT standards only when the measures do not breach the BIT standards. If this is the interpretation of the general exceptions clause then it is difficult to envisage the conformity of climate change precautionary measures to this model.

Both, the unrestricted model of general exceptions and the model that conditions their efficacy to the consistency with BITs, face contradictions if they are interpreted literally. The first model seems to undermine the rationale of the investment treaty law to a point where the system loses its functionality as an instrument to promote the flow of transnational investment. The second model appears to condemn to failure any attempt to invoke regulatory measures as exceptions. It is therefore

229 Qatar-Argentina BIT (2016).

230 *Ibid.*

231 TPP (2018); PACER (2017).

232 Qatar-Argentina BIT (2016).

233 TPP (2018); PACER (2017).

imperative to interpret the text according to the purpose and the context (Article 31 VCLT reappears as a guidance). In this respect, I argue that other alternatives of 'exceptions with restriction' may enlighten the interpretation of these two models: the *chapeau*-restriction should enlighten the interpretation of the unrestricted model and the ponderation-restriction should enlighten the consistency model. This would make it possible to harmonise climate change precautionary measures with BITs without distorting the investment treaty law system.

The *chapeau*-model consists of limiting the scope of the open exceptions through—the usual[234]—threshold of discrimination, arbitrariness and disguised restrictions. That is the case of the Iran-Slovakia BIT, which Article 11 establishes as an exception 'adopting or enforcing measures to ensure compliance with laws and regulations'[235] conditioned 'to the requirement such measures are not applied in a manner that would constitute arbitrary or unjustifiable discrimination (..) or disguised restriction on the international trade or investment'.[236] Under this model, precautionary climate change measures would rank as general exceptions as long as they do not trespass the *chapeau.*

Since the *chapeau* model maximises the chances of regulatory measures to conform with the general exceptions, the interpretation of the *unrestricted model* should also obey this threshold. In fact, prohibiting the host state from taking arbitrary, discriminatory and/or disguised restrictive measures is the basic minimum element that defines the identity of the BIT in accordance with its rationale. Indeed, an instrument without these prohibitions may not be conceived as an investment agreement. Consequently, if the *unrestrictive model* needs to be limited to suit the system, it makes sense that the interpretation is conducted by deploying the *chapeau,* because this threshold purports the maximum radio of regulatory powers that the system withstands concerning general exceptions.

In addition to the *unrestricted model*, the *consistency model* and the *chapeau model*, the Morocco-Nigeria BIT represents a fourth model, which seems to embrace the scholarship's trend of recognising the need to ponder host states' public interest against the investors' interest in investment treaty law:

> Except where the rights of Host State are expressly stated as an exception to the obligation of this Agreement, a Host State's pursuit of its rights to regulate shall be understood as embodied within a balance of the rights and obligations of Investors and Investments and Host States, as set out in the Agreement.[237]

This article conceives two levels of exceptions. While the first level corresponds to general exceptions that are explicit in the text of the agreement, whose effectiveness is incontrovertible, the second level is reserved for regulatory measures that are not within the explicit exceptions, whose effectiveness is subject to the result of pondering the public interest of the host state against the investor's interests (*the*

[234] *See Notes* 222 and 223.

[235] Iran-Slovakia BIT (2016) Art. 11.

[236] *Ibid.*

[237] Morocco-Nigeria BIT (2016) Art. 23 (2).

ponderation model). This second level certainly poses interest interpretative challenges that lead to methods and concepts such as proportionality and reasonability, which certainly involve a higher threshold than the *chapeau*. In this context, precautionary climate change measures have the chance to be adapted to the general exceptions, but the success of the conformity depends on the particular circumstances showing that the climate interest trumps over the other competing interests.

The *ponderation-model* could also serve as an instrument to interpret the *consistency model* and overcome the contradiction that the latter involves. This is so because the *ponderation-model* represents the minimum level of regulatory power that the efficacy of a general exception could admit. Requiring the host state to prove the consistency of the regulatory measure beyond concepts such as reasonability or proportionality is not only against the high level of deference that is inherent to BITs but it would also lead to confusion between the general exception and the international defence of necessity, which are different in nature, as explained above.[238] If the *consistency-model* needs to be moderated to avoid contradictions, it is reasonable to adopt an interpretation that profits from the ponderation of interests, because this methodology entails the minimum radio of regulatory powers that the system withstands concerning general exceptions.

Without intending to undermine the importance of the preservation of the regulatory powers to the operation of the investment treaty law, it could be said that the absence of open exceptions in most BITs is grounded on the excess of their counter rationale. In fact, the system is, in principle, structured to encourage the flow of investment by protecting the investor, thus open exceptions are not only counter rational—as any exception would be—but their openness could lean the balance of interests too far in favour of the investor. Admittedly, the models that limit the open exceptions lessen to a great extent the distorting effects that they may have on the system. Nevertheless, most BITs opt for restricting general exceptions to particular, legitimate objectives that represent individualised interests.

While do not impose a major interpretative challenge to harmonising with climate change precautionary measures, models that adopt an exhaustive list of interest—such as security, peace, health, environment, etc.—require an additional interpretative effort to adapt climate change precautionary measures to the general exceptions clause. This is the subject of the next subsection.

9.2.2.2 Exhaustive Exceptions

Precautionary climate change measures might match the interests that are listed in the general exceptions clause. In some cases, the link between climate change and the measure excepted is clear, such as when general exceptions comprehend

[238] *Supra* Sect. 9.2.1.

measures intended to protect human, animal, plant life or health;[239] the environment,[240] to promote sustainable development[241] or that comply with international law obligations.[242] On other occasions, the link requires a further elaboration that could be questioned for taking too far the rebound effect that climate change may have on the interests that are listed as general exceptions, such as the public order,[243] the essential security interest[244] or peace.[245]

Vadi and Sands argue that it is possible to defend that climate change measures suit the general exceptions as a result of the effect that climate change may have on the essential security interest, public order and peace.[246] The authors support this assertion with scientific evidence that envisages different threats to these interests that include international wars motivated by the water scarcity and massive migrations;[247] stagnation of the economic development of most vulnerable countries;[248] redrawing of national borders;[249] armed conflicts over the Arctic;[250] agricultural and food crises resulting from desertification;[251] infrastructure destruction resulting from rising water levels and other natural disasters;[252] deaths and diseases related to the

[239] ECT (1994) Art. 24 (2) (b)(i); Hong Kong-ASEAN (2017) Art. 9(b), CETA (2014) Art. 28.3 2 (b); Israel-Japan BIT (2017) Art. 15 (a); Turkey-Rwanda BIT (2016) Art. 5 (1)(a), Canada-Hong Kong BIT (2016) Art. 17(b), Japan-Iran (2016) Art. 13 (a); Iran-Slovakia BIT (2016) Art. 11(b); Singapore-Turkey (2015) Art. 12.24 (b); Australia-China FTA (2015) Art. 9.8. (a), 9.11 (4).

[240] NAFTA (1994) Art. 1114(1); CAFTA (2004) Art. 11.16 (1), PACER (2017) Art. 19 (2), TPP (2018) Art. 9.16., CETA (2014) Art. 24.4.; Morocco-Nigeria BIT (2016) Art. 13.4.; Turkey-Rwanda BIT (2016) Art. 5 (1)(a); Singapore-Turkey (2015) Art. 12.24 (b); Australia-China FTA (2015) Art. 9.8. (2), 9.11 (4).

[241] Morocco-Nigeria BIT (2016) Art. 23 (1).

[242] *Ibid.*

[243] ECT (1994) Art. 24 (3) (c); Hong Kong-ASEAN (2017) Art. 9(a); TPP (2018) Art. 29.2. (b), CETA (2014) Art. 28.3 2 (a); Iran-Slovakia BIT (2016) Art. 11(a); Singapore-Turkey FTA (2015) Art. 12.24 (a); Australia-China FTA (2015) Art. 9.11 (4).

[244] ECT (1994) Art. 24 (3) (a); China-Japan-Korea BIT (2012) Art. 18; TPP (2018) Art. 29.2. (b); CETA (2014) Art. 28.3 2 (a); Iran-Slovakia BIT (2016) Art. 11(a); Australia-China FTA (2015) Art. 9.11 (4).

[245] China-Japan-Korea BIT (2012) Art. 18; Singapore-Turkey FTA (2015) Art. 12.24.

[246] Vadi (2015), p. 1285. Sattorova adds that host state's measures dealing with renewable energy suit the general exception because energy security could be understood within the concept of essential security: Sattorova (2012). The IISD study on Climate Change and International Investment Agreements argues that the exception of essential security interest would be likely to succeed manly when it is invoked by 'more vulnerable countries'. Marshal (2010b), pp. 52–60.

[247] Vadi (2015), p. 1347.

[248] Johnson (2009), p. 11154.

[249] *Ibid.*

[250] *Ibid.*, p. 11155.

[251] *Ibid.*

[252] *Ibid.*

extreme heat;[253] as well as authoritarianism, radicalisation and extremism because of governments' failure to deal with the extreme conditions.[254]

Even if the dystopian scenario illustrated in the last paragraph has become more realistic in the last 30 years, the bond between the climate change interest and peace, security and public order interests seems to lack an element of proximity. This is so because the plethora of factors that need to coincide for global warming to result in war, social unrest or public chaos blur their immediate connection with the climate interest. The handling of the economy, diplomatic relations and cooperation between countries, adaptative measures to climate change, among others, emerge in-between an allegedly cause-effect relationship.

For example, the degree of proximity of a regulatory measure establishing limitations to CO2 emissions with the effect of impeding a war over water is very weak, in comparison to a regulatory measure banning the import of a substance needed to process cocaine to prevent drug cartel wars in countries such as Colombia or Mexico. Despite concluding that the climate change threat to peace among the international community for the scarcity of water is reasonable, it needs to rely on too many uncertain assumptions[255] other than global warming to be supported and, thus, the direct link between climate change and war appears to be unclear.

Contrarily, climate change precautionary measures and the measures adopted to protect the human, animal, plant life or health appear to have a more direct link, which is an expression of the impact that global warming has on ecosystems. The effects of global warming on animal, plant and human health are already evident—famine among polar bears, disappearing corals, deaths caused by heatwaves and the fires in the Amazon Forest are some publicly known examples.[256] Against this background, one may think that precautionary climate change measures fall within this exception. This would be even clearer for BITs that expressly establish that '[t]he Parties understand that the measures referred [. . .] include **environmental measures** necessary to protect human, animal or plant life or health'[257] [emphasis added].

If identifying precautionary climate change measures with a *human, animal, plant life or health* exception is possible, the nexus between the latter and the environmental exception is uncontested. Furthermore, one can affirm that any harm to climate is *per se* a harm to the environment, so any climate change precautionary measure is, in fact, an environmental precautionary measure. Thus, there would be no problem of conformity for climate change precautionary measures if the catalogue of general exceptions includes environmental measures.

[253] *Ibid.* p., 11156.

[254] *Ibid.*

[255] For example, from the warming to desertification, from desertification to health and agricultural problems, from agricultural and health problems to political pressure, from political pressure to social unrest, from social unrest to civil war, from civil war to international relationships, etc.

[256] Time, Special Climate Issue (2019).

[257] Singapore-Turkey FTA (2015) Art. 12.24 (b).

The assimilation of precautionary climate change measures to the concept of sustainable development seems to be as clear as the connection to the environment. The whole idea of sustainable development involves the challenging of the excessive voracity in pursuing economic interests and conceiving the notion of development as inseparable from social and environmental interests. Preventing, mitigating and adapting to global warming is within the notion of sustainable development in the perspective that the measures to fulfill the needs of today require us to consider their impact on the world of tomorrow. In this context, precautionary climate change measures suit general exceptions when sustainable development measures are included in the text.

While including measures protecting human, animal and plant life or health interests within the general exceptions is common, the inclusion of environmental and sustainable development measures is sporadic. However, it is possible to argue that there is a trend in the new generation of BITs to highlight the importance of environmental interests and, consequently, conceiving environmental measures as exceptions. NAFTA and CAFTA are precursors in this regard by establishing as an exception—in an independent article—the host state's measures that are directed '*to ensure that investment activity in its territory is undertaken in a manner sensitive to environmental concerns*'.[258] Yet, the effectiveness of this exception could be at odds because the text appears to require the measures to be consistent with the agreement by using the language 'otherwise consistent with this Chapter [investment chapter]'.[259] This kind of limitation to the environmental exception—as explained above concerning open exceptions—should be understood as adopting a *ponderation model.*[260]

Another instrument that could be regarded as a predecessor of the trend of including the environmental interest within the general exceptions is the ECT:

> In pursuit of sustainable development and taking into account its obligations under those international agreements concerning the environment to which it is party, each Contracting Party shall strive to minimize in an economically efficient manner harmful Environmental Impacts occurring either within or outside its Area from all operations within the Energy Cycle in its Area, taking proper account of safety. In doing so each Contracting Party shall act in a Cost-Effective manner. In its policies and actions each Contracting Party shall strive to take precautionary measures to prevent or minimize environmental degradation.[261]

Although this article does not express that environmental measures are general exceptions to the compliance with the ECT's obligations, it does compel host states to make efforts to minimise environmental impacts as well as to take precautionary

---

[258] NAFTA (1994) Art. 1114(2); CAFTA. (2004) Art. 11.16 (2).

[259] NAFTA (1994) 1114 1(1).

[260] *See* above Sect. 9.2.2.1; Bjorklund highlights an 'alternative interpretation' to overcome the contradiction of Article 1114 that 'would attach more importance to the provision and suggests that measures must meet the requirements of Chapter 11—e.g., environmental measures must be enacted on a non-discriminatory basis in accordance with Articles 1102 and 1103': Bjorklund (2013), p. 498.

[261] ECT (1994) Art. 19(1).

measures. Moreover, it contextualises these obligations within the notion of sustainable development and the obligations under international environmental agreements. The extent to which this obligation constitutes an exception could be debatable, but we should recognise—at least—an interpretative grey zone on the legitimacy of the host state environmental (or climate) precautionary measures that requires us to assume a position.

Thus—to provide any efficacy to the article and coherence to the agreement as a whole—one may think that, when the host state harms the investor by taking environmental precautionary measures, a ponderation between the investor's interest and the climate interest may be appropriate to define whether an obligation has been breached. The logic of relying on an exercise of balance in this case, is reinforced by the fact that the article itself fosters an interpretation that harmonises the agreement with international environmental obligations as well as with sustainable development—and, no less important, that the preamble of the ECT redirects to the international climate change legal framework.

Contrary to the difficulties that the precautionary climate change principle may face to suit the general exceptions of NAFTA, CAFTA or ECT, there is evidence that a significant portion of recently adopted agreements would embrace the climate interest within their exceptions. Thus, some BITs just plainly establish environmental measures within the exceptions without the limitation that expressions such as 'otherwise consistent' may entail.[262] Other BITs include the environmental exception within a context of environmental protection and sustainable development.[263] An archetype of this model is PACER that not only establishes environmental measures as exception[264] but also makes direct reference to the 'aspiration to promote high standards of environmental and labour protection and, to uphold these in the context of sustainable development',[265] and grants the host states the 'right to regulate and […] to preserve set legislative and regulatory priorities, safeguard public welfare, and protect legitimate public welfare objectives, such as […] the environment'.[266]

The CETA agreement seems to go beyond the PACER and the NAFTA in setting a more progressive climate change, environmental and sustainable development context that could boost the chances for climate change precautionary measures to be categorised within the general exceptions. Unlike PACER and NAFTA, the CETA contains a whole chapter on 'Trade and Environment', whose Article 24.2 (object and context) provides that the 'environment is a fundamental pillar of sustainable development and recognize the contribution that trade could make to

---

[262] Turkey Rwanda (2016) Art. 5 (1)(a); Singapore-Turkey (2015) Art. 12.24 (b); Australia-China FTA (2015) Art. 9.8. (2), 9.11 (4).

[263] Within the same category could be included the TPP (2018) Article 9.16 and Preamble.

[264] 19(2).

[265] PACER (2017) Preamble.

[266] *Ibid.*

sustainable development [...]'.[267] Accordingly, the chapter highlights the right of the host state to regulate environmental matters,[268] burdens the host state with the duty to promote environmental protection[269] and prohibits the host state from relaxing environmental regulations to encourage investment[270]—among other provisions promoting the environment interest.[271] Furthermore, CETA also contains a whole chapter on 'Trade and Sustainable Development', whose Article 22.1 (context and objective) redirects to several international instruments that highlight the relevance of combatting global warming;[272] e.g., Agenda 21 on Environment and Development of 1992 that highlights the impact of climate change in desertification [273] and fresh water;[274] the Johannesburg Declaration on Sustainable Development of 2002 that recognises that 'the adverse effects of climate change are already evident, natural disasters are more frequent and more devastating';[275] the 2002 World Summit on Sustainable Development Implementation Plan that establishes a 'Change in the Earth's climate and its adverse effects are a common concern of humankind'.[276]

However—on the adaptation of climate change measures to general exceptions—the most innovative aspect of CETA is defining environmental measures as 'those that aim at the prevention, abatement or control of the release, discharge, or emission of pollutants or environmental contaminants',[277] as well as consecrating the measures 'taken pursuant to multilateral environmental agreements to which they are party'[278] as general exceptions.

On the one hand, defining environmental measures through the example of tackling the 'release, discharge, or emission of pollutants or environmental contaminants' automatically evokes the idea of measures regulating CO2 emissions and addressing global warming. This reading of the text—that leads to climate measures being perceived as environmental measures—becomes more palpable, if it is contextualised through other articles of the agreement that directly highlight the relevance of climate change within the scope of the environment and sustainable

---

[267] CETA (2014) Art. 24.2.

[268] *Ibid.*, Art. 24.3.

[269] *Ibid.*

[270] *Ibid.*, Art. 24.5.

[271] *Ibid.*, Chapter 24.

[272] 'The Parties recall the Rio Declaration on Environment and Development of 1992, the Agenda 21 on Environment and Development of 1992, the Johannesburg Declaration on Sustainable Development of 2002 and the Plan of Implementation of the World Summit on Sustainable Development of 2002(...)': *Ibid.*, Art. 22.1.

[273] United Nations (1992), no. 12.

[274] *Ibid.*, no. 18.

[275] World Summit on Sustainable Development, & United Nations (2003), no. 13.

[276] *Ibid.*, no. 38.

[277] CETA (2014) Art. 24 (1).

[278] *Ibid.*, Art. 24.4(4).

development[279] or that indirectly incorporate international instruments that embrace this linkage.[280] The assimilation of climate change measures with environmental measures in CETA clears out any doubts about the legitimacy of precautionary climate change measures as general exceptions.

On the other hand, the consecration in Article 24.4 of the exceptional character of measures 'taken pursuant to multilateral environmental agreements to which they are party'[281] also leads to conceive precautionary climate change measures within the realm of general exceptions. This article must be interpreted in concordance with Article 23.3. which reiterates that the host state has the 'right [. . .] to adopt or modify its laws and policies accordingly and in a manner consistent with the multilateral environmental agreements'.[282] Jumping from the—fairly progressive—effort of CETA to harmonise with international environmental agreements, to the conclusion that the United Nations Framework Convention on Climate Change is within the realm of the general exceptions is not a difficult interpretative task, since in essence, climate change international instruments are environmental instruments.

The initiative of embracing the measures complying with international agreements as general exceptions is also an innovative feature of the Morocco-Nigeria BIT, which precedes the CETA.[283] In Article 23—on the right to regulate—the agreement emphasises that '[f]or greater certainty, non-discriminatory measures taken by a State Party to comply with its international obligations under other treaties shall not constitute a breach of this Agreement'.[284] Contrary to CETA, this article allows a direct integration of the United Nations Framework Convention on Climate Change without the need to rely on association with the environment or sustainable development. Thus, the relevance of precautionary climate change measures as a basis for justifying the noncompliance of BIT obligations is straight-forward in the Morocco-Nigeria BIT.

In this subsection, I presented the formulas that determine the reception of climate change precautionary measures in investment treaty law through general exceptions. All the possible combinations between the catalogue of *subject complements* that

[279] For example, Article 24.9(2): '1. The Parties are resolved to make efforts to facilitate and promote trade and investment in environmental goods and services (. . .). The Parties shall, consistent with their international obligations, pay special attention to facilitating the removal of obstacles to trade or investment in goods and services of particular relevance for climate change mitigation and in particular trade or investment in renewable energy goods and related services'; Article 24.12. 1.: 'The Parties recognise that enhanced cooperation is an important element to advance the objectives of this Chapter, and commit to cooperate on trade-related environmental issues of common interest, in areas such as efficiency and the development and deployment of low-carbon and other climate-friendly technologies'. *Ibid.*

[280] *Supra* text to *Notes* 272–276.

[281] *Ibid.*, Art. 24.4.

[282] *Ibid.*, Art. 23.3.

[283] While CETA entered into force on 21 September 2017 (2014), the Morocco-Nigeria agreement did so on 3 December 2016 (2016).

[284] Morocco-Nigeria BIT (2016) Art. 23.

link the measure with the interest protected—which can range from *self-judging clauses* to *necessity* clauses—and the catalogue of the public interests measures that the exception aims to cover through open statements or exhaustive lists—which often include references to health, the environment or sustainable development—suggest that it is likely to envisage the legitimacy of climate change measures as a justification to prevent liability in investment treaty climate change-related disputes.

9.3 Concluding Remarks on the Direct Application of the Climate Change Precautionary Principle to Investment Treaty Law Disputes: Framing Necessity as an Instrument for Weighting Global Public Interests and the Exceptionality of Its Application

This chapter posits the application of the climate change precautionary principle to investment treaty-climate change disputes, by envisaging the preclusion of wrongfulness as a consequence of the host government's necessity to take precautionary measures—in line with the Article 25 DASR—as well as through the operation of the BIT's general exceptions clauses.

To prove that the preclusion of wrongfulness based on the precautionary principle matches the operation of the investment treaty law system, I highlighted the fact Article 25 DASR has been deployed in investment treaty disputes to prevent liability for measures promoting global public interests (i).

Thus, I describe the way in which the evolution of case law on the disputes arising from the Argentinian Crisis of the early 2000s demonstrates that the operation of the international defence of necessity (Article 25 DASR) embraces global public interests that are different from the state's existence, self-preservation or self-defence; e.g., national security, political and economic stability, ecology, environment, health and human rights (ii).

Furthermore, I show that the expansion of the notion of necessity to annulment decisions and dissenting opinions on the Argentinian crisis awards, as well as awards related to disputes that arise from other matters, reiterates—on the one hand—the connection between the defence of necessity and global public interests (iii) and recognises—on the other hand—that the defence of necessity could be deployed to solve tensions between compliance with host states' obligations under BITs and compliance with other obligations from international law (iv) such as promoting anti-discriminatory action, preventing risks to health, protecting the autonomy and equality of indigenous communities, ensuring the provision of water and energy services to citizens, and preventing human rights violations and damages to the ecology and the environment.

After demonstrating that the operation of the international defence of necessity has the potential to embrace the global climate interest and to recognise the relevance of climate change international obligations, I argued that some of the requirements of

Article 25 DASR needed to be interpreted in a flexible manner—beyond the literal interpretation—if one intends to affirm that climate change precautionary measures trigger the defence (v).

While asserting that a precautionary climate change measure suits the requirement of representing an *essential interest* does not involve a considerable interpretative effort (vi)—because the pliability of the text and the intention of the drafters of the DASR are clear in including international global interests within the scope of Article 25 (1)(a)—conceiving that such a measure responded to an *imminent peril*, was *the only way to safeguard the interest* and that the state did not *contribute to the state of necessity*, needs further elaboration.

In this respect, I revealed that tribunals have mastered an interpretation of Article 25 DASR that overcomes the strict-literal interpretation of the requirements, allowing precautionary climate change measures to suit the defence of necessity (vii). First, precautionary climate change measures might conform to the requirement of *imminence,* since the evidence that is *reasonably available* in an investment climate change-related dispute would probably show that the industrial (or non-industrial) activity of the investor who affected by the measure, represents an actual risk to climate (viii). Second, a precautionary climate measure might be considered the *only measure available,* as long as there is evidence that there were no alternatives that would have either not breached international obligations or breached international obligations in a less grave manner, and which were equally or more effective than the measure that was taken (ix). Third, a precautionary climate change measure could fulfill the requirement of *non-contribution to the situation of necessity* either when the host state has not materially propelled climate change—which is the case of least developed countries—or when the effects of global warming were not foreseeable for the host state, which is the case for contributions before the 1990s (x).

Fourth, to boost the suitability of precautionary climate change measures to *essentiality* and *imminence*, I posited that the climate interest constitutes a global public interest in itself (xi). This would allow Article 25 DASR to embrace the climate interest within its scope without the need to resort to the—often—difficult maneuver of envisaging the essentiality of climate change under the umbrella of neighboring interests—such as health, the environment or human rights—(xii) and without the need to determine *imminence* by reference to the rebound effects of global warming over these interests (xiii).

Fifth, I suggested that the *onus probandi* should be eased and dynamised when concerning the proof of the *only measure available* and the *non-material contribution* to global warming (xiv). This proposal responds to a host state's impossibility to prove that uncountable alternatives within the myriad of climate change measures were less reasonable or less effective than the one implemented; and from the difficulties in proving that the state complied with the expectations of the community concerning the—often abstract—climate change obligations. In both cases—following the reasoning of *Continental*—it was said that the host state has the duty to prove *prima facie*—in a general manner—that the measure was the only one available to protect the climate interest within the reasonable alternatives that the host state contemplated (xv), and that it did not materially contribute to global warming by complying with the international climate change obligations (xvi). However, it was also clarified that the *onus probandi* should be assessed considering that—in practice—the verification of Article 25 DASR requirements is the result of the evaluation of the proofs that both parties provide to the tribunal and those that the tribunal has at hand—rather than the result of the mechanical application of the *onus probandi* (xvii).

Having clarified that the precautionary climate change principle finds a space within investment treaty law through the deployment of Article 25 DASR, the second part of the chapter postulates that climate change precautionary measures might also be justified through the application of general exceptions (xviii). This claim involves moving back from law ascertainment to the text-interpretation of BITs and, accordingly, looking for harmonisation through the balance between the investor's and the host state's interest as entrenched in the BIT, this being governed by Article 31 VCLT.

In this respect, it was revealed that climate change precautionary measures need to be confronted against the interests that the text of general exceptions aims to protect and the link between the measure and the interest protected (xix). The variations on the combination of these two conditions might lead to different tailor-made formulas that involve more or less stringent requirements.

It was found convenient to distinguish the link (subject complement) of general exceptions in two categories. While one category embraces the group of clauses that prevent liability when the measure is 'necessary for' the protection of the interest, the other group encompasses clauses that contain expressions that are not related to *necessity*, such as 'directed to', 'appropriate to', 'designed and applied for', etc. (xx).

Concerning the first group—those exceptions that contain the expression 'necessary for' or equivalent—I described that the interpretation of tribunals pendulates between conflating the general exception with the international defence of necessity and conceiving a *lex generalis-lex specialis* or a *primary-secondary application* relationship between the general exception and the international defence (xxi). I suggested that the method that best adapts to the harmonisation of the precautionary climate change principle and the investment treaty law was the *primary-secondary application* (xxii). This method makes it possible to detach the general exception from the strict requirements of Article 25 DASR and, simultaneously, leaves open the possibility to apply the international defence of necessity in the cases where the text of general exceptions rules out the climate change interests from its scope (xxiii).

On the second group—those expressions that are not directly related to the expression necessity—I affirmed that the chances for climate change precautionary measures to prevent liability were greater than through clauses that belong to the group that qualifies the exception with the expression 'necessary for' or equivalents (xxiv). Contrary to the expression 'necessary to', I argued that the expressions in the second group do not seem to condition the success of the exception to the inexorability of the climate change measure in light of the power of the situation (xxv). In this sense, expressions such as 'directed to', 'in pursuance of', 'considers appropriate for', 'intended to', 'relating/ed to' 'in pursuing', 'imposed to', 'that provide for' or 'applied and designed to' imply recognising that the host state has some level of discretion in adopting climate change precautionary measures.

The analysis of the clauses that include *subject complements* that are not related to necessity also revealed that, while some of these condition the legitimacy of the precautionary climate change measure—as a general exception—to the (objective) verification that the measure actually promotes the climate interest, others redirect the examination of the legitimacy of the measure to the host state's subjective considerations (self-judging clauses), such as 'considers appropriate' or 'considers necessary' (xxvi).

Against this background, I concluded that the spectrum of clauses that embrace climate change precautionary measures is characterised, on one extreme, by *self-judging* clauses that confer the host state the widest discretion to take climate change precautionary measures without facing liability, and, on the other extreme, by necessity clauses that restrict the discretion of the host state to the inexorability of the climate change precautionary measures (xxvii). Between these two extremes there would be formulas that attribute a different degree of discretion to the host state for choosing to adopt and design climate change measures, which are not easy to define in abstract terms or by reference to each other, such as 'directed to', 'in pursuance of', 'intended to', 'relating/ed to', 'in pursuing', 'imposed to', 'that provide for' and 'applied and designed to' (xxviii). There would also be formulas that establish a special limit or condition to the legitimacy of climate change precautionary measures. Limits would include aspects such as arbitrariness, abuse, discrimination and disguised restrictions, while conditions would include aspects such as extreme emergency, serious threat or not avoiding international obligations (xxix).

After having examined the suitability of precautionary climate change measures with the general exceptions stemming from the link between the measure and the interest protected, in the last subsection, I analysed suitability based on the interests that the exception aims to cover. I found two basic models that define the conformity of climate change precautionary measures to general exceptions: *open exceptions* and *exhaustive exceptions* (xxx). While the first group appeared to embrace climate change precautionary measures by including any public interest measure within its scope through a general statement in that sense, or through a list of non-exhaustive measures characterised with the phrase 'such as' (xxxi), the second group conditions the suitability of a climate change precautionary measure to a closed-list of specific interests—such as environment, health or sustainable development—that may be affected as a rebound effect (xxxii).

Concerning *open exceptions*, I mentioned that they might be limited –expressly or implicitly– either by the threshold (*chapeau*) of discrimination, arbitrariness or disguised restrictions or by the ponderation between the climate interest and the investor's interests (xxxiii). While the implicit limit to the open exception would be the *chapeau* when the objective of the exception clause seems to reflect an expansive approach towards the protection of global public interests (*unrestricted model*) (xxxiii), the implicit limit would be the ponderation when the exception suggests embracing an equilibrium between the public interest and the protection to the investor, which is the case of the clauses that condition the exception to consistency with BITs (*consistency model*) (xxxiv).

More challenging than defining climate change precautionary measures as open exceptions, is to determine their conformity to exhaustive exceptions. While the conformity of precautionary to the exceptions of public order, essential security interest and peace was contested because the casual link between the effects of climate change and these interests is very weak (xxxv), the relationship between climate change and the exceptions of 'human, animal, plant life or health', the environment and sustainable development seems realistic enough to suggest that precautionary climate change measures could justify the breach (xxxvi). It was concluded that BITs' trend of including this last set of interests within the general exceptions, in addition to the increasing relevance of climate change, the environment and sustainable development in BITs—as well as the relevance of international treaties promoting these interests in BITs—evinces that the recognition of precautionary climate change measures within the scope of general exceptions is gaining ground within investment treaty law (xxxvii).

Although the considerations that I have made in this chapter to this point evince that the precautionary climate change principle is applicable to investment treaty-climate change disputes, through either the international defence of necessity or the operation of general exceptions, it is worth explaining—before finalising the chapter—the space that these instruments occupy in the whole body of my work by contrasting them. Although the function of the international defence of necessity and the general exceptions is hindering the rationale of the investment treaty law system by introducing climate change considerations into investment treaty disputes (Sect. 9.3.2), they are independent operation mechanisms (Sect. 9.3.1).

9.3.1 Independence of the International Defence of Necessity and General Exceptions in Investment Treaty-Climate Change Disputes: Conceptual, Dimensional and Methodical Differences

Necessity and general exceptions in investment treaty-climate change disputes are two different operation mechanisms of harmonisation (Sect. 9.3.1.1) that address different dimensions of conflict of interests and suppose different methods to solve the tensions (Sect. 9.3.1.2); while the application of the precautionary principle in the international defence of necessity (Art. 25 DASR) is the result of an operation of law ascertainment that solves the tension between global public interests through a systematised-ponderation exercise, the application of the principle in general exceptions is the result of an operation of text-interpretation that solves the tension between host states' interests and investors' interests by examining conformity to BITs' exceptions.

9.3.1.1 Harmonisation Through Different Operational Mechanisms: The Verification of General Exceptions Determines the Absence of the Breach and the Verification of the International Defence of Necessity Determines the Preclusion of Wrongfulness of the Breach

Although it may be difficult to categorise primary and secondary rules in international law—in understanding the notions of the international defence of necessity and general exceptions—it is useful to distinguish between the rules concerning the determination of the obligations (primary law) from the rules determining the consequences of the breach (secondary law). While the international defence of necessity concerns liability, the general exceptions concern the content of the obligations.[285] More specifically, while the verification of general exceptions determines the absence of breach of BIT standards, the international defence of necessity determines the preclusion of wrongfulness of the breach.

Accordingly, when the invocation of climate change precautionary measures prevents liability as a result of their conformity to the general exceptions, the conclusion is that the host state's conduct conforms to the content of BIT obligations. In the same vein, when climate change measures prevent liability as a result to their suitability with the requirements of Article 25 DASR the conclusion is that, even if the conduct did not conform to BIT obligations, the host state is excused from liability because the breach is considered to be legitimate.

As shown in this chapter, the conceptual independence between *necessity* and general exceptions has practical consequences. It defines to what extent the requirements of Article 25 DASR may (or may not) condition the application of general exceptions in investment treaty-climate change disputes. Thus, I highlighted that opting for a complete differentiation between the two concepts through the model of *primary-secondary application*, rather than adopting a conflation or *lex specialis* model, optimises the chances of applying the climate change precautionary principle in investment treaty disputes. Indeed, the *primary-secondary model*, on the one hand, detaches the general exception from the stringentness of Article 25 DASR's requirements and, on the other, makes it possible to apply Article 25 DASR in case the general exception fails to cover the climate change interest within its scope.

Moreover, conceiving general exceptions and the international defence of necessity as primary law and secondary law, respectively, also has an impact on the framing of the conflict of interests that each of these mechanisms represents and the method for solving these conflicts. While the international defence of necessity makes it possible to adopt a heterodox approach to harmonise the investment treaty law with the climate change international legal framework, general exceptions entail the deployment of an orthodox approach.

---

[285] Cassella (2011), pp. 72–73.

9.3.1.2 Direct Application of Precautionary Climate Change Measures Through Article 25 DASR (Heterodox Approach) and the Conformity of Climate Change Precautionary Measures to General Exceptions (Orthodox Approach)

Although this chapter was initially only intended to address the direct application of the climate change precautionary principle through the international defence of necessity, the overlapping in functional terms of the international defence with BITs' general exceptions led me to conclude that it was convenient to study both instruments together. In fact, both instruments are counter rational to the system and have the potential to prevent liability when the host state takes climate-change precautionary measures. Furthermore, it was also useful to include both instruments in the same chapter to analyse the possible conflation between them.

Nevertheless, I must highlight that the study of general exceptions as modulators of the content of BIT obligations, could perfectly have suited the Part II of this book, which deals with the orthodox approach to investment treaty-climate change-related disputes. In fact, the method used to study the general exceptions in this chapter was the same as the one I deployed in Part II to study the expropriation and the FET standards. This method focuses on harmonising the climate change international legal framework and international investment law through BIT content-interpretation, and is limited by the principles of Article 31 VCLT (*orthodox approach*), rather than through law ascertainment (*heterodox approach*).

Bearing this in mind, it is possible to affirm that, while Article 25 DASR could be deployed as a rule of law ascertainment, the interpretation of general exceptions is limited to unearthing the balance between the host state and the investor's interest as it is entrenched in the text of the BIT. This involves a different framing of the interests at hand when dealing with the international defence of necessity than when dealing with the general exceptions in investment treaty-climate change disputes.

Article 25 involves weighing all the global interests that may enter into conflict with the climate change interest to ascertain the application of an international law instrument that represents the global interest that prevails (the public dimension). On first instance, one might envisage the evident conflict between the global interest of transnational investment and the global climate interest. However, the climate interest might also enter into conflict in investment treaty disputes with other global interests, such as human rights or social rights that may align under certain circumstances with the interest of the flow of investment.

I have shown in this chapter that the pondering of the climate change interest against other interests that underlie BIT effectiveness (principally, interest in the flow of investment and *pacta sunt servanda)* involves overcoming the strict requirements of Article 25 DASR. However, tribunals also have the responsibility to choose which interest should prevail considering the circumstances. In this sense, once the precautionary measure fulfills the requirements of *the essentiality, the only way to safeguard the interest, the imminence and the non-contribution of the host state,* the tribunal will still have a wide measure of discretion (responsibility) to establish whether the climate change interest should prevail over the other global interests. This exercise of ponderation is translated in Article 25 (1) (b), which requires, in this case, that the precautionary climate change measure 'does not seriously impair an essential interest of the State or States towards which the obligation exists, or of the international community as a whole'.[286]

Only the circumstances of the case at hand can suggest which interest should prevail. However, if the conflict is framed between the global climate interest and the international flow of investment only, it would be difficult to argue that the legitimacy of a precautionary measure in a given case could impair the global flow of investment in such a magnitude that the interest of the global community would result affected on a larger scale. The same could be said about the affectation to the other states that are parties to a BIT. In principle, the host state's precautionary climate change measures would not have the potential to threaten the essential interests of the other states that are party to the treaty.

The decision on which interest prevails would be followed by the application of the international instrument that represents that interest. If the climate interest prevails, then the climate change precautionary principle is going to be ascertained as the applicable rule over the *pacta sunt servanda* or over other rules of international law that represent other conflicts of interests. In such an event, the dispute can be solved through the direct application of the precautionary climate change principle.

Contrarily, the operation of general exceptions does not, in principle, involve an exercise of ponderation between global interests, nor the ascertainment of international law instruments. The interests that conflict when interpreting general exceptions are the investor's and the host state's interests (*the private dimension*). In the case of the investor, these interests are translated into the protection of its assets and the freedom to operate in the host state's market. In the case of the host state, these interests are translated into its right to regulate the market and to pursue its public policy. The focus of enquiry when interpreting general exceptions defines the space of the host state's right to regulate against the extent to which the investor's assets are protected, i.e. which is the balance between the parties' interests according to the BIT.

I am not claiming that the interest of the parties when interpreting general exceptions could not be aligned with or translated into global public interest. As I described in this chapter, general exceptions could, through their text, integrate global public interests such as the environment, health and security, among others, and even comprehend international obligations within the definition of the exceptions. However, these interests and international obligations play the role of filling the content of the general exceptions and, broadly speaking, of filling the content of the host state's obligations.

Consequently, the international instruments that represent global interests do not intervene directly when interpreting general exceptions. They are simply the result of the application of the BIT and, accordingly, they do not involve a tension between

[286] International Law Commission (2001).

principles of different international law subsystems; they do not, in principle, suppose any law ascertainment activity. Admittedly, this reasoning could be at odds when the exception itself calls for the ponderation of global public interests against the investor's interests, as in the model that was described above as the *ponderation-model* of general exceptions. Nevertheless, under this model, the ponderation still provides a way to define the content of BIT obligations and the application of the BIT (*pacta sunt servanda*) does not conflict with other international instruments.

In this vein, the application of the precautionary climate law principle through general exceptions is nothing else but the result of the application of BIT—the effectiveness of the *pacta sunt servanda*—after verifying the conformity of the host state's precautionary measure to their text. Conversely, the precautionary principle may directly determine the outcome of the dispute by prevailing over the application of the BIT, as a result of the intervention of the international defence of necessity, understood as a systematised-ponderation rule of law ascertainment.

9.3.2 The Counter Rationale of the International Defence of Necessity and the General Exceptions in Investment Treaty-Climate Change Disputes

The application of the climate change precautionary principle through the international defence of necessity and the general exceptions in climate change-related disputes harmonises the investment treaty law with the climate change international legal framework. In fact, the success of the international defence or the *general exceptions* in preventing liability, when dealing with the legitimacy of climate change measures, entails the prevalence of the climate interest in investment treaty disputes and, accordingly, incentivises host states to take climate change measures.

However, the rationale of the investment treaty law is to promote the flow of transnational investment and the global economic growth by protecting the investor. Thus, the possibility of justifying climate change measures, either through the defence of necessity or the general exceptions, reveals that these instruments could operate as counter rationale to the system. The idea of limiting the efficacy of BIT obligations and the host state's responsibility with the introduction of climate change considerations in investment treaty disputes, certainly, discourages investments.

The counter rationale of *necessity* and the general exceptions explain why these instruments operate only under extraordinary circumstances and subject to strict conditions. Thus, the requirements of Article 25 DASR that were described in this chapter serve as a guarantee that the defence of necessity remains exceptional, and is *essentiality, the only way to safeguard the interest, immediacy and non-contribution of the host state.* Likewise, tailor-made general exceptions were interpreted in this chapter as limited by the *chapeau-model* or the *ponderation model*, apart from the limitations that are inherent to the subject complement—necessary to, directed to, in pursuance of, etc. [. . .]—and to the scope of the interests that the exception covers.

The application of the climate change precautionary principle through *necessity* or general exceptions in investment treaty disputes shows that the operation of the investment treaty law system could be permeated by climate change interests. The counter rationale that this permeation represents for the system could be translated into the need for adaptation mechanisms that make it possible to soften the voracity of the economic interests that the investment treaty law intends to promote. As Cassella puts it, 'the state of necessity [and the general exceptions] permits to avoid an excessive social cost, functioning as integrator of divergent interests'.[287]

Although the penetration of the precautionary climate change principle is against the rationale of the system, it shapes the operation of the investment treaty law and, furthermore, enhances its survival. As long as the aim of boosting foreign investments stays in balance with climate interests, countries will continue to rely on the investment treaty law system. Otherwise, the system would lose its capability of adaptation to the changing circumstances that global warming supposes and, consequently, would lose the support of host states. Yet, the penetration of climate change interests—or any other global public interest—through Article 25 DASR or general exceptions should be kept within the limits of exceptionality to maintain the general unity and identity of the investment treaty law system.

References

Bjorklund A (2013) NAFTA Chapter 11. In: Brown C (ed) Commentaries on selected model investment treaties. OUP, p 498

Cassella S (2011) La Nécessité en Droit International. Nijhoff, pp 168–169

Crawford J (1999) Second report on State responsibility. A/CN.4/498 and Add.1–4, pp. 72–73

De Brabandere E (2018) Human rights and international investment law. Grotius Centre Working Paper 2018/75-HRL, Leiden Law School Research Paper 19. https://ssrn.com/abstract=3149387. Accessed 02 August 2022

Di Benedetto S (2013) International investment law and the environment. Edward Elgar, Cheltenham, p 175

Heathcote (2010) Circumstances precluding wrongfulness in the ILC articles on state responsibility: necessity. In: Crawford J and others (eds), The Law of International Responsibility, OUP, 491

Johnson L (2009) International investment agreements and climate change: the potential for investor-state conflicts and possible strategies for minimizing it. ELR 39:11147, 11150

Kurtz J (2010) Adjudging the exceptional at international investment law: security, public order and financial crisis. Int Comp Law Q 59:325–371

Marshal F (2010a) Climate change and international investment agreements: obstacles or opportunities? IISD Trade, Investment and Climate Change Series 53. http://www.fao.org/fileadmin/user_upload/rome2007/docs/Climate%20Change%20and%20International%20Investment%20Agreements.pdf. Accessed 02 August 2022

[287] Cassella (2011), p. 517.

Marshal F (2010b) Climate change and international investment agreements: obstacles or opportunities? IISD Trade, Investment and Climate Change Series 52–60. http://www.fao.org/fileadmin/user_upload/rome2007/docs/Climate%20Change%20and%20International%20Investment%20Agreements.pdf. Accessed 02 August 2022

Monebhurrun N (2013) Is investment arbitration an appropriate venue for environmental issues? A Latin American perspective. Braz J Int Law 10(1):195. https://ssrn.com/abstract=2672945. Accessed 02 August 2022

Paddeu F (2018) Justification and excuse in international law. CUP, pp 404–409

Sattorova M (2012) International investment law and renewable energy: enabling national policy-making? Conference Paper 8–10. https://papers.ssrn.com/sol3/papers.cfm?abstract_id=2170667. Accessed 02 August 2022

Vadi V (2015) Beyond known worlds: climate change governance by arbitral tribunals? Vanderbilt J Transnatl Law 48:1285

Viñuales J (2015) Foreign investment and the environment in international law. CUP, pp 384–389

Decisions

ICSID (2004) *LG&E Energy Corp., LG&E Capital Corp., and LG&E International, Inc. v. Argentine Republic*, Case No. ARB/02/1, Decision on the Objections to Jurisdiction https://www.italaw.com/cases/621. Accessed 02 August 2022

ICSID (2005) *CMS Gas Transmission Company v. The Republic of Argentina,* Case No. ARB/01/8 https://www.italaw.com/cases/288. Accessed 02 August 2022

ICSID (2007) CMS Gas Transmission Company v. The Republic of Argentina, Case No. ARB/01/8, Decision of the ad hoc Committee on the Application for Annulment https://www.italaw.com/cases/288. Accessed 02 August 2022

ICSID (2007) *Enron Corporation and Ponderosa Assets, L.P. v. Argentine Republic, Case No. ARB/01/3 https://www.italaw.com/cases/401.* Accessed 02 August 2022

ICSID (2007) *LG&E Energy Corp., LG&E Capital Corp., and LG&E International, Inc. v. Argentine Republic*, Decision of Liability, Case No. ARB/02/1 https://www.italaw.com/cases/621. Accessed 02 August 2022

ICSID (2007) Sempra Energy International v. The Argentine Republic, Case No. ARB/02/16. https://www.italaw.com/cases/1002. Accessed 02 August 2022

ICSID (2008) *Continental Casualty Company v. The Argentine Republic*, Case No. ARB/03/9 https://www.italaw.com/cases/329. Accessed 02 August 2022

ICSID (2010) *Enron Corporation and Ponderosa Assets, L.P. v. Argentine Republic,* Case No. ARB/01/3, Decision of the ad hoc Committee on the Application for Annulment. http://www.italaw.com/cases/401. Accessed 02 August 2022

ICSID (2010) *Sempra Energy International v. The Argentine Republic,* Case No. ARB/02/16, Decision of the ad hoc Committee on the Application for Annulment. https://www.italaw.com/cases/1002. Accessed 02 August 2022

ICSID (2010) Suez, Sociedad General de Aguas de Barcelona S.A., and InterAguas Servicios Integrales del Agua S.A. v. The Argentine Republic, Case No. ARB/03/17 Decision on Liability

ICSID (2010) Suez, Sociedad General de Aguas de Barcelona, S.A. and Vivendi Universal, S.A. v. Argentine Republic, Case No. ARB/03/19, Decision on Liability https://www.italaw.com/cases/309. Accessed 02 August 2022

ICSID (2010) *Total S.A. v. The Argentine Republic,* Case No. ARB/04/01. https://www.italaw.com/cases/1105. Accessed 02 August 2022

ICSID (2011) *Continental,* Decision on the Application for Partial Annulment, Case No. ARB/03/9. https://www.italaw.com/cases/329. Accessed 02 August 2022

ICSID (2011) *Impregilo S.p.A. v. Argentine Republic,* Case No. ARB/07/17. https://www.italaw.com/cases/554. Accessed 02 August 2022
ICSID (2012) *EDF International S.A., SAUR International S.A. and León Participaciones Argentinas S.A. v. Argentine Republic,* Case No. ARB/03/23 https://www.italaw.com/cases/372. Accessed 02 August 2022
ICSID (2015) *Bernhard von Pezold and Others v. Republic of Zimbabwe,* Case No. ARB/10/15 https://www.italaw.com/cases/1472. Accessed 02 August 2022
ICSID (2016) Urbaser S.A. and Consorcio de Aguas Bilbao Bizkaia, Bilbao Biskaia Ur Partzuergoa v. The Argentine Republic, Case No. ARB/07/26 https://www.italaw.com/cases/1144. Accessed 02 August 2022
ICSID (2017) Bear Creek Mining Corporation v. Republic of Peru, Case No. ARB/14/21 https://www.italaw.com/cases/2848. Accessed 02 August 2022
ICSID (2017) *Bear Creek Mining Corporation v. Republic of Peru,* Case No. ARB/14/21, Dissenting Opinion https://www.italaw.com/cases/2848. Accessed 02 August 2022
ICSID (2018) *Unión Fenosa Gas, S.A. v. Arab Republic of Egypt,* Case No. ARB/14/4 https://www.italaw.com/sites/default/files/case-documents/italaw10061.pdf. Accessed 02 August 2022
ICSID (2011) *El Paso Energy International Company v. The Argentine Republic,* Case No. ARB/03/15. https://www.italaw.com/cases/documents/383. Accessed 02 August 2022
UNCITRAL (2013) *CC/Devas (Mauritius) Ltd., Devas Employees Mauritius Private Limited and Telecom Devas Mauritius Limited v. India,* PCA Case No. 2013-09 https://www.italaw.com/cases/1962. Accessed 02 August 2022
UNCITRAL (2017) *Deutsche Telekom v. India,* PCA Case No. 2014-10, Interim Award https://www.italaw.com/cases/2275. Accessed 02 August 2022

International Legal Instruments

Agreement between Japan and the Islamic Republic of Iran on Reciprocal Promotion and Protection of Investment (2016). https://investmentpolicy.unctad.org/international-investment-agreements/treaties/bilateral-investment-treaties/3625/iran-islamic-republic-of%2D%2D-japan-bit-2016-. Accessed 02 August 2022
Agreement between Japan And The Kingdom Of Cambodia for the Liberalization, Promotion And Protection of Investment (2007). https://investmentpolicy.unctad.org/international-investment-agreements/treaty-files/3116/download. Accessed 02 August 2022
Agreement between The Argentine Republic and Japan for the Promotion and Protection of Investment (2018). https://investmentpolicy.unctad.org/international-investment-agreements/treaty-files/5799/download. Accessed 02 August 2022
Agreement between the Government of Canada and the Government of the Hong Kong Special Administrative Region of the People's Republic of China for the Promotion and Protection of Investments Canada Hong (2016). https://investmentpolicy.unctad.org/international-investment-agreements/treaty-files/5094/download. Accessed 02 August 2022
Agreement Between the Government of the Republic of Korea and the Government of Japan for the Liberalisation, Promotion And Protection of Investment (2002). https://investmentpolicy.unctad.org/international-investment-agreements/treaty-files/1727/download. Accessed 02 August 2022
Agreement between The Government of the Republic of Rwanda and The Government of The Republic of Turkey concerning the Reciprocal Promotion and Protection of Investments (2016). https://investmentpolicy.unctad.org/international-investment-agreements/treaty-files/5415/download. Accessed 02 August 2022
Agreement between The Government of the Republic of Singapore and The Government of the Republic of Peru on the Promotion and Protection of Investments (2003). https://

investmentpolicy.unctad.org/international-investment-agreements/treaty-files/2160/download. Accessed 02 August 2022
Agreement between The Republic of Hungary and The Republic of India for the Promotion and Protection of Investments (2003). https://investmentpolicy.unctad.org/international-investment-agreements/treaty-files/1523/download. Accessed 02 August 2022
Agreement between the Slovak Republic and the Islamic Republic Of Iran for The Promotion And Reciprocal Protection of Investments (2016). https://investmentpolicy.unctad.org/international-investment-agreements/treaty-files/3601/download. Accessed 02 August 2022
Agreement Between the State of Israel and Japan for the Liberalization, Promotion and Protection of Investment (2017). https://investmentpolicy.unctad.org/international-investment-agreements/treaty-files/5849/download. Accessed 02 August 2022
Agreement on Investment among the Governments of the Hong Kong Special Administrative Region of the People's Republic of China and the Member States of the Association of Southeast Asian Nations (2017). https://investmentpolicy.unctad.org/international-investment-agreements/treaties/treaties-with-investment-provisions/3793/asean%2D%2D-hong-kong-china-sar-investment-agreement-2017-. Accessed 02 August 2022
Bilateral Agreement for the Promotion and Protection of Investments between The Republic of Colombia and_ Colombian Model (2007). https://www.italaw.com/documents/inv_model_bit_colombia.pdf. Accessed 02 August 2022
Canada-Colombia Free Trade Agreement (2008). https://www.international.gc.ca/trade-commerce/trade-agreements-accords-commerciaux/agr-acc/colombia-colombie/fta-ale/22.aspx?lang=eng. Accessed 02 August 2022
Canada-Peru Free Trade Agreement (2008). https://www.international.gc.ca/trade-commerce/trade-agreements-accords-commerciaux/agr-acc/peru-perou/fta-ale/22.aspx?lang=eng. Accessed 02 August 2022
Central American Free Trade Agreement (2004). http://www.sice.oas.org/TPD/USA_CAFTA/Jan28draft/CAFTAind_e.asp. Accessed 02 August 2022
Comprehensive Economic and Trade Agreement (CETA) between Canada, of the one Part, and The European Union, of the other part (2014). http://trade.ec.europa.eu/doclib/docs/2014/september/tradoc_152806.pdf. Accessed 02 August 2022
Draft Agreement for the Promotion and Protection of Investment between the Republic of Austria and for the promotion and protection of investments (2010). https://pca-cpa.org/wp-content/uploads/sites/6/2016/11/Agreement-Between-the-Republic-of-Austria-and-the-for-the-Promotion-and-Protection-of-Investments-2010.pdf. Accessed 02 August 2022
ECCOWAS Energy Protocol (2003). http://www.ecowapp.org/sites/default/files/eprotocol.pdf. Accessed 02 August 2022
Energy Charter Treaty (ECT) (1994). https://www.europarl.europa.eu/meetdocs/2014_2019/documents/itre/dv/energy_charter_/energy_charter_en.pdf. Accessed 02 August 2022
Free Trade Agreement between The Government of Australia and The Government of the People's Republic of China (2015). https://dfat.gov.au/trade/agreements/in-force/chafta/official-documents/Documents/chafta-agreement-text.pdf. Accessed 02 August 2022
Free Trade Agreement Between the Republic of Singapore and the Republic of Turkey (2015). https://www.enterprisesg.gov.sg/-/media/esg/files/non-financial-assistance/for-companies/free-trade-agreements/TRSFTA/Legal-text/Chapter-18-Institutional-General-and-Final-Provisions. Accessed 02 August 2022
International Law Commission (1981) *Draft Articles on State Responsibility*, *Report of the Commission to the General Assembly on the work of its thirty-second.* Yearbook of the International Law Commission 1980, A/CN.4/SER.A/1980/Add.l (Part 2) 34 https://legal.un.org/ilc/publications/yearbooks/english/ilc_1980_v2_p2.pdf. Accessed 02 August 2022
International Law Commission (2001) Draft articles on Responsibility of States for Internationally Wrongful Acts, with commentaries 2001. Yearbook of the International Law Commission 2, Part 2, 83, pars. 1-2 https://legal.un.org/ilc/texts/instruments/english/commentaries/9_6_2001.pdf. Accessed 02 August 2022

International Law Commission (2001) Draft Articles on Responsibility of States for Internationally Wrongful Acts. Supplement No. 10 (A/56/10)

Investment promotion and protection agreement between Japan, Republic of Korea and China (2012). https://investmentpolicy.unctad.org/international-investment-agreements/treaty-files/2633/download. Accessed 02 August 2022

North American Free Trade Agreement (NAFTA) (1994). https://www.nafta-sec-alena.org/Home/Texts-of-the-Agreement/North-American-Free-Trade-Agreement. Accessed 02 August 2022

Pacific Agreement on Closer Economic Relations (PACER) (2017). https://dfat.gov.au/trade/agreements/not-yet-in-force/pacer/Pages/pacific-agreement-on-closer-economic-relations-pacer-plus.aspx. Accessed 02 August 2022

Reciprocal Investment Promotion and Protection Agreement Between The Government of the Kingdom of Morocco And The Government of the Federal Republic of Nigeria (2016). https://investmentpolicy.unctad.org/international-investment-agreements/treaty-files/5409/downloada. Accessed 02 August 2022

Republic of Mauritius and The Government of The Republic of India for The Promotion and Protection of Investments (1998). https://dea.gov.in/sites/default/files/Mauritius.pdf. Accessed 02 August 2022

The Reciprocal Promotion and Protection of Investments between The Argentine Republic and The State Of Qatar (2016). https://investmentpolicy.unctad.org/international-investment-agreements/treaties/bit/3706/argentina%2D%2D-qatar-bit-2016. Accessed 02 August 2022

Trans-Pacific Partnership (2018). https://ustr.gov/sites/default/files/TPP-Final-Text-Investment.pdf. Accessed 02 August 2022

Treaty Between United States of America and the Argentine Republic concerning the Reciprocal Encouragement and Protection of Investment (1991). https://investmentpolicy.unctad.org/international-investment-agreements/treaty-files/127/download. Accessed 02 August 2022

United Nations (1969) *Vienna Convention on the Law of Treaties*. Treaty Series, vol. 1155, p. 331 https://www.refworld.org/docid/3ae6b3a10.html. Accessed 02 August 2022

United Nations (1992) Agenda 21, Rio Declaration, Forest Principles. https://sustainabledevelopment.un.org/content/documents/Agenda21.pdf. Accessed 02 August 2022

United Nations (1992) *Rio Declaration on Environment and Development 1992* (31) ILM, p. 874 https://www.jus.uio.no/lm/environmental.development.rio.declaration.1992/portrait.a4.pdf. Accessed 02 August 2022

United Nations Conference on Trade and Development (2009) The Protection of National Security In IIAs. UNCTAD Series on International Investment Policies for Development, p. 91-92

United Nations Framework Convention on Climate Change (1992) FCCC/INFORMAL/84 GE.05-62220 (E) 200705 https://unfccc.int/resource/docs/convkp/conveng.pdf. Accessed 02 August 2022

World Summit on Sustainable Development, & United Nations (2003) *Johannesburg Declaration on Sustainable Development and Plan of Implementation of the World Summit on Sustainable Development* https://ec.europa.eu/environment/archives/wssd/documents/wssd_pol_declaration.pdf. Accessed 02 August 2022

Chapter 10
Direct Application of the Common But Differentiated Responsibilities and Respective Capabilities Principle (CBDRRC) in Investment Treaty-Climate Change Disputes: Vindicating a Differential Approach to Compensation

The 'common but differentiated responsibilities and respective capabilities' principle represents the global consensus on the differential role that states occupy in the climate change international legal framework. Thus, from the implementation of the UNFCCC in 1992[1] to the adoption of the Paris Agreement,[2] the CBDRRC principle has been central to the discussions that led to the different climate change international legal instruments as well as to their text.[3] In some of these instruments, the principle has been captured through a rigid binary-differentiation model that distinguishes between the responsibilities and capabilities of developing and developed countries.[4] Other instruments incorporate a more dynamic conception of the principle, acknowledging that the *developing-developed* countries dichotomy needs to be supported by a permanent review of the particular circumstances of each country.[5] Regardless of the rigid or dynamic conception of the instrument, the CBDRRC principle bestows legal relevance to the fact that countries have different conditions and, accordingly, that these differences should be reflected in the text and the operation of the climate change international legal framework.

---

[1] United Nations Framework Convention on Climate Change (1992).

[2] United Nations Paris Agreement (2015).

[3] United Nations Paris Agreement (2015) Arts. 2 (1), 3, 4(1), 4 (3), 4(4), 4(5), 4(6), 4(15), 4(19), 5(2), 6(6), 6(8), 7(2), 7(6), 8(3), 9(1) and 9(3); Arts. 2 (1) (a), 2 (4), Art. 13 (4) (c), Art. 13 (4) (d), 10, Kyoto Protocol to the United Nations Framework Convention on Climate Change (adopted 11 December 1997, entered into force 16 February 2005), p. 2303; United Nations Framework Convention on Climate (1992) Arts. 3.1., 4.1.

[4] United Nations Framework Convention on Climate Change (1992) preamble, Arts. 3.1., 4.1.; Kyoto Protocol to the United Nations Framework Convention on Climate Change (1997) Annexes A and B.

[5] United Nations Paris Agreement (2015) Arts. 2 (1), 3, 4(1), 4 (3), 4(4), 4(5), 4(6), 4(15), 4(19), 5(2), 6(6), 6(8), 7(2), 7(6), 8(3), 9(1) and 9(3); Arts. 2 (1) (a), 2 (4), Art. 13 (4) (c), Art. 13 (4) (d), 10.

T. Restrepo Rodríguez, *Investment Treaty Law and Climate Change*,
https://doi.org/10.1007/978-3-031-18655-4_10

I argue in this chapter that the application of the CBDRRC principle may influence the assessment of damages in investment treaty-climate change disputes. However, BITs do not have specific provisions transporting this principle to the assessment of the compensation. Furthermore, generally—with the exception of lawful expropriation—BITs do not contain specific provisions for establishing compensation. This absence leads to the conclusion that an *orthodox approach* (content-interpretation) is insufficient to bring the CBDRRC principle to the investment treaty law system. Thus, a *heterodox approach* is needed to give weight to the special needs and circumstances of host states before climate change in the assessment of the compensation.

The application of the CBDRRC principle to the compensation in investment treaty-climate change disputes has not been proposed in any other study. Studies on the relationship between environmental law and investment treaty law have not taken this approach either. However, the solidity of the 'all or nothing' paradigm (either full compensation or no compensation) in the cases of lawful expropriation has been disputed by Kriebaum. She posits the adoption of a proportionality test that includes the evaluation of whether a special (overriding) public interest—such as health, human rights or the environment—justifies the reduction of the full compensation of the investment that was expropriated.[6] Kulick takes a similar view but extends the reasoning to the FET.[7] He proposes a three-tier methodology, whose third layer consists of exercising a proportionality *stricto sensu* test—that is subsequent to the verification of the scope of the investor's right (first layer) and the infringement of the standard (second layer).[8] The exercise that he proposes consists of weighting the global public interest against the investor's interests by analysing '(1) necessity; (2) gravity of the infringement; (3) legitimate expectations; and (4) seriousness of pursuing the global public interest'.[9] In his perspective, if the result of the test reveals that the defence of global public interest is successful, the legal consequence could be the reduction of the compensation.[10]

While these studies admit that the principle of full compensation could be superseded when the infringement is justified by the global public interest, I contend that in the particular case of investment treaty-climate change disputes compensation could be proportional to the responsibilities and capabilities of the host state when the defence of necessity fails. In other words, while Kulick and Kriebaum associate the proportionality of the compensation to a sort of (global) public interest defence, this research grounds the proportionality of the compensation as a result of the direct application of the CBDRRC (and climate justice) rather than on the success of necessity or any general style defence. Furthermore, I contend that in the event that a breach is justified by necessity or a general defence, the application of the

[6] Kriebaum (2007), pp. 739–742.

[7] Kulick (2012), pp. 205–208.

[8] *Ibid.*, pp. 195–197.

[9] *Ibid.*, p. 214.

[10] *Ibid.*, p. 209.

CBDRRC principle may suggest that the imposition of compensation is still possible to promote investments in the green industry.

Applying a differential treatment to the assessment of compensation instead of the full compensation in investment treaty law faces two clear challenges. First, it must be proven that elements that are different from the magnitude of the victim's prejudice influence the determination of the compensation, i.e. it must be proven that the full compensation principle is not absolute (Sect. 10.1). Second, it must be clarified that a principle that appears to regulate cooperation between countries—such as the CBDRRC—could be transferred to a system of liability that is connected to the investor's protection such as the investment treaty law. In other words, the second challenge requires envisaging the operationalisation of the CBDRRC in investment treaty-climate change disputes (Sect. 10.2). On the first challenge, I contend that tribunals establish compensation, not only in based on compensatory justice, but also—occasionally—deploying elements that represent retributive and distributive justice. Furthermore, I describe how *distributive justice* has been deployed to determine compensation according to a differentiation approach that considers the particular conditions of the host state. Concerning the second challenge, I claim that the CBDRRC principle is not devoid of liability features, nor is the investment treaty law system devoid of cooperation features. Further, I show that climate change law and investment treaty law coincide in recognising an instrumental position of the investor in relation to the rationale of each system. The chapter concludes that harmonisation is possible without transgressing the operation of the investment treaty law system, through the exceptional application of the CBDRRC, and discards the deployment of a sort of environmental general defence—connected to the examination of the proportionality of the measure—as a means to modulate compensation (Sect. 10.3).

10.1 Suitability of the CBDRRC to the Practice of the Investment Treaty Law Tribunals: Demystifying the Full Compensation Principle in Investment Treaty Law

The full compensation principle is the most refined expression of compensatory justice. The victim's prejudice is the basis for measuring the compensation. Therefore, neither a compensation failing to repair the entire loss, nor a compensation superseding the extent of the loss, pleases the ideal of compensatory justice. Contrarily, the application of the CBDRRC to the compensation in investment treaty law would be seen—at least on a first approach—as the consecration of distributive justice. Accordingly, the host state's capabilities, vulnerabilities and responsibilities before climate change would be the basis for assessing the compensation rather than the magnitude of the loss. In the end, tribunals dealing with compensation in investment treaty-climate change disputes should address the disjunctive of choosing which model of justice—compensatory or distributive—prevails.

Unlike the CBDRRC, the application of the full compensation principle is a practice that is firmly embedded in investment treaty law. Thus, Arbitral tribunals frequently refer to Article 31 DASR, which establishes that '[t]he responsible State is under an obligation to make full reparation for the injury caused by the internationally wrongful act';[11] and/or invoke the *Chorzow factory award,* which expresses that the reparation should 'wipe out all the consequences of the illegal act and re-establish the situation which would, in all probability, have existed if that act had not been committed'.[12]

However, the principle of full compensation is not absolute. The (ordinary) text of the (lawful) expropriation standard limits the compensation to the 'market value', the 'fair market value' or any similar expression (Sect. 10.1.1). Additionally, elements of retributive and distributive justice can be traced in some of the secondary rules that investment treaty tribunals deploy to determine compensation (Sect. 10.1.2). The embracing of justice models different from compensatory justice reveals that the CBDRRC principle is not alien to investment treaty law. Furthermore, examples of tribunals exceptionally adopting a differentiation model in light of the particular conditions of the parties to the dispute suggests that the CBDRRC principle suits the investment treaty law system.

10.1.1 Limitations to the Full Compensation Principle Arising from the Text of the (Ordinary) Expropriation Standard: Paving the Way for the Application of the CBDRRC Through the Identification of Retributive Justice Elements in the Compensation

As described in the chapter dedicated to the orthodox approach to the expropriation standard,[13] the ordinary text permits an expropriation as long as the measure has been taken 'for a public purpose', 'in a non-discriminatory manner', 'in accordance with due process' and 'upon payment of prompt, adequate and effective compensation'.[14] Furthermore, treaties generally establish that the compensation should be

[11] International Law Commission (2001).

[12] Court of International Justice, *Case concerning the factory at Chórzow (2017)*, p. 47.

[13] *Supra* Chap. 6.

[14] '"Prompt" means compensation must be paid without unreasonable delay, "adequate" means that it must be equal to the "fair market value" of the property taken immediately before the taking' and '"effective" means that it must be made in a freely transferable currency': Sabahi and Birch (2010), p. 760.

objectively assessed by reference to the 'market value'[15] of the investment at the time of the expropriation. The (ordinary) text has generated two crucial debates. The first concerns the question of whether the limitation to the compensation is applicable to any expropriation infringing any of these requirements or, contrarily, whether the limitation is restricted to those expropriations that only lack the payment of compensation. The second involves wondering whether subjecting compensation to the 'market value' constitutes a limitation to the full compensation principle at all.

For the first question—whether the limitation to the compensation is applicable to any expropriation—there are two approaches. While one posits that the threshold of 'market value' applies indistinctly to any expropriation lacking one of the conditions [16]—whichever the condition is—the other conceives that the threshold corresponds exclusively to lawful expropriations, i.e. expropriations that fulfill all the conditions but the payment of compensation. [17] Although the ordinary text of the expropriation standard does not clearly express that the compensation based on the 'market value' only belongs to lawful expropriations—or refers to the dichotomy lawful-unlawful expropriation—it could be said that the prevailing approach is that the compensation for lawful expropriation is different from the compensation for unlawful expropriations;[18] while the former is the 'market value' of the investment, the latter should be determined by reference to the full compensation principle.

The basis for the differentiation in the compensation between lawful and unlawful expropriations—in the words of *AMOCO v. Iran*—is grounded on the fact that 'the legal bases for the two concepts are totally different and, logically, the practical methods to be used in order to derive the amount due should differ'. [19] The same view is found in *UP v. Hungary,* where the tribunal expressed that '[i]t is by no

[15] Albeit a variety of expressions qualify the 'market value' in different investment treaties, the variation seems no to be relevant: 'fair market value', 'genuine value', 'appropriate market value' are taken in general as synonyms. In the perspective of *CMS* 'These concordant provisions are variations of an agreed, essential theme, namely, that when a state takes foreign property, full compensation must be paid': *CMS Gas Transmission Company v. The Republic of Argentina (2005)*, para. 498.

[16] *CME Czech Republic B.V. (The Netherlands) v. The Czech Republic (2003)*; *Wena Hotels Ltd. v. Arab Republic of Egypt (2000)*; *Metalclad Corporation v. The United Mexican States (2000)*; *Middle East Cement Shipping and Handling Co. S.A. v. Arab Republic of Egypt (2002).*

[17] *Compañía de Aguas del Aconquija S.A. and Vivendi Universal S.A. v. Argentine Republic (2007)*, para. 8.2.10.; *ADC Affiliate Limited and ADC & ADMC Management Limited v. The Republic of Hungary* (2006), para. 481.; *Karkey Karadeniz Elektrik Uretim A.S. v. Islamic Republic of Pakistan (2017)*, para. 662.; *Siemens A.G. v. The Argentine Republic (2007)*, para. 352.; *UP (formerly Le Chèque Déjeuner) and C.D Holding Internationale v. Hungary (2018)*, para. 511. Supporters of this approach: Marboe (2015), pp. 1033–1034.

[18] 'With regard to unlawful expropriations early cases treated it identically to lawful expropriation, generally applying the BIT provisions on compensation for legal expropriation in question, regardless of whether the expropriation met the provisions' other requirements (...) Such treatment of unlawful expropriation through treaty provisions for lawful expropriations was the norm until the decision in *ADC v. Hungary* in 2006 (...) Recent awards follow the division of lawful and unlawful expropriation proposed by *ADC v. Hungary*': Sabahi and Birch (2010), pp. 168–769.

[19] *AMOCO International finance Corporation v. Islamic Republic of Iran (1987),* para. 191.

means "*logical*" that a breach of Art. 5(2) [expropriation standard] by an unlawful dispossession should not result in any wider liability than a lawful dispossession'. [20] This appears to signify that a glimpse of retributive justice irradiates the assessment of the expropriation, since the wrongfulness of the expropriatory measure is a factor that defines whether the host state should pay more or less;[21] whether it must pay 'market value' compensation or a 'full compensation'.

However, the practicality of the distinction between the effects of lawful and unlawful expropriations has been disputed by some commentators[22] and rulings. [23] This concerns the second question—whether subjecting compensation to the 'market value' constitutes a limitation to the full compensation principle at all. The doubts arise from the difficulties in distinguishing between 'market value' and full compensation. It appears that the method for assessing the 'market value' of a company or property encapsulates the loss of profit.[24] If that is the case, there would be no material difference between the two concepts; both would include *damnum emergens* and *lucrum cessans*. Furthermore—apart from the events of lawful/unlawful expropriations—the 'market value' has been even considered to be appropriate to measure the compensation for the breach to the FET.[25] Considering that the standard of compensation for the breach to the FET is the principle of full compensation under international customary law (*Chorzow Factory* and Art. 31 ILC), the plausibility of melting the concepts of 'market value' and full compensation seems to be a common practice of investment tribunals.

Nevertheless, the unification of the concepts of 'market value' and full compensation is not always possible in the case of unlawful expropriation. The full

---

[20] *UP (formerly Le Chèque Déjeuner) and C.D Holding Internationale v. Hungary (2018)*, para. 511.

[21] 'There can be no doubt about the vitality of this statement of the damages standard under customary international law, which has been affirmed and applied by numerous international tribunals as well as the PCIJ's successor, the International Court of Justice. It is also clear that such a standard permits, if the facts so require, a higher rate of recovery than that prescribed in Article 5(2) for lawful expropriations': *Compañía de Aguas del Aconquija S.A. and Vivendi Universal S.A.* (2007), para. 8.2.5.

[22] Sabahi and Birch (2010), p. 785.

[23] *Bernardus Henricus Funnekotter and others v. Republic of Zimbabwe (2009)*, paras. 112, 115.; *Waguih Elie George Siag and Clorinda Vecchi v. The Arab Republic of Egypt (2009)*, para. 541.; *Marion Unglaube v. Republic of Costa Rica (2012)*, paras. 305–307.

[24] 'The recourse to a discounted cash flow analysis for expected future revenue/profit of the project is aimed at ascertaining a present market value for the Property and Project in 1996, and the calculation produces a result not materially different from the alternative basis upon which compensation was calculated, namely an assessment of the market value of the land at the time of the expropriation': *Siag* (2009), para. 541; *Unglaube* (2012), para. 307.

[25] 'As pointed out by the Tribunal in CMS Gas Transmission Co. v. Argentine Republic, a "fair market value" standard might also be appropriate for other breaches which result in long-term losses. The Azurix tribunal also concluded that it could properly resort to fair market value to compensate breaches other than expropriation – in particular the fair and equitable standard. In its award, it noted the particular relevance of the government having taken over the concession': *Compañía de Aguas del Aconquija S.A. and Vivendi Universal S.A.* (2007), para. 8.2.10.

compensation can still reflect a higher compensation than the 'market value' when the value of the assets or the company considerably increases between the date of the expropriation and the date of the judgement. In this case, the date of valuation of the compensation for unlawful expropriation would be the date of the award rather than the date of the expropriation, while for lawful expropriations, it would be the date of expropriation, as established in the text of the ordinary standard. [26] A compensation higher than the 'market value' would also be granted when the compensation comprehends moral damages[27] or consequential damages.[28]

Moreover, although it is true that the valuation methods that are often deployed in the cases of lawful expropriation,[29] particularly the Discounted Cash Flow (DCF) method, include the calculation of profits,[30] other approaches that have been adopted by tribunals to perform this valuation do not consider the revenues, e.g. the 'book value',[31] 'the investment value' [32] or 'sunk value' (Asset Value or Cost Benefit Approach[33]). Admittedly, the DCF method is the most popular among the tribunals, firms and academics. Nevertheless, the adoption of other methods to assess compensation that exclude the loss of profit in cases of lawful expropriation shows that occasionally there are differences between 'market value' and the determination of full compensation.

Against this background, it could be said that the (ordinary) text of the expropriation standard establishes a limitation to the full compensation principle on lawful expropriations. This limitation is not only conceptual; it also has practical consequences. In conceptual terms, the full compensation of international customary law

[26] *ADC* (2006), para. 497; *Ioannis Kardassopoulos v. The Republic of Georgia (2010)*, para. 514. *Phillips Petroleum case, 'the lawful/unlawful* taking distinction (...) is relevant only to two possible issues: whether restitution of the property can be awarded and whether compensation can be awarded for increase of the value of the property between the date of the taking and the date of the judicial or arbitral decision awarding compensation'.

[27] Sabahi and Birch (2010), p. 771.

[28] Marboe (2015), p. 1039.

[29] Apart from the DCF method, tribunals may include the compensation for loss of profit through the comparable transaction approach or market approach [CMS v. Argentina (2005), paras. 411–417, CME v. Czech Republic (2003), para. 264, Ros Invest v. Russia (2013), para. 675, Valores Sicav v. Russia, 675], the differential method ('but for') [AMOCO (1987), para. 255].

[30] 'The DCF approach has become widely accepted because it is the best method for valuing lost profits. Critics who complain that it produces excessive levels of compensation are perhaps not paying sufficiently enough attention to how the method is operated in practice. The DCF method, properly employed, reduces speculation because it forces claimants to explain and quantify each individual area of their claim. Difference with parties can thus be highlighted and speculative valuations can be dismissed': Mclachlan et al. (2017), p. 434.

[31] APP v. Sri Lanka para. 103. *Siemens v. argentina* (2007), para. 355; *Señor Tza Yap Shum v. The Republic of Peru (2011)*, paras. 264–267.

[32] *Azurix Corp. v. The Argentine Republic (2006)*, para. 374.; *Metalclad* (2000), paras. 114–125; *Wena Hotels (*2000), para. 122.; *Suez, Sociedad General de Aguas de Barcelona, S.A. and Vivendi Universal, S.A. v. Argentine Republic (2010)* Decision on Liability, para. 8.3.13.

[33] Marboe (2015), pp. 1076, 1079–1080; *Siemens v. Argentina* (2007), para. 386, *Siag v. Egypt* (2009), para. 587.; *Funnekotter v. Zimbabwe* (2009), para. 138.

is applicable to unlawful expropriations, while the 'market value' of the expropriation standard is applicable to lawful expropriations. This distinction is intended to impose a more severe compensation to be paid by the host state that infringes the due process, discriminates the investor or does not follow the public interest, in comparison to the host state that only misses paying the compensation to the investor. In practical terms, the full compensation exceptionally differs from the 'market value', since the latter is usually assessed by reference to the loss of profit among other factors.

The difference in the compensation between lawful and unlawful compensation seems to represent a punitive character—as Reisman and Slone affirm[34]—or at least a variation to the principle that the prejudice is the measure of the compensation. Such consideration reveals that the principle of full compensation is not absolute, and that retributory justice could also play a role in the assessment of damages. Furthermore, accepting that the compensation for expropriation might be driven by these factors, begins to pave the way for accepting that a differential approach to compensation—such as the CBRRDC's approach—may be applicable to investment treaty-climate change disputes.

10.1.2 Limitations to the Full Compensation Principle Arising from Secondary Rules: Confirming the Suitability of the CBDRRC to the System's Operation Stemming from a Differentiation Approach to Compensation

In addition to the limitation to the full compensation that is embedded in the text of the expropriation standard, the system's operation occasionally admits variations to this principle. While some of these variations seem to reflect retributive justice, other limitations appear to stem from distributive justice considerations. The influence of the contributory negligence and the moral damages in the determination of the compensation appears to reflect retributive justice elements. Contrarily, the recognition that the investor should bear a portion of the legal and regulatory risk and the principle of equity seems to reveal distributive justice considerations.

On contributory negligence, Article 39 DASR establishes that the reparation is not only tied to the magnitude of the loss but also to the contribution of the victim's 'willful or negligent action or omission'[35] to the injury. This rule has been adopted by several investment treaty tribunals. In *Oxy v. Ecuador*, for example, the tribunal established that the 'the Claimants should pay a price for having committed an

[34] Reisman and Sloane (2004), p. 138.

[35] International Law Commission (2001), Art. 39.

unlawful act which contributed in a material way to the prejudice'. [36] In its perspective, the investor's celebration of a farmout agreement without the host state's approval constitutes a 'material and significant wrongful act [that] contributed to the extent of 25% to the prejudice'.[37] Similarly, in *Yukos v. Russia* the tribunal considered that the compensation should be reduced by 25% [38] because the claimant 'should pay a price (...)' for the 'abuse of the low-tax regions by some of its trading entities (...)'[39]—which was regarded as a 'material and significant misconduct'.[40] As illustrated by these cases, the determination of the contributory negligence is not only a matter of causation; it requires the victim's contribution to be judged as wrongful. Consequently, it seems correct to aver that the reduction of the compensation in these events has, in its core, a sense of retributive justice.[41]

Retributive justice is also embedded in the determination of the compensation for moral damages when this is imposed to of punish the host state's behaviour. In principle, the degree of severity of the host state's fault, negligence or willful misconduct should not affect the assessment of the compensation. However, some tribunals have deployed moral damages as punitive damages. This is the case of *von Pezold v. Zimbabwe*, where the tribunal awarded moral damages in addition to expropriatory damages in part because the 'threats of, and actual, physical violence and detainment that Heinrich [the investor] reported clearly contravene the conduct expected of states'. [42] This ruling was explicit about the punitive instrumentalisation of moral damages by affirming that '[i]n some sense'[43] the compensation of moral damages 'serves not only the function of repairing intangible harm, but also of condemning the actions of the offending State'.[44] The punitive character of moral damages was also explicit in *Lemire v. Ukraine*, where the tribunal asserted that moral damages can be—exceptionally [45]—awarded when 'the ill-treatment [to the investor] contravenes the norms according to which civilized nations are expected to act'.[46] Similarly, in *Desert Line v. Yemen,* the compensation of moral damages was

[36] *Occidental Petroleum Corporation and Occidental Exploration and Production Company v. The Republic of Ecuador (2012),* para. 680.

[37] *Ibid.*, para. 687.

[38] *Yukos Universal Limited (Isle of Man) v. The Russian Federation (2014),* para. 1637.

[39] *Ibid.*, para. 1634.

[40] *Ibid.*, para. 1637.

[41] The same conclusion could be extracted from the reasoning of: *MTD Equity Sdn. Bhd. And MTD Chile S.A. v. Republic of Chile (2004)*, para. 242.; *Yuri Bogdanov and Yulia Bogdanova v. Republic of Moldova (2013)*, para. 84.; *RosInvestCo UK Ltd. v. The Russian Federation (2010).*

[42] *Bernhard von Pezold and Others v. Republic of Zimbabwe (2015),* para. 920.

[43] *Ibid.*, para. 916.

[44] *Ibid.*

[45] *Joseph Charles Lemire v. Ukraine (2011),* para. 333.

[46] *Ibid*. Although in Gavazzi v. Romania moral damages were rejected because the damage was not significant enough, the Tribunal cited and followed the reasoning of *Lemire v. Yemen* on the plausibility of punishing through moral damages: *Marco Gavazzi and Stefano Gavazzi v. Romania (2017)*, paras. 292–293.

justified on the grounds that the host state's behaviour was 'malicious (...) fault-based'.[47]

Contrary to the retributive justice quality of compensation in the event of contributory negligence and the moral (punitive) damages, the duty of the investor to bear part of the regulatory risk is grounded on distributive justice considerations. The proportionality of the compensation to the risk that the investor assumes is not really an expression of contributory negligence since the lawfulness of the investor's behaviour is not determinant in the reduction. The justification for the reduction in these cases will be grounded on the fairness of distributing the negative externalities of public interest regulations between the investor and the host state.

In the *MTD v. Chile* case, for example, the investor took the risk of investing in a construction project, despite the fact that the location was earmarked as agricultural land. Even if the national authorities promised the investor that the regulation was going to change to allow urbanisation projects, the tribunal ruled that the investor should bear 50% of the damages[48] because claimants 'incurred costs that were related to their business judgment irrespective of the breach',[49] and 'made decisions that increased their risks in the transaction and for which they bear responsibility, regardless of the treatment given by Chile'.[50]

Admittedly, the limits between the investor's assumption of an extraordinary risk and negligence could be blurred, as could distinguishing between reducing the compensation because of contributory negligence and reducing it because of transferring the risk to the investor.[51] Nevertheless, it is important to highlight that even if accompanied by negligence or willful misconduct by the investor, the apportionment of the risk is a matter of distributive justice. Thus, in *Yukos v. Russia*, it was expressed that a tribunal is entitled to reduce compensation 'if the victim of the wrongful act of the respondent State also committed a fault'[52] and if—under its 'discretion'[53]—the Tribunal 'considers the claiming party should bear some responsibility'.[54]

Similarly, the difference between a full compensation based on compensatory justice and a compensation that attributes part of the risk to the investor based on distributive justice could be diffused. This concerns the cases where the assessment

---

[47] *Desert Line Projects LLC v. The Republic of Yemen (2008)*, para. 290.

[48] *Ibid.*, para. 243.

[49] *Ibid.*, para. 242.

[50] *Ibid.*

[51] It is not uncommon that in defining the risk that the investor takes, the Tribunal refers to the due diligence of the investor. Thus, in *MTD v. Chile* the Tribunal asserted that the 'investor accepted to pay a price for the land with the Project without appropriate legal protection. A wise investor would not have paid full price up-front for land valued on the assumption of the realization of the Project; he would at least have staged future payments to project progress, including the issuance of the required development permits': MTD (2004), para. 242.

[52] *Yukos* (2014), para. 1633.

[53] This margin of discretion was also highlighted in *Oxy v. Ecuador. Oxy* (2012), para. 678.

[54] *Ibid.*

of the market value of the investment includes a premium weighting regulatory and legal risk. In *Tidewater v. Venezuela*, the determination of the market value of the oil company that was expropriated comprehended a risk premium of 14.75%. The tribunal chose to adopt this approach of weighting 'the risk associated with such [investor's] cash flow under realistic circumstances' [55] because, even if 'the Treaty seeks to encourage investment between the Contracting States [...] it is not an insurance policy or guarantee against all political or other risks associated with such investment'.[56] Importantly, the distribution of the risk through a premium is not exclusive for the compensation of expropriation. It has also been adopted in other breaches that rely on the market value to determine compensation. The tribunal in *Valores v. Venezuela*, for example, applied a 4% country risk premium to the compensation for the violation of the FET and the free transfer of funds. [57]

Distributive justice could also influence the determination of the compensation through the plain application of the principle of equity. The movement for the nationalisation of oil reserves in the 60s and 70s provided tribunals with the opportunity to challenge the principle of full compensation through equity considerations.[58] In this respect, the tribunal in *LIAMCO* awarded the investor a lump sum of US$ 66,000,000 as 'equitable compensation' [59] and was explicit about the general application of the principle of equity to the compensation for oil nationalisations:

> (...) the rule of "full and prior" compensation is no more imperative, and that only "convenient and equitable" compensation is required in cases of nationalisation. This trend has been justified by the necessity of taking into consideration not only the interest of the owner of the property nationalised, but also those of Society (Community) and of the nationalising State (...)[60]

This formulation is certainly harmonising with the general trend of international theory and practice on the concepts of sovereignty, destination of national wealth and natural resources, nationalistic motivations in the attitude and behaviour of 'Third World' nations, the lawfulness and frequency of nationalisations, and the recent declaration affirmed in successive United Nations Resolutions by the majority members of the General Assembly.[61]

---

[55] *Tidewater Inc., Tidewater Investment SRL, Tidewater Caribe, C.A., et al. v. The Bolivarian Republic of Venezuela (2005)*, para. 185.

[56] *Ibid.*, para. 184.

[57] *Valores Mundiales, S.L. and Consorcio Andino S.L. v. Bolivarian Republic of Venezuela (2017)*, para. 791.

[58] Stewart (1991), pp. 122–123.

[59] *Lybian American Oil Company v. The Government of Libyan Arab Republic* (LIAMCO) (1977), para. 160.

[60] *Ibid.*, para. 145.

[61] *Ibid.*, paras. 150–151. Cfr. *Government of the State of Kuwait v. American Independent Oil (AMIOIL) (1982)*, para. 144. In AMIOIL, arbitrators brought into consideration the *United Nations General Assembly Resolution n degrees 1803, XVII of 14 December 1962*, which establishes 'appropriate compensation' to redress the victim of expropriation. According to the Tribunal, appropriate compensation should be assessed on a case-by-case basis, taking into account the

Likewise, in *Shahin Shane Ebrahimi v. Iran,* it was held that the prevailing standard in the practice of international law was 'appropriate compensation' rather than 'prompt, adequate and effective' compensation.[62] According to the tribunal, an 'appropriate compensation' requires assessment and reparation 'in a flexible manner, that is, taking into account the specific circumstances of each case'.[63] The radio of flexibility that the tribunal conceived embraced partial compensations, without excluding the possibility of adopting a full compensation.[64]

On the compensation arising from the breach of contract, in *Sapphire International Petroleums Ltd. v. National Iranian Oil,* it was held as a principle that the investor is entitled to be placed in the same pecuniary position as it would if the contract had been properly performed by the host state. [65] In the tribunal's view, the principle described derives from the principle of *pacta sunt servanda,* since the compensation is the substitute of the obligation promised.[66] According to this decision, the compensation for the breach of a contract comprehends *damnum emergens* and *lucrum cessans.* However, as the outcome of the operations (oil operations in this case) was not certain, the tribunal decided to order the compensation of the loss of profit as 'loss of opportunity'.[67] Since the assessment of the 'loss of opportunity' cannot be established with absolute certainty, it was held that the compensation for this category should be evaluated *ex aequo et bono.*[68]

The application of the principle of equity stemming from the particular circumstances of the dispute has also found some acceptance outside the conflicts concerning the nationalisation of the oil industry. In this respect, *Santa Elena v. Costa Rica* directly quotes *Philips v. Iran*—a case on oil nationalisation—to express that 'the determination of [market]value by a tribunal must take into account all relevant circumstances, including equitable considerations'. [69] Even if this tribunal did not conceive equity as a limitation to the full compensation principle, the compensation was determined in a peculiar manner that shows traces of distributive justice. The method consisted of adding the claimant's estimation of the market value (U.S. $6,400,000) to the respondent's estimation of the market value (U.S. $1,700,000) and dividing this into two (U.S. $4,150,000).[70] The tribunal concluded that '[i]n the circumstances of this case the sum of U.S. $4,150,000 constitutes a reasonable and fair approximation of the value of the Property

particular circumstances, instead of assessing compensation by resourcing to abstract theoretical discussions.

[62]Cited in Highet et al. (1995), pp. 385, 386–387.

[63]*Ibid.*, p. 387.

[64]*Ibid.*

[65]*Sapphire International Petroleums Ltd. v. National Iranian Oil (1963)*, para. 8.

[66]*Ibid.*, para. 7.

[67]*Ibid.*

[68]*Ibid.*, para. 15; Cfr. *Gold Reserve Inc. v. Bolivarian Republic of Venezuela (2014)*, para. 686.

[69]*Compañia del Desarrollo de Santa Elena S.A. v. Republic of Costa Rica (2000)*, para. 92.

[70]*Ibid.*, para. 93.

(...)'.[71] In *CME v. Argentina,* it was established that the investor must bear—'in a reasonable manner' [72]—part of the costs of the Argentinian 2001 economic crisis with the same argument that was used in LIAMCO: that the investment protection under international investment law does not 'amount to an insurance policy against business risk'.[73]

The deployment of the principle of equity closest to the notion of distributive justice that is embedded in the CBDRRC is found in *American Manufacturing v. Zaire.* This award concerned the failure of the government to provide *full protection* to the investor against riots that ended up destroying property. [74] The tribunal openly expressed that compensation could be determined based on 'equitable principles'.[75] Furthermore, the tribunal considered the 'existing conditions of the country' and was able to not make the mistake of grounding its decision on 'abstraction[s] based on a criterion for the assessment which does not correspond to all reality, nor the current happenings in Zaire, nor indeed to the commercial and industrial activities of the claimant'.[76] In addition, it is important to highlight that—similar to the rationale of the CBDRRC principle—the tribunal emphasised the importance of differentiation under extreme conditions:

> AMT would have liked to adopt a method of calculating compensation including interests practicable in normal circumstances prevailing in an ideal country where the climate of investment is very stable, such as Switzerland or the Federal Republic of Germany. The Tribunal does not find it possible to accede to this way of evaluating the damages with interests in the circumstance under consideration, in which is apparent that the situation remains precarious [...] It would be neither impractical nor reasonable to apply the method of assessment of compensation in a way so far removed from the striking reality of the current situation.[77] [...] Preferably, the Tribunal will opt for a method that is more plausible and realistic in the circumstances of the case, while rejecting all other methods of assessment which would serve unjustly to enrich an investor who, rightly or wrongly, has chosen to invest in a country such as Zaire, believing that by so doing the investor is constructing a castle in Spain or a Swiss chalet in Germany without any risk, political or even economic or

---

[71] *Ibid.* A similar approach was taken in *ADC v. Hungary*: 'The Tribunal is of course grateful to the experts on both sides for their enormous help on the issue of damages. However the Tribunal feels bound to point out that the assessment of damages is not a science. True it is that the experts use a variety of methodologies and tools in order to attempt to arrive at the correct figure. But at the end of the day, the Tribunal can stand back and look at the work product and arrive at a figure with which it is comfortable in all the circumstances of the case'. *ADC v. Hungary* (2006), para. 521.

[72] *CME* (2003), para. 248.

[73] *Ibid.* A similar approach was taken in *Siag v. Egypt*: '[...] there were undoubtedly considerable risks associated with the further investment required to bring the Project to fruition and to realise the potential which the Property and the Project seemed to hold, and it is trite to observe that success in business ventures can never be assured. The Tribunal bears these risks in mind in reaching the final level of compensation which is set out further below': *Siag* (2009), para. 543.

[74] *American Manufacturing & Trading, Inc. v. Republic of Zaire (1997)*, para. 6.02–6.13.

[75] *Ibid.*, para. 7.16.

[76] *Ibid.*, para. 7.13.

[77] *Ibid.*, para. 7.14.

> financial or any risk whatsoever [...][78] For practical reasons founded on equitable principles, the Tribunal finds that the Republic of Zaire which is responsible in international law, is under a duty to compensate AMT for the very losses[79][...] In effect, the Republic of Zaire has pleaded in its rejoinder that "[n]o one on earth could ignore the fact that for the past four years, the Republic of Zaire has been going through a most painful and unfortunate period in its history". Zaire continues, "This requires a benevolent and compassionate attention on the part of all our partners, even those who have encountered unfortunate and disastrous consequence, for there was a time when the same persons where enjoying the benefit of the good situation of Zaire."[80]

The full compensation principle is not absolute. I have shown in this section that the magnitude of the prejudice is not the only standard used for assessing compensation. Retributive justice and distributive justice also play a part in the determination of damages. Even if it is not always clear in the rulings, the complete independence of distributive justice considerations from compensatory justice—such as in the cases where the business risk and the regulatory risk influence the market value of the investment—the full compensation principle accepts exceptions. This conclusion opens the gate to the CBDRRC principle to enter into investment treaty law. Moreover, the system's operation—exceptionally—embraces a *differentiation* approach to determine compensation in light of the particular circumstances of the parties to the dispute—such as the relevance of the position of 'Third World' nations concerning natural resources in LIAMCO, or the striking reality of the Zairean civil war in *American Manufacturing v. Zaire*. However, what remains puzzling is how a rule that in principle concerns cooperation could operate in a system of liability. This is the topic of the next section.

10.2 The Operationalisation of the CBDRRC in Investment Treaty-Climate Change Disputes

The application of the CBDRRC is generally conceived as a cooperation rule. Broadly, this principle posits that the responsibility concerning the adverse effects of climate change corresponds to the whole global community, but propends for a differentiation in the burden that developed countries should assume in comparison to developing and least developed countries. The distributive justice logic of the CBDRRC seems easy to simplify: wealthy countries (northern hemisphere) should assume higher costs than non-wealthy ones (southern hemisphere).

Several provisions of the climate change international legal framework reflect the CBDRRC.[81] For example, the Paris Agreement, not only recognises in its recitals

[78] *Ibid.*, para. 7.15.

[79] *Ibid.*, para. 7.16.

[80] *Ibid.*

[81] *Supra Note* 3.

and in Article 2(2) the principle as a guiding principle,[82] but it also incorporates rules allocating most of the costs of climate change to developed countries as well as rules considering the particular conditions of developing and least developed countries.[83] In this respect, Article 9 (1) provides a good picture that summarises the relationship between developed and developing countries that is entrenched in the concept of CBDRRC.

> Developed country Parties shall provide financial resources to assist developing country Parties with respect to both mitigation and adaptation in continuation of their existing obligations under the Convention.[84]

At first glance, it seems that the CBDRRC principle does not concern liability, i.e. it does not directly regulate the response to non-compliance of climate change obligations, nor the compensation for damages caused by the host state's climate change measures, nor does the investor seem to be the addressee of the CBDRRC principle.

However, I argue that a deeper analysis that considers BITs' cooperation character and the liability elements of the CBDRRC, as well as the position of the investor in both systems (investment treaty law and climate change law) leads to the outcome that the principle might exceptionally intervene in the process of determining compensation (Sect. 10.2.1).

Furthermore, I propose that the operationalisation of CBDRRC in investment treaty law would benefit from the application of some secondary rules of liability to fully integrate the principle into the system—namely the principle of equity, the 'polluter should pay', and Article 27 DASR—as long as the conditions of the parties are weighted under a dynamic approach to the principle (Sect. 10.2.2).

[82] 'In pursuit of the objective of the Convention, and being guided by its principles, including the principle of equity and common but differentiated responsibilities and respective capabilities, in the light of different national circumstances'; '[...]noting the importance for some of the concept of "climate justice" when taking action to address climate change': United Nations Paris Agreement (2015) Preamble.

[83] Thus, Article 3 highlights 'the need to support developing country Parties' in achieving the objectives of the treaty; Article 4 (1) establishes 'that peaking [of greenhouse gas emissions] will take longer for developing country Parties', Article 4 (6) provides that the planning and strategies of least developed countries to lower gas emissions should be prepared 'reflecting their special circumstances'; Article 4 (15) mandates the parties to consider in the implementation of the agreement 'the concerns of Parties with economies most affected by the impacts of response measures, particularly developing country Parties'; Article 6 (6) compels to direct a 'share of the proceeds from activities under the mechanism [for mitigation] (...) is used (...) to assist developing country Parties that are particularly vulnerable to the adverse effects of climate change to meet the costs of adaptation': United Nations Paris Agreement (2015). *See* also, Arts. 2 (1), 3, 4 (3), 4(4), 4(5), 4(19), 5(2), 6(8), 7(2), 7(6), 8(3) and 9(3) United Nations Paris Agreement.

[84] United Nations Paris Agreement (2015) Art. 9.

10.2.1 Exceptional Application of the CBDRRC in Investment Treaty-Climate Change Disputes: Topical Connection Between Investment Treaty Law and the Climate Change International Legal Framework

There are reasons to contend that the CBDRRC exceptionally applies to investment treaty-climate change disputes. Even if the full compensation principle undoubtedly remains the general rule—according to the abundant case-law that relies on the *Chorzow factory award* and Article 31 DASR—a differentiation approach in line with the global climate interest might be justified in light of the circumstances of the parties to the dispute. The contribution to and vulnerability before climate change of host states might influence the compensation, just like the position of third world countries in *LIAMCO* or the Zairean civil war in *American Manufacturing* was determinant in its assessment.[85]

However, it is difficult to conceive the operation of the CBDRRC in investment treaty law without facing an evident topical concern. It could be argued against its application that the investment treaty law system and the system of climate change law—including the CBDRRC principle—have been erected for different purposes, have different addressees, and have a different structure. One may think that, on the one hand, the rationale of the climate change international legal framework is the levelling of global warming, the addressees of this system are the States-Parties, and that the structure of the treaties consists of distributing the burdens between them; on the other hand, the rationale of BITs is the flow of transnational investment, the foreign investors—apart from the States-Parties—are among the addressees of these treaties, and the structure consists of establishing obligations for host states to protect investors.

The possibility of exceptionally adopting the CBDRRC in investment treaty-climate change disputes requires the adopting of an interpretation that topically identifies the CBRDDC with investment treaty law. I contend that even if both systems obey to different rationales, they have points of connection that justify the application of the CBDRRC to the compensation in investment treaty-climate change disputes. This involves understanding the CBDRRC as a principle that is capable of burdening the investor with climate change costs (Sect. 10.2.1.1) and envisaging both systems (international climate change law and international investment law) equally as a hybrid between cooperation and liability (Sect. 10.2.1.2).

[85] *Supra* text to *Notes* 59–60, 74–80.

10.2.1.1 The Instrumental Role of the Investor and the Effectiveness of the CBRDDC in Investment Treaty Law

The first point of connection is the analogous instrumental position of the investor in the two systems. Indeed, while in investment treaty law the protection to the investor is instrumental to promoting the flow of international investment, in international climate change law the position of the investor is instrumental to levelling of global warming. Moreover, none of the systems impose obligations to the investors—at least not in the technical sense of the term obligation.

However, the investor is more notorious in investment treaty law than in climate change law, since it is the investor that sets the whole system in motion by relying on international arbitration when its interests are under threat. As creditor and claimant, it is certainly recognised with personhood in investment treaty disputes.

In the international climate change legal framework, contrarily, the investor takes a secondary place—but still a place. The notion of global warming as a 'common concern of humankind'[86] and the acknowledgement that the climate action requires the 'engagements of (...) various actors'[87] to place the investor within the scope of the treaty, in the sense that the fight against climate change is only possible with the contribution of the private sector. Moreover, Article 6 of the Paris Agreement calls the attention on the enhancement of the 'private sector participation in the implementation of nationally determined contributions'[88] as an element of 'integrated, holistic and balanced non-market' [89] approaches.

In fact, there are some initiatives that already incorporate investors' action. For example, the Global Adaptation and Resilience Investment Working Group (GARI), that was launched during COP21 in conjunction with the Climate Resilience Initiative A2R, incorporates more than 300 private investors engaged in aligning investment to climate change.[90] Caring for Climate is another initiative led by the UN—since 2007—that gathers almost 500 companies committed to climate action.[91] Also, the Adaptation Private Sector Initiative (PSI) is a UN programme for engaging private investors, by providing a platform for creating partnerships between companies, governments, and non-governmental organisations to innovate and produce technologies for adaptation to climate change.[92]

Therefore, claiming that the addressee of both systems is different is not sufficiently solid to sustain that the CBDRRC is not applicable to investment treaty-climate change disputes. Admittedly, the investor is a protagonist in the investment treaty law system, but its role is still instrumental, and—similar to the position of the

---

[86] United Nations Paris Agreement (2015) Preamble.

[87] *Ibid.*

[88] *Ibid.*, Art. 9.

[89] *Ibid.*

[90] Global Adaptation and Resilience Investment Working Group.

[91] United Nations Global Compact (2020).

[92] United Nations Framework Convention on Climate Change.

investor in the climate change international legal framework—it is not subject to obligations under the BIT.

Furthermore, the absence of investors' climate change obligations in both systems does not mean that it is not burdened with the duty to contribute to the worldwide goal of levelling global warming. Investment treaty arbitration—via compensation—seems to be a suitable scenario for the effectiveness of the duty to internalise part of the costs of climate change. This would not be strange to the system's operation. As I have shown above, the occasional moderation of compensation in investment treaty law disputes may involve the investor assuming part of the costs of public policies that embrace different interests—e.g., agricultural, fiscal, economic, exhaustible resources, and environmental interests.[93]

In this vein, the distinction between duty and obligation is relevant to the argument that the CBDRRC is applicable to compensation. Contending that neither of the systems imposes 'obligations' in the technical sense of the word, means that investors are not legally bound to perform a specific action in favour of the state or the global community under the text of the treaties and, accordingly, that they cannot be declared liable to compensate for the non-performance of an obligation that does not exist. In other words, if investors do not contribute to the levelling of global warming, they cannot be brought to international tribunals to be held responsible under investment treaty law or under the international climate change legal framework.

Contrarily, affirming that investors have the duty to contribute to levelling global warming entails that they are called to participate in the solution to the 'common concern' and, consequently, that apportioning some burden to them appears valid. Indeed, since the climate change international legal framework is not equipped with any adjudging authority to enhance its effectiveness, investment treaty arbitration provides one of the very few chances to transfer some of the costs of global warming to the investor. As described above, this is possible since the determination of compensation in investment treaty law has served as a channel to distribute the costs of global public policies between the investor and the host state through the reduction compensation.

Bearing that in mind, it is correct to affirm that the personhood of the investor under investment treaty law and its role as claimant, are crucial for the effectiveness of the investor's duty to assume the costs of climate regulation and the application of the CBDRRC. The determination of compensation provides an opportunity to burden the investor with part of the costs of climate change.

Paradoxically, the similarities and differences in the role of the investor when comparing both systems—international climate change law and investment treaty law—suggest the application of the CBDRRC to compensation. While the similar instrumental status of the investor in both systems is one of the connections that legitimises the application of the CBDRRC principle in investment treaty law, the difference in how the investor intervenes in the arbitration proceeding contrary to the

93 *Supra* Sect. 10.1.2.

absence of a special proceeding for the climate change international legal framework, reinforces the idea of its application to enhance the effectiveness of the principle.

10.2.1.2 The Cooperative-Liability Nature of the Investment Treaty Law and Climate Change Law

The second point of connection is the analogous cooperative-liability nature of the investment treaty law and climate change law. Indeed, neither is investment treaty law only a system of international civil liability for the protection of the investor, nor is the climate change international legal framework is exclusively a cooperation mechanism. While the former is also a cooperation mechanism for the promotion of transnational investment, the latter is also impregnated with the idea of attributing to developed countries the responsibility for the causation of global warming and the correlative duty to compensate.

Again—on the cooperative nature of the investment treaty law—it is important to reiterate the instrumentality of the investor to the system. The compensation to the investor is only central as an inherent condition for the effectiveness of the cooperation in promoting the flow of transnational investment. The cooperative side of the investment treaty law highlights the common interest of the parties to dynamise and grow their economies through bidirectional investment, leading to a win-win situation of mutual collaboration. If the compensation imposed for taking a public interest measure in a particular case is so expensive that it jeopardises the country's stability, the collaborative relationship between the parties is broken, as it would no longer be a win-win situation when considering the big picture. Thus, the intention of the parties to a BIT is clearly not to sacrifice the stability of their economies for compensating the investor's interest or for promoting the global interest of transnational investment. The cooperative nature of the investment treaty law justifies a differential approach in the marginal cases where the application of the full compensation principle would imply a ruinous burden to host states for taking public interest measures.

The cooperative nature of the investment treaty law and the primary importance of the host state's welfare are explicit in the recitals and the objectives of many BITs. The NAFTA's preamble, for example, not only declares that the treaty intends to 'strengthen the special bonds of friendship and cooperation among their nations'[94] but also to 'preserve their flexibility to safeguard the public welfare'.[95] The preamble also highlights the protection of non-economic values such as sustainable development and the environment.[96] Likewise, the ECT expresses in the preamble that the treaty intends to 'implement and broaden their co-operation' and in Article 2 that the

[94] North American Free Trade Agreement (NAFTA) (1994), Preamble.
[95] *Ibid.*
[96] *Ibid.*

purpose is to 'promote long-term cooperation in the energy field, based on complementarities and mutual benefits'.[97] Furthermore, the preamble recognises the 'increasingly urgent need for measures to protect the environment'[98] and underlines the importance of international treaties that represent non-economic interests, such as those embedded in the treaties on the Proliferation of Nuclear Weapons; air pollution; the environment; and, more importantly, for this work, the United Nations Framework Convention on Climate Change.[99]

Apart from recognising economic cooperation within a frame of sustainable development, in its preamble, the 'PACER plus'[100] goes beyond the ECT and the NAFTA by incorporating an explicit reference to the importance of differentiation. Thus, the treaty stresses:

> Taking into account the significant differences in the size of the Parties; the unique and particular vulnerabilities and development challenges of developing country and least-developed country Parties in the region; the need to build their capacities with a view to fostering their enhanced participation in international trade and investment; and that many of the Parties are not Members of the World Trade Organization[101]

The differentiation approach is also tangible in the TPP, which acknowledges 'the different levels of development and diversity of economies'[102] as well as 'the flexibility of the parties to set legislative and regulatory priorities, safeguard public welfare, and protect legitimate welfare objectives'.[103] Within these priorities, the text provides a non-exhaustive list of public interests that includes health, safety, environment, conservation of natural resources, financial stability and public

---

[97] Energy Charter Treaty (ECT) (1994), Preamble.

[98] *Ibid.*

[99] *Ibid.*

[100] 'Mindful of the links between economic development, social development and environmental protection and the important role of development and economic cooperation in promoting sustainable development, the Parties resolve that PACER Plus must be a development tool for the Forum Island Countries to, *inter alia*, increase their production capacity and exports; support their economic and trade diversification, and competitiveness efforts; attract investment to them; and lead to the creation of jobs; (...)'. Pacific Agreement on Closer Economic Relations (PACER) (2017), preamble. A similar approach: Reciprocal Investment Promotion and Protection Agreement Between The Government of the Kingdom of Morocco And The Government of the Federal Republic of Nigeria (2016); Investment Promotion and Protection Agreement between The Government of the Federal Republic of Nigeria and The Government Of The Republic Of Singapore (2016); Agreement for The Promotion And Protection Of Investment between The Government of the Republic Of Austria and The Government Of The Kyrgyz Republic (2016); Agreement between the Government of Canada and the Government of the Hong Kong Special Administrative Region of the People's Republic of China for the Promotion and Protection of Investments Canada Hong (2016); Agreement between the Slovak Republic and the Islamic Republic Of Iran for The Promotion And Reciprocal Protection of Investments (2016).

[101] PACER (2017) Preamble.

[102] Trans-Pacific Partnership (2018) Preamble.

[103] *Ibid.*

morals.[104] The same differentiation approach[105] as well as the host state's flexibility to define the regulatory priorities[106] is found in the preambular text of several BITs.

Just like the international civil liability mechanism of the investment treaty law is influenced by the cooperative nature of BITs, the cooperative mechanism of the climate change international legal framework and the CBDRRC principle are influenced by international civil liability considerations. The differential approach of the CBDRRC is not plainly grounded on distributive justice (the wealthy pays more) but also on the narrative that developed countries should bear the costs climate change because global warming could be attributed to them (the polluter pays).[107] The preamble of the UNFCCC is stark, thus:

> Noting that the largest share of historical and current global emissions of greenhouse gases has originated in developed countries (...) [and] [r]ecognizing also the need for developed countries to take immediate action in a flexible manner on the basis of clear priorities (...) with due consideration of their relative contributions to the enhancement of the greenhouse effect (...)[108]

Providing that developed countries should take action in proportion to their contribution to climate change equates in liability terms to imposing a duty to repair the prejudice to the global climate interest. Evidently, the possibility to reverse climate change and its effects as if the damage has never occurred is an impossible task, but bearing the costs to remediate and adapt to climate change as well as the channeling of funds from developed countries to developing countries could certainly be perceived as reparations.

---

[104] *Ibid.*

[105] For example, Agreement on Trade in Goods under The Framework Agreement on Comprehensive Economic Cooperation between The Republic Of India and The Association Of Southeast Asian Nations (2009).

[106] For example, Comprehensive Economic and Trade Agreement (CETA) between Canada, of the one Part, and The European Union, of the other part (2014).

[107] The debate on which type of 'justice' represents the country's direct responsibility proportional to its emissions remains open (Distributive Justice, retributive justice or compensatory justice?) however there seems to be an agreement that it constitutes a liability mechanism: 'Notably the UNFCCC and the Kyoto Protocol hold that developed countries are principally responsible for the current high level of GHGs in the atmosphere because it is they that allowed the emission of those gases over more than 150 years of industrial activity. Both the UNFCCC and the Protocol place on them a burden of mitigating those emissions. They do so on the principle of 'common but differentiated responsibilities'. On this position, the denotation of 'responsibility' is retributive. That is developed countries are being 'made to pay' for what 'they have done': Maniruzzaman and Al-Tayer (2011), p. 651. 'Different principles could come into play to address these issues. Some of them relate to theoretical notions of distributive justice, such as causal and moral responsibility. The former refers to the responsibility for contributing to climate change via emissions of GHGs, the latter to the responsibility for solving the problem – noting that these two aspects are not always connected. Other principles invoke compensatory justice, such as the polluter-pays principle, the community-pays principle or the beneficiary-pays principle; and a third set involves procedural justice based on the way in which outcomes are brought about': 'Dynamic Differentiation': Voigt and Ferreira (2016), p. 288.

[108] United Nations Framework Convention on Climate Change (1992) Preamble.

Even if the Paris Agreement is less clear than the UNFCCC in pointing the finger at developed countries as responsible parties that are obligated to repair the damages related to the global climate interest, political discussions between developed and developing countries throughout the development of the climate change international legal framework concern to this issue.[109] The background of any climate change protocol or agreement is set with developing countries contending that developed countries have the historical debt of repairing what they have done to climate.[110] This background reiterates that the cooperative character of the CBDRRC is not devoid of reparation features that are typical of the reasoning of international liability.

Having clarified that the topical boundaries between the international climate change law and investment treaty law are not so pronounced when they regard to the liability-cooperation dichotomy, and that the connection between the systems not only legitimises the application of the CBDRRC in investment treaty climate change but also enhances its effectiveness through the possibility of transferring some of the costs of climate change to the investor, the next section proposes deploying international civil liability devices to facilitate the translation of the principle to investment treaty law and its limits.

10.2.2 Translation of the CBDRRC Principle to Investment Treaty Law Through International Civil Liability Devices and Dynamic Analysis of the Conditions of the Parties

Although the evidence that differentiation is within the operation of the investment treaty law and the topical coincidence (liability-cooperation) between systems should be enough to legitimise the exceptional-direct application of the CBDRRC to investment treaty-climate change disputes, there are international civil liability rules which facilitate the operationalisation of the principle. These rules of liability could be understood as expressions of the CBDRRC that intervene exceptionally in investment-climate change disputes to alter the general principle of full compensation.

---

[109] 'As a political reality, however, developing countries are reluctant to pay a high price or undermine their economic growth in other to combat climate change; they expect any actions they take to be heavily subsidized or compensated. For many, this is a question of fairness: The industrialized countries are historically responsible for a large majority of greenhouse gas emissions and has much higher emissions per capita on average, and yet it is less developed nations that will be harmed the most by climate change. In this context that North-South finance has emerged as such a critical issue. Any grand bargain on climate change will hinge on an agreement over specific obligations and mechanisms to promote resource flows from more to less developed countries': Thompson (2016), pp. 138–139.

[110] *Ibid.*, pp. 138–160.

The principle of equity, the polluter should pay, and the Article 27 DASR could be viewed as instruments that are familiar to the practice of international responsibility that consider the special conditions of the parties to assess compensation (Sect. 10.2.2.1). However, the use of these instruments needs to be justified through a dynamic approach to the CBDRRC principle that grasps differentiation stemming from the particular situation of the parties at the time of the dispute rather than relying on the static dichotomy of developed-developing countries (Sect. 10.2.2.2).

10.2.2.1 Application of International Civil Liability Provisions that Represent the CBDRRC Principle to Investment Treaty Law: The Principle of Equity, the Polluter Should Pay and the Article 27 DASR

The devices that facilitate the operationalisation of the CBDRRC are nothing but rules that are common in the practice of international civil liability. The purpose of presenting these devices is not to deny the direct application of the CBDRRC but to translate the principle into other rules that are more familiar to the practice of international investment law. This translation, then, should enhance the effectiveness of the principle's application in investment treaty-climate change disputes.

While the principle of equity—as shown above[111]—has been deployed by investment treaty tribunals, and the Article 27 DASR—as part of the DASR—belongs to the investment treaty law system; the *polluter should pay* has not been deployed by the investment treaty tribunals. This outsideness should not result in inhibiting its application, not only because the system exceptionally admits differentiation in the determination of the compensation by reference to the conditions of the parties, but also because investment treaty law is a system of international civil liability.

The principle of equity and the CBDRRC are reciprocally dependent by mutually defining each other in investment treaty-climate change disputes. While—as described above[112]—equitable considerations legitimise the application of the CBDRRC as the source for *differentiation* in the compensation, the application of the equity principle in investment treaty-climate change disputes determines its content by reference to the CBDRRC. Consequently, the application of differentiation in the compensation as a result of the deployment of the principle of equity in climate change-related disputes should obey—in principle—to the division between developed countries, developing countries, and least developed countries.

The principle of equity—under the key of the CBDRRC—is determinant for the effectiveness of the other rules that I propose to apply in investment treaty-climate change disputes. In a certain way the *polluter should pay* and Article 27 DASR would not apply to a particular case if not based on equity reasons. In other words,

[111] *Supra* Sect. 10.1.2.

[112] *Ibid.*

the notion of equity justifies the choice of applying these rules over the general principle of full compensation. On the other hand, these rules are more concrete than the equity principle and simplify its application to the particular cases. Also, a decision sourced on more concrete rules rather than the general principle of equity gives the impression of being more systematic.

The Application of the Principle of Equity

The cases referred to above on the nationalisation of the oil industry and civil unrest represent the influence of the special vulnerability of host states in the reduction of compensation.[113] This consideration is equally applicable to the cases where climate change measures are at the centre of an investment treaty dispute. In fact, the notion of equity under the key of the CBDRRC compels arbitrators to consider that the position of the parties before climate change may suggest a differential approach to compensation.

Importantly, the CBDRRC principle is an expression of the concept of climate justice, which in turn, evokes the principle of equity. Not in vain does the Paris Agreement highlight 'the importance for some of the concept of 'climate justice', when taking action to address climate change'.[114] This concept focuses on the inequalities between developed and developing countries in their contribution to and vulnerability against climate change as well as calls for positive discrimination measures in favour of developing countries.[115]

It is commonly accepted that, while developed countries have achieved considerable industrialisation thanks to the burning of fossil fuels and, coincidentally, they are less vulnerable to the effects of global warming in comparison to developing and least developed countries—in proportion to developed countries—have not contributed materially to climate change. They have only incidentally benefited from emission activities and they are more vulnerable to the consequences of global warming. Moreover, developing and least developed countries are not in the capacity to assume the costs of levelling global warming to the same degree as developed countries.

---

[113] *Ibid.*

[114] United Nations Paris Agreement (2015) Preamble.

[115] '[...] the moral dimensions of climate change cannot be ignored—in particular as those nations and individuals who benefit the least from industrialization and development are often the most affected by the adverse impacts of climate change. Beyond the scope of positive rules, here exist general principles underpinning international law, such as the principle of responsibility and some principles governing remedial obligations, from which legitimate expectations arise as the outcome of political negotiations, and which therefore should not arbitrarily be disregarded when responses to new issues such as climate change are being imagined. Only if it appears fair to most peoples around the globe can a global climate agreement trigger the costly measures necessary to mitigate climate change. International law, despite all of its flaws and biases, is certainly a strong reflection of broadly accepted moral principles': Mayer (2017), p. 186.

The imbalance between developed and developing countries concerning the contribution to and vulnerability before climate change as well as the incapacity of developing countries to afford the costs of climate change justifies a differential treatment. The imposition of full compensation in investment treaty-climate change disputes, disregarding this inequality would go against the principle of equity.

The application of the principle of equity to investment treaty-climate change disputes under the key of the CBDRRC could manifest itself in two ways. On the one hand, when a developing or least developed country is declared liable for taking measures against global warming—such as phasing out coal industries or fighting deforestation—the compensation could be reduced or even disappear. On the other hand, when a developed country has committed a breach by taking a measure against the climate interest—such as cutting RESSs—but it was justified by the international defence of necessity, compensation or partial compensation may still be possible. These solutions not only weight the conditions of the host state before climate change, but also encourage the protection and promotion of the climate interest as a common concern of humankind, and they grant the investor a share of the costs of global warming.

The effects of the application of the equitable principle—as an expression of the CBDRRC—in the determination of the compensation could be further concretised through rules of liability. The *polluter should pay*, which attributes part of the costs of climate change to the investors that contributed to global warming; and Article 27 DASR, which serves as a source to impose compensation to developed host states for taking measures that are against the climate interest, even if the wrongfulness of the breach has been excluded.

The Polluter Should Pay

To begin with the *polluter should pay* was conceived as an economic rule whereby the environmental costs of pollution for the public at large were internalised by private actors as part of their operative or production costs.[116] The rationale behind the rule at that time was to tackle pollution—without distorting the market with state aid—by encouraging polluters to optimise emissions instead of subsidising green

[116]Chemain (2010), pp. 877–879. 'The principle to be used for allocating costs of pollution prevention and control measures to encourage rational use of scarce environmental resources and to avoid distortions in international trade and investment is the so-called "Polluter-Pays Principle". This principle means that the polluter should bear the expenses of carrying out the above-mentioned measures decided by public authorities to ensure that the environment is in an acceptable state. In other words, the cost of these measures should be reflected in the cost of goods and services which cause pollution in production and/or consumption. Such measures should not be accompanied by subsidies that would create significant distortions in international trade and investment': OCDE, Recommendation of the Council on Guiding Principles concerning International Economic Aspects of Environmental Policies (1972).

industries.[117] However, soon after that, the *polluter should pay* principle was transmuted into environmental liability codifications[118] and even reached the status of guiding principle in the Rio Declaration:

> National authorities should endeavour to promote the internalization of environmental costs and the use of economic instruments, taking into account the approach that the polluter should, in principle, bear the costs of pollution, with due regard to the public interest and without distorting international trade and investment[119]

In addition to the Rio Declaration, several international conventions, [120] as well as national and regional legislations[121] have adopted the principle. Although the polluter should pay principle lacks tools for its implementation and the drafting techniques of its adoption are based on a 'should' rather than on a 'must', investment treaty-climate change disputes provide an opportunity to put the principle into action.[122]

If the *polluter should pay* is understood from a CBDRRC perspective, it may have a double-consequence in investment treaty-climate change disputes. The investor that produces high emissions may be obliged to internalise part of the costs of climate change mitigation measures, even if the host state has breached a BIT standard. Accordingly, developing or least developed countries may have the chance to obtain a reduction or cancellation of the compensation, even if they are responsible for breaching a BIT.

This proposal considers the specific conditions of the investor and the host state. On the one hand, the host state must be a developing country—acknowledging that the *polluter should pay* is an expression of the CBDRRC that makes it possible to

[117] Chemain (2010), pp. 877–879.

[118] *Ibid.*

[119] United Nations, *Rio Declaration on Environment and Development 1992* (1992) Principle 16.

[120] *See* Khan (2015), pp. 638–653.

[121] *Ibid.*

[122] Wagner envisaged this opportunity on expropriatory environmental regulation: 'Given the undeniable impact of human activities on environmental health, and the equally obvious relationship between environmental health and human welfare, it is clear that regulations to protect the environment are at least as legitimate as regulations to address any of these other governmental concerns. Furthermore, the nature of environmental protection indicates that environmental regulations should be accorded particular protection from compensation claims. Such regulations are an important method of internalizing environmental costs, which is an essential component of environmental protection and "a general principle of international law." As the OECD has explained, internalizing environmental costs depends in large part on requiring the *polluter* to "bear the expenses of carrying out [pollution prevention) measures decided by public authorities to ensure that the environment is in an acceptable state." Requiring the government to provide compensation for the economic impact of environmental regulations does exactly the opposite, shifting the cost of complying with environmental measures, and thereby the cost of environmental harm, onto the general public. Such a shift constitutes a significant obstacle to environmental protection. For these reasons, the polluter pays principle and the multitude of international agreements for the protection of the environment require that governments generally be free to implement environmental regulations without having to pay to do so': Wagner (1999), pp. 528–529.

lessen the application of the full compensation principle. On the other hand, the investor must be a high emitter, given that the *polluter should pay* attributes liability by reference to the duty of the investor to internalise (or repair) the costs that its emissions entail for the entire society.

Article 27 (b) of the DASR

Article 27 (b) DASR establishes that '(t)he invocation of a circumstance precluding wrongfulness (...) is without prejudice to (...) the question of compensation for any material loss caused by the act in question'.[123] This article appears to incur a systemic-logical contradiction, since the natural consequence of precluding liability—either by the intervention of necessity, distress or force majeure—is the cancelation or suspension of the obligation to compensate.[124] Yet, it seems that the intention of the drafters to leave open the alternative of imposing reparation in the cases of preclusion of wrongfulness was not accidental. Contrarily, it appears to conform to the aim of providing some flexibility under exceptional circumstances that may justify a compensation, even if the breach is not wrongful. According to the commentaries to the DASR:

> [...][A]rticle 27 is a reservation as to questions of possible compensation for damage in cases covered by chapter V [circumstances for preclusion of wrongfulness] [...] the article [...] is concerned with the question whether a State relying on a circumstance precluding wrongfulness should nonetheless be expected to make good any material loss suffered by any State directly affected [...]
>
> Subparagraph (b) does not attempt to specify in what circumstances compensation should be payable. Generally, the range of possible situations covered by chapter V [circumstances for preclusion of wrongfulness] is such that to lay down a detailed regime for compensation is not appropriate.[125]

The commentaries to the DASR not only acknowledge the possibility to exceptionally grant compensation when wrongfulness is excluded but also to expressly abstain from establishing which would be the conditions of these exceptional circumstances.

The issue on the imposition of compensation under Article 27 (b) has not been ignored by investment treaty tribunals. Generally, one could say that case law in this respect recognises that the imposition of compensation is facultative to the tribunals rather than mandatory. In *LG&E v. Argentina*—one of the cases that concerns the

---

[123] International law Commission (2001) Art. 27 (b).
[124] Forteau (2010), p. 887.
[125] International law Commission (2001), p. 86.

Argentinian 2000-2001 economic crisis—the tribunal verified the state of necessity[126] and expressly discarded the application of Article 27 (b).[127] In its perspective, the article does not provide that a compensation must be paid to the investor and, moreover, decided that under the circumstances of the case 'the damages suffered during the state of necessity should be borne by the investor'.[128] A similar conclusion could be found in *EDF v. Argentina,* where the tribunal affirmed that 'the successful invocation of the necessity defense does not per se preclude payment of compensation'. [129]

Conversely, in *CMS v. Argentina*—also on the Argentinian crisis—the tribunal considered that it was obligatory to impose compensation in any a case. [130]Yet, this conclusion was rejected by the annulation committee of this decision because

> Article 27 itself is a "without prejudice" clause, not a stipulation. It refers to "the question of compensation" and does not attempt to specify in which circumstances compensation could be due, notwithstanding the state of necessity [131]

Given the flexibility to impose compensation in the case of preclusion of wrongfulness, I argue that the special conditions of developed countries before climate change, legitimise the compensation for material losses in investment treaty-climate change disputes. This is a solution that seems to adapt to the higher proportion of costs of climate change that developed countries should bear according to the CBDRRC principle and equity. Additionally, this compensation for material losses should only be possible when such measures have the effect of lessening the climate interest, such as ruling out RESSs. A compensation as such should discourage developed countries from taking measures against the goal of levelling global warming.

In brief, if the polluter should pay principle allows developing host states to transfer part of the costs of climate change to the investor that produces high emissions, Article 27 (b) serves as an instrument to allocate to developed host states the burden of compensating investors in the green industries that have been affected by measures that—although justified by the circumstances precluding

[126] *LG&E Energy Corp., LG&E Capital Corp., and LG&E International, Inc. v. Argentine Republic,* Decision of Liability (2007), para. 259.

[127] *Ibid.*, paras. 260, 264.

[128] *Ibid.*, para. 264.

[129] *EDF International S.A., SAUR International S.A. and León Participaciones Argentinas S.A. v. Argentine Republic (2012)*, para. 1177.

[130] *CMS (*2015), paras. 383–384. This ruling based its position on several rulings of the PCJ: 'this conclusion finds support in the Gabcikovo-Nagymaros case, where the Court noted that "Hungary expressly acknowledged that, in any event, such a state of necessity would not exempt it from its duty to compensate its partner. [...]" This criterion was also the basis for the decisions in earlier cases, such as the Compagnie Générale de l'Orinocco case and the Properties of the *Bulgarian Minorities* in Greece case invoked by the Claimant, or the Orr & Laubenheimer case'.

[131] *CMS Gas Transmission Company v. The Republic of Argentina,* Decision of the ad hoc Committee on the Application for Annulment of the Argentine Republic (2007), para. 147.

wrongfulness—run counter to the climate interest.[132] This proposal intends to reflect the CBDRRC principle by considering the dichotomy of developed-developing countries. However, this approach is not complete if this categorisation is not taken with some reserves for the purposes of compensation. The CBDRRC principle compels a dynamic analysis of the distinction. This is explained in the he next section.

10.2.2.2 Dynamic Approach to the CBDRRC Principle in Investment Treaty-Climate Change Disputes: Differentiation Stemming from the Particular Circumstances of the Host State at the Time of the Award

The Paris Agreement establishes, in the third recital of the preamble, that it is guided by 'the principle of equity and common but differentiated responsibilities and respective capabilities'[133] and complements these principles with the expression 'in the light of different national circumstances'.[134] From there on, this last sentence is included each time that the CBDRRC principle is—either explicitly or implicitly—invoked in the text of the convention.[135] This addition is not casual; it stems from the tension between developed and developing countries in reaching a consensus on the static or dynamic notion of differentiation.[136]

In fact, the text of the 1992 Framework Convention on Climate Change (FCCC) did not mention the 'changing circumstances of the parties' or any other analogical expression suggesting the adoption of a dynamic approach to differentiation. Therefore, not only did the convention allocate the biggest portion of the costs of climate change to developed countries,[137] it also disregarded the relevance of the potential natural, social, and economic vicissitudes that both groups—developed and developing countries—could experiment.[138]

---

[132] 'Art. 27 could even play a role in making necessity more appropriate when reconciling foreign investors right and environmental protection. Indeed, because its function is different, this kind of compensation should be assessed according to parameters different to those typical ILL. This would involve, in particular, avoiding the criterion of full compensation which turns on the notion of fair market value. In this way, ILC 27, far from preventing the concrete application of necessity in "environmental disputes" before investment tribunals, would lead to a balanced result by means of discerning and prudent application of criteria to assess compensation': Di Benedetto (2013), p. 175.

[133] United Nations Paris Agreement (2015) Preamble.

[134] *Ibid.*

[135] *Ibid.*, Arts. 2 (2), 4(3), 4 (4), 4(6), 4 (19), 13 (3), 13 (12), 15 (2).

[136] *See* Voigt and Ferreira (2016), p. 291.

[137] 'The 1992 Framework Convention on Climate Change is unabashedly favourable to developing countries from the recognition in its preamble that the share of emissions from developing countries will grow to meet their social and development needs to annex-based differentiation that exempts developing countries from emission reduction obligations': Rajamani (2016), pp. 493, 506.

[138] 'As a growing number of developing countries become industrialized and increase their pressure on the global environment along with their capabilities, the expectation grows that they also assume

Although the 1997 Kyoto Protocol includes in its Article 10, a reference to 'regional development priorities, objectives and circumstances'[139]—in addition to the CBDRRC—and refers to the circumstances of the parties[140] in various provisions, its design clearly results in a *de facto* regime which only imposes concrete obligations to developed countries. While only the countries listed in Annexes A and B were committed to pursue and achieve targets of limiting and reducing emissions of greenhouse gases,[141] the great majority of developing or least developed countries[142] were only engaged by 'good intentions'. In the end, the Kyoto Protocol replicated the static approach of the FCCC.

The negotiations that preceded the 2015 Paris Agreement were characterised by two opposing positions. On the one hand, the block of developing countries intended to keep the static structure of the FCCC and defended the position that the Paris Agreement had to be understood under its authority.[143] On the other hand, developed countries defended the position that only the objective of the FCCC—the levelling of global warming—had to be understood as part of the Paris Agreement and differentiation should reflect the economic reality of the countries.[144] The final synthesis of this pulse between developed and developing countries was the adoption of the ambiguous text of Article 2[145] and the addition of the expression 'in the light of different national circumstances'[146] to the CBDRRC principle,. This infused the content of the CBDRRC principle with the changing circumstances, as a response to the claim of developed countries to modify the static dichotomy of developed-developing countries.

greater responsibilities in international environmental law. The question is thus how to design legal instruments that can reflect the different "situations" of states in an equitable and dynamic fashion, as they develop overtime. The experience from 1992 to 2016 tells us that such differentiation cannot be static; it needs to allow the structure and content of international agreements to evolve dynamically': Voigt and Ferreira (2016), p. 287.

[139]Kyoto Protocol to the United Nations Framework Convention on Climate Change (1997) Art. 10.

[140]*Ibid.*, Art. 2 (1) (a), 2 (4), Art. 13 (4) (c), Art. 13 (4) (d).

[141]*Ibid.*, Annex A, Annex B.

[142]Countries that were 'undergoing the process of transition to a market economy' in Europe also committed to achieve targets, even if their condition as developed countries was very doubtful. *Ibid.*, Art. 3.5, Annex A, Annex B.

[143]*See* Rajamani (2016), p. 494.

[144]*Ibid.*

[145]'The phrase "in enhancing the implementation of the Convention, including its objective" could be read as an acceptance of the developing country position that this Agreement will enhance the implementation of the Convention. This would permit the entirety of the Convention to be engaged in interpreting the provisions of the Paris Agreement. The phrase could also be read as a statement of fact that the Agreement, such as it is, enhances the implementation of the Convention. Thus limiting, by implication, the engagement of the entirety of the Convention in interpreting the Paris Agreement. In this reading, the Convention is engaged only in so far as it is specifically referred to in the Paris Agreement.' *See* Rajamani (2016), p. 507.

[146]United Nations Paris Agreement (2015).

The switch from a static to a dynamic approach to the CBDRRC principle has an impact on my proposal of applying this principle to the compensation in investment treaty-climate change disputes. The rules for the internalisation of the costs of climate change in the compensation that stems from the bifurcation—developed-developing countries—need to operate in harmony with the social-economic circumstances that the host state experiences at the time of the award.

Consequently, the application of the polluter should pay principle would legitimise a reduction in (or the cancelation of) the compensation under two conditions. First, that the hypothetical imposition of the full compensation would risk the developing host state's economic or social stability; and second, that the developing country has not increased its emissions in a disproportionate manner. While the first condition acknowledges that under a dynamic approach the economic conditions of developing countries could improve, the second condition recognises that the responsibilities of a developing country could become broader as a result of increasing its contribution to climate change.

Analogously, Article 27 (b) DASR would not be applicable as a source for granting compensation if compensation jeopardises the economic or social stability of the developed host state. This acknowledges that, under a dynamic approach, the economic conditions of developed countries are relevant when determining the burden that they should bear for the objective levelling of global warming. Furthermore, the imposition of compensation under Article 27 could be rejected on the basis that the developed country has made a considerable effort in reducing emissions. In this case, the non-application of Article 27 (b) obeys a dynamic approach in the sense that a reduction in the contribution to climate change could lead to the shrinking of the host state's responsibilities.

Moreover, the CBDRRC, understood as an expression of the principle of equity, suggests that the adoption of a static approach could be inappropriate. One of the most important roles of equity in law consists precisely of providing some flexibility to rules that may appear inadequate in particular circumstances. In this vein, conceiving that the condition of developed or developing countries determines the exceptional reduction or imposition of compensation in investment treaty-climate change disputes, would be contradictory to equity, if it disregards the dynamic of the changing economic and social circumstances of host states.

The operationalisation of the CBDRRC in investment treaty-climate change disputes through the principle of equity, *the polluter should pay,* and the Article 27 DASR enhances the harmonisation of the investment treaty law system with the climate change international legal framework. The reduction or imposition of the compensation stemming from the dichotomy of developed-developing country needs to be determined without ignoring the circumstances that the host state is experiencing at the time of the award. As a consequence, the polluter should pay principle would legitimise a reduction in the compensation as long as the application of the full compensation principle would threaten the developing host state's stability and the developing host state has not increased emissions considerably. Likewise, Article 27 DASR would be applicable as long as the developed host state's stability is not at odds for the imposition of compensation and the country has not considerably reduced emissions. This dynamic approach provides the CBDRRC with the flexibility that the principle of equity demands.

10.3 Concluding Remarks on the Direct Application of the CBDRRC to Compensation: Inadequacy of a *Global Public* (Environmental/Climate) Interest Defence and the Systematicity of a Differential Approach

In this chapter, I proposed to challenge the full compensation principle in investment treaty-climate change disputes by envisaging the reduction or the imposition of the compensation in proportion to the conditions of vulnerability, responsibility, and capability of the host state before climate change. In short, I proposed the exceptional application of the CBDRRC to compensation.

To demonstrate that the reduction or imposition of compensation based on the CBDRRC principle suits the operation investment treaty law, I revealed that the principle of full compensation accepts restrictions stemming from retributory and distributive justice considerations (i).

In this respect, I explained that the compensation for lawful expropriation was lesser than the compensation for unlawful expropriation. While the compensation for lawful expropriation according to the (ordinary) standard equals to the market value, unlawful expropriation is assessed by reference to the full compensation principle (ii).

Additionally, I unearthed how some secondary law adopted by tribunals to assess liability, leads to a compensation that is measured by reference to elements that are different from the magnitude of the prejudice to the investor (iii). While the examination of decisions on compensation in the events of contributory negligence and moral damages revealed traces of retributive justice (iv), it was found that distributive justice influences compensation in the cases where tribunals determine that the investor should bear part of the country/regulatory risk, or where the principle of equity suggests that the circumstances of the case requires the exclusion of the application of the full compensation principle (v). Furthermore, distributive justice considerations that take into account the particular conditions of the host state were relevant in the cases of oil nationalisations—*LIAMCO*—and social unrest—*American Manufacturing v. Zaire*—as application of the principle of equity (vi).

After demonstrating that distributive justice elements and the principle of equity could limit the principle of full compensation and, thus, that the CBDRRC could be conceived within the investment treaty law system, I assumed the challenge of envisaging the operationalisation of the CBDRRC in investment treaty law. Thus, I first argued that the fact that the CBDRRC principle is generally conceived as a rule of cooperation and that the investment treaty law is regarded as a system of liability does not entail a topical problem for its application (vii). And second, I showed that the operationalisation of the principle would benefit from the translation of its terms into rules that are typical of international civil liability (viii).

On the alleged difficulties for filtering a cooperation principle that belongs to international climate change law into investment treaty law, I highlighted that the two systems coincide in both the instrumental role that the investor represents in them and in their common bidimensional liability-cooperative character.

In fact, I posited that in both systems investors are relevant actors, that in both systems investors are not burdened with direct obligations and that in both systems the role of investors is of secondary importance to the rationale—either the flow of transnational investment or the stabilisation of global warming (ix). Moreover, I mentioned that the relationship between the systems was complementary because investment treaty law is one of the few channels through which the CBDRRC principle can be effective and may facilitate transferring part of the costs of climate change to investors (x).

On the topical coincidence on the bidimensional liability-cooperative nature of both systems, it was demonstrated that, although investment treaty law is a system of liability, BITs are adopted for the purpose of cooperating in the flow of investment (xi). Thus, a full compensation that destabilises the host state's situation and ignores its particular conditions may go against the system; it appears to break the equilibrium that is expected from a cooperation relationship. Analogously, I found that although the climate change international legal framework is a cooperative mechanism, the CBDRRC is partly based on international liability considerations (xii), since the responsibility of the state under the *CBDRRC* is proportional to its contribution to climate change. Therefore, the application of the CDBRRC as a catalyser for exceptionally reducing or imposing compensation has no inconvenience for operating in a liability system, such as investment treaty law.

The evidence that the system admits exceptions to the full compensation principle grounded on distributive justice and that the topical coincidence between systems should be enough to justify the application of the CBDRRC in investment treaty law climate change-related disputes. However, I mentioned some international liability devices that could enhance the effectiveness of the principle in investment treaty-climate change disputes: the principle of equity, the *polluter should pay* and Article 27 (b) DASR.

While the principle of equity defines the effectiveness of the CBDRRC—since the exceptional application of the former is accepted in the practice—the CBDRRC shapes the content of the equity principle in investment treaty-climate change disputes with special regard to the concept of climate justice (xiii). Thus, it was argued that the application of the principle of equity in these disputes incarnates the differential treatment between developed and developing countries when determining compensation. In this vein, the application of the principle of equity—under the influence of the CBDRRC—might result in the exceptional reduction of compensation grounded on the host state's condition as developing country, as well as in the imposition of compensation based on the host state condition as developed country (xiv).

The other two devices that facilitate the application of the CBDRRC to investment treaty law could be regarded an expression of the principle of equity in the context of climate change. However, its application is still convenient since they are rules that are more concrete than the principle of equity. While the *polluter should pay* makes it possible to transfer part of the costs of climate change from the developing host state to the investor that produces high emissions through the reduction or cancellation of compensation (xv), Article 27 (b) makes it possible to allocate—partly or fully—the burden of compensating investors in the green industries to developed host states, even if the wrongfulness of the breach was excluded (xvi).

However, I also mentioned that the differentiation in compensation based on the CBDRRC needs to be tempered through a dynamic approach, in accordance with the notion of the CBDRRC principle in the Paris Agreement. This means that the determination of compensation grounded on the bifurcation of developed-developing country cannot ignore the actual circumstances of the host state at the time of the award (xvii). Therefore, the *polluter should pay* should only reduce compensation in the event that the application of the full compensation principle would jeopardise the social or economic stability of the developing country and given that the host state has not considerably increased its contribution to climate change (xviii). Likewise, the application of Article 27 (b) as a source for imposing compensation should only be valid if the imposition of compensation would not threaten the developed country's social or economic stability and if the country has not considerably reduced its contribution to climate change (xix).

Although the considerations that I have made in this chapter to this point provide evidence that the application of the heterodox approach to the compensation harmonises the climate change international legal framework with the investment treaty law without transgressing the system's operation, it is worth presenting two arguments that lead me to reiterate the place that my approximation to compensation occupies in the whole body of this book. My first argument has to do with the inadequacy of Kulick's proposal of deploying a sort of environmental general defence—connected to the examination of the proportionality of the measure—as a means to modulate compensation (Sect. 10.3.1). My second argument is that the direct application of the CBDRRC over the full compensation principle overcomes the practical disadvantages of formal equality (Sect. 10.3.2).

10.3.1 *The Rejection of a Global Public (Environmental/Climate) Interest Defence to Modulate Compensation in Investment Treaty-Climate Change Disputes: Emphasising the Importance of Distinguishing Between the Proportionality of the Measure and the Proportionality of the Compensation*

Kulick proposes reducing compensation as a result of the success of a global public (environmental/climate) general defence, that stems from testing the proportionality of the public interest measure against the private interest of the investor.[147] His proposal to modulate compensation to reach a middle point between the investor's interests and the global public (environmental/climate) interest follows a twofold analysis. In the first level of the analysis, he depicts a three-step method for weighting the global public interest against the investor's interests: 'verifying the scope of the investor's right', the 'infringement of the standard' and the 'proportionality *stricto sensu*'.[148] While *verifying the scope of the investor's right* concerns the definition of whether the investment is covered under the BIT,[149] the determination of the *infringement of the standard* examines whether the host state's measure conforms to the standard.[150] If the investor's right is within the *scope of the investment* and if the standard was *infringed* by the host state, then the second level of the analysis enters into action: proportionality *stricto sensu*.[151]

The analysis of the proportionality *stricto sensu* in Kulick's theory is not much different from the proportionality analysis that is generally proposed by different scholars for verifying the breach of the standards, when appropriate.[152] In other words, the examination is aimed at finding whether the measure was taken within limits that constitute 'safeguards against the abuse by the host state'.[153] In his view, the elements that determine the proportionality of the measure *stricto sensu* are '(1) necessity; (2) gravity of the infringement; (3) legitimate expectations; and (4) seriousness of pursuing the global public interest'.[154]

However, unlike the mainstream approach to proportionality in investment treaty law,[155] Kulick postulates that the determination of proportionality should be

---

[147] Kulick (2012), p. 209.
[148] *Ibid.*, pp. 195–197.
[149] *Ibid.*
[150] *Ibid.*
[151] *Ibid.*
[152] *Supra* Chap. 7, Sect. 7.3.1.
[153] Kulick (2012), p. 214.
[154] *Ibid.*
[155] *Supra* Chap. 7, Sect. 7.3.1.

undertaken after the infringement is verified[156] and he comprises within the balance exercise the 'analysis whether there exists a particular interest in reducing the amount of compensation [...]'.[157] He concludes that '[o]nly if the host State paying the full amount of compensation would result in undermining the Global Public Interest as such, will a reduction be justified'.[158]

Kulick is correct in positing that it is possible to overcome the 'all or nothing' paradigm (either full compensation or no compensation). Yet, his approach might be difficult to implement, not only for systematic reasons that characterise the operation of the investment treaty law, but also because the connection between the proportionality of the measure and the reduction of compensation that he proposes seems complicated to grasp.

First, it is extremely difficult to envisage how the proportionality *stricto sensu* of the measure could be assessed independently from the determination of the infringement of the standard. The relationship between the infringement and the analysis of proportionality is inseparable. While the finding of a disproportional measure indicates that there is a breach, a proportional measure indicates that the host state's behaviour conforms with the standard.[159] Indeed, most of the elements of Kulick's proportionality test appear to be directed precisely towards unravelling whether the measure of the host state infringes the standard: legitimate expectations, necessity, seriousness of pursuing the global public interest.

Admittedly, Kulick does not equate infringement to breach. In his perspective—inspired on German law—'infringement does not necessarily mean a violation'.[160] He adds that 'in a third step [proportionality *stricto sensu*], the judge must analyze whether there is any basis for justification of the infringement'.[161] Yet, this complicates the understanding of his theory even more, since difficulties unearthing whether the author is giving a role to *proportionality* in the determination of the violation or whether he refers to a general defence that is similar to the international defence of necessity. If the case is the latter, one would not understand, for example, the function of the investor's legitimate expectations in examining such a general defence, given that the legitimate expectations have a crucial role in the determination of the obligations and the conformity of the host state's behaviour with these obligations, i.e. the analysis of legitimate expectations can only take place in the analysis of the violation (breach).

Second, if the examination of the proportionality of the measure is interpreted either as an element of the violation (breach) or as an element of a general defence, there is no systematic explanation about why this should result in the reduction of the compensation. In the case that proportionality is interpreted as an element of the

[156] Kulick (2012), p. 196.

[157] *Ibid.*, p. 206.

[158] *Ibid.*

[159] *Supra* Sect. 7.1.2.

[160] Kulick (2012), p. 196.

[161] *Ibid.*

violation, the legal consequence of finding disproportionality would be the attribution of the violation (breach) to the host state and, accordingly, the legal consequence of the proportionality of the measure would be the absence of violation to the obligation of the host state. In other words, the proportionality of the measure would define whether the measure conforms to the obligation or not, but nothing more.

Likewise, when it comes to understanding the proportionality of the measure as an element of a general defence that is similar to the defence of necessity, the legal consequence of its verification would be the preclusion of the wrongfulness rather than the reduction of compensation. Thus, the proportionality of the measure, neither as an element for assessing the violation (breach) nor as an element for determining a general defence leads to the legal consequence of reducing compensation—contrary to Kulick's posture on the matter.[162]

Third, regardless of the layer where the proportionality test is placed in Kulick's approach—breach or general defence—the author does not provide any explanation on how the reduction of compensation is connected to the proportionality of the measure. The proposal to include a balance exercise within the determination of the proportionality of the measure that aims to establish 'whether there exists a particular interest in reducing the amount of compensation [...]'[163] has no connection to the assessment of the lawfulness of the measure at all.

While the determination of the proportionality of the measure is intended to establish whether the behaviour of the host state is allowed or forbidden, the analysis of whether 'the host State paying the full amount of compensation would result in undermining the Global Public Interest'[164] is related to the proportionality of the compensation. Briefly, Kulick's methodology of wrapping the proportionality of the compensation in the same box that covers the legitimacy of the measure results in contradiction. This suggests that the analysis of the proportionality of the compensation—whether the full compensation undermines the promotion of the *global public* (environmental/climate) *interest*—should be independent from the analysis of the proportionality of the measure—whether the host state's measure is legitimate even if it affects the investor.

It is worth highlighting that Kulick's approach has the virtues of unearthing that the global public (environmental/climate) interests could be affected by the application of the full compensation principle, and that the reduction of the compensation could be a solution that simultaneously recognises an infringement and promotes the global public interest—a kind of middle point between the investor's interests and the global community. My proposal of exceptionally adopting a differential

[162] 'What should be the actual consequence of introducing defense by the host state and balancing them with the investor rights according to proportionality analysis? The simple answer is: The—legitimate—defense might find reflection in the amount of compensation and damages': *Ibid.*, p. 209.

[163] *Ibid.*, p. 206.

[164] *Ibid.*

approach to compensation in investment treaty-climate change disputes embraces these virtues of Kulick's thinking and aligns them within the operation of the investment treaty law system, as I describe in the last section.

10.3.2 The Systematicity of the Exceptional Application of a Differential Approach to Compensation in Investment Treaty-Climate Change Disputes: The Proportionality of Compensation as Rationale and Counter Rationale

The proportionality of the compensation in investment treaty-climate change disputes stemming from a differential approach achieves the objective of harmonising the investment treaty law with the climate change international legal framework. This harmonisation is twofold. On the one hand, the direct application of the CBDRRC principle in investment treaty disputes is per se evidence of harmonisation. On the other hand, the reduction of compensation—in cases where the developing host state's measure aims to contribute to levelling global warming—or the imposition of compensation—in cases where the developed state's measure affects the investor in the green economy—encourages host states to adopt climate change measures and promote investments in climate friendly industries, respectively.

The harmonisation of the investment treaty law and international climate change law at this stage, constitutes the last resort to break the fragmentation between the systems. The direct application of the CBDRRC would not be necessary, if the interpretation of the text—of the FET or the expropriation standard in context—already provides a balance between the interest of the host state and the investor's interests that matches the global climate interest, as seen in Part II (private dimension).[165] Also, the direct application of the principle would be useless, if the success of the international defence of necessity legitimises the host state's measure to promote the climate, as shown in Chap. 9.[166]

Contrary to Kulick's approach, harmonisation through the application of the CBDRRC is achievable without incurring the confusion of squeezing the determination of the proportionality of the compensation into the analysis of the proportionality of the measure. The methodology I have adopted makes it possible to analyse the proportionality of the compensation with more freedom and without contradictions with the determination of the breach or its wrongfulness. Under this approach, the proportionality of the compensation is viewed not as an issue of BIT interpretation (orthodox approach) but as a matter of law ascertainment (heterodox approach).

---

[165] *Supra* Part II.

[166] *Supra* Chap. 9.

In fact, BITs generally do not contain provisions addressing the determination of compensation, except for establishing the market value as a standard of compensation for lawful expropriations. Therefore, the determination of compensation is an operation of law ascertainment rather than an operation of text interpretation, i.e. an operation of choosing the applicable rule to the dispute. The heterodox approach makes it possible to adopt a solution to fragmentation through the direct application of climate change rules when it is decided that the climate interest should prevail over other global interests.

Investment treaty tribunals commonly opt to apply the full compensation principle, but—as explained above—they also exceptionally apply rules of equity that represent retributive or distributive justice. In this context, the application of the CBDRRC as an expression of equity competes against the full compensation principle. I argued that the difference between developing and developed countries—in their capability, vulnerability, responsibility—before climate change, and in light of the circumstances, may justify the application of the differential approach to compensation.

However, the tension between the full compensation principle and the CBDRRC does not always represent a struggle between rational and counter rational considerations in investment treaty law. While the CBDRRC intervenes against the rationale of the investment treaty law in the cases where its application generates the reduction of the compensation, the CBDRRC conflates with the rationale of the investment treaty law in the cases where the result of its application is the imposition of a compensation.

Indeed, the flow of transnational investment might be affected by reduced compensation in investment treaty-climate change disputes because investors will see themselves lacking the full protection of their interests. In this case, the prevalence of the global climate interest over the global interest of economic growth is counter rational to the system. Contrarily, the flow of transnational investment might be fostered by the imposition of compensation in these disputes, because investors will find themselves protected even if the wrongfulness of the measure was precluded. In this case, the global climate interest is in line with the global interest of economic growth.

The paradoxical dynamic of the CBDRRC of simultaneously acting as rationale and counter rationale of the investment treaty law—depending on the case—also explains how the tension between global public interests is organised when a tribunal addresses compensation in investment treaty-climate change disputes.

When the host state takes a measure in favour of levelling the effects of global warming that affects investors, the tension would be between the global climate interest against the global economic growth interest. If in this conflict of interests the tribunal chooses to privilege the global climate interest against the rationale of the investment treaty law, then its decision should be to apply the CBDRRC principle and reduce compensation.

Contrarily, when the host state takes a measure against levelling the effects of global warming that affects investors in the green industry, the tension would be between the global climate interest and global economic growth interest, on the one hand, against the other global public interests such as sovereignty, social development, labor protection, human rights, etc., on the other hand. If in this conflict of interests the tribunal chooses to privilege the global climate interest and the flow of investment against the other global interests, then its decision should be to apply the CBDRRC and impose compensation in proportion to the circumstances, even if the wrongfulness of the measure has been precluded.

Although the application of the CBDRRC to compensation inclines the balance towards the global climate interest, it should be noted that the principle—moderately—reaches a middle point in the tension between global interests in investment treaty-climate change disputes. On the one hand, it provides the opportunity to reduce compensation to promote the global climate interest rather than cancelling compensation, thus the flow of investment will still be promoted to a certain extent. On the other hand, it provides the opportunity to impose compensation in proportion to the circumstances of the parties to promote the global climate interest rather than imposing full compensation, so that other global interests are not completely undermined.

This middle point between global public interests is certainly an advantage of overcoming the 'all or nothing' approach inherent to the full compensation principle and embracing instead a compensation proportional to the conditions of the parties before climate change.

References

Chemain R (2010) The 'Polluter Pays' Principle. In: Crawford J et al (eds) The law of international responsibility. OUP, pp 877–879

Di Benedetto S (2013) International investment law and the environment. Edward Elgar, Cheltenham, p 175

Forteau M (2010) Reparation in the event of a circumstance precluding wrongfulness. In: Crawford J et al (eds) The law of international responsibility. OUP, p 887

Khan M (2015) Polluter-Pays-Principle: the cardinal instrument for addressing climate change. Laws, pp. 638–653. https://www.mdpi.com/2075-471X/4/3/638. Accessed 03 August 2022

Kriebaum U (2007) Regulatory takings: balancing the interests of the investor and the state. J World Invest Trade 8(5):717, 739–742

Kulick A (2012) Global public interest in international investment law. CUP, pp 205–208

Maniruzzaman A, Al-Tayer A (2011) Addressing the global climate change problem in GATT/WTO law: the vision of a new international climate law based on international distributive justice. J World Invest Trade 12(5):651. https://ssrn.com/abstract=1969845. Accessed 03 August 2022

Marboe I (2015) Restitution, damages and compensation. In: Bungenberg M et al (eds) International investment law. C.H. BECK. Hart, Nomos, pp 10333–11034

Mayer B (2017) Climate change reparations and the law and practice of state responsibility. Asian J Int Law 7:186. https://www.researchgate.net/publication/296692772_Climate_Change_Reparations_and_the_Law_and_Practice_of_State_Responsibility. Accessed 03 August 2022

Mclachlan C et al (2017) International investment arbitration, substantive principles, 2nd edn. OUP, p 434

Rajamani L (2016) Ambition and differentiation in the 2015 paris agreement: interpretative possibilities and underlying politics. Int Comp Law Q 65(2):493–514. https://doi.org/10.1017/S0020589316000130

Reisman W, Sloane R (2004) Indirect expropriation and its valuation in the BIT generation. Faculty Scholarship Series, Paper 1002, p. 138. http://digitalcommons.law.yale.edu/fss_papers/1002. Accessed 03 August 2022

Sabahi B, Birch N (2010) Comparative compensation for expropriation. In: Schill S (ed) International investment law and comparative public law. OUP, p 760

Stewart D (1991) Philips Petroleum Co. Iran. v. Islamic Republic of Iran. Am J Int Law 85:122–123. http://www.jstor.org/stable/2203573. Accessed 03 August 2022

Thompson A (2016) The global climate regime for climate finance, political and legal challenges. In: Carlarne C et al (eds) The Oxford handbook of international climate change law. OUP, pp 138–139

Voigt C, Ferreira F (2016) The principles of CBDR-RC, progression and highest possible ambition in the Paris Agreement. Transnatl Environ Law 5(02):288. https://www.researchgate.net/publication/309437478_%27Dynamic_Differentiation%27_The_Principles_of_CBDR-RC_Progression_and_Highest_Possible_Ambition_in_the_Paris_Agreement. Accessed 03 August 2022

Wagner J (1999) International investment, expropriation and environmental protection. Golden Gate Univ Law Rev 465:528–529. http://digitalcommons.law.ggu.edu/ggulrev/vol29/iss3/6. Accessed 03 August 2022

Decisions

American Society of International Law *(1977) Lybian American Oil Company v. The Government of Libyan Arab Republic* (LIAMCO) (1), p. 160 http://www.jstor.org/stable/20692245. Accessed 03 August 2022

American Society of International Law (1982) *Government of the State of Kuwait v. American Independent Oil (AMIOIL)* (21) http://www.jstor.org/stable/20692245. Accessed 03 August 2022

American Society of International Law *(1987) AMOCO International finance Corporation v. Islamic Republic of Iran*, Partial Award http://www.jstor.org/stable/20693263. Accessed 03 August 2022

Court of International Justice (1927) *Case concerning the factory at Chórzow*. Publications of the Permanent Court of International Justice, Series A. -No. 17, September 13th, 1928, p. 47. http://www.icj-cij.org/pcij/serie_A/A_17/54_Usine_de_Chorzow_Fond_Arret.pdf. Accessed 03 August 2022

Highet K et al. (1995) Shahin Shane Ebrahini v. the Government of the Islamic Republic of Iran. Am J Int Law 89:385, 386–387

ICISID (2007) *LG&E Energy Corp., LG&E Capital Corp., and LG&E International, Inc. v. Argentine Republic*, Decision of Liability, Case No. ARB/02/1 https://www.italaw.com/cases/621. Accessed 03 August 2022

ICSID (1997) *American Manufacturing & Trading, Inc. v. Republic of Zaire*, Case No. ARB/93/1 http://www.italaw.com/cases/76#sthash.v8UbOlhb.dpuf. Accessed 03 August 2022

ICSID (2000) *Compañia del Desarrollo de Santa Elena S.A. v. Republic of Costa Rica*, Case No. ARB/96/1 https://icsid.worldbank.org/apps/icsidweb/cases/Pages/casedetail.aspx?caseno=ARB/96/1. Accessed 03 August 2022

ICSID (2000) Metalclad Corporation v. The United Mexican States, Case No. ARB(AF)/97/1. https://www.italaw.com/cases/671. Accessed 03 August 2022

ICSID (2000) *Wena Hotels Ltd. v. Arab Republic of Egypt*, Case No. ARB/98/4. https://www.italaw.com/cases/1162. Accessed 03 August 2022

ICSID (2002) Middle East Cement Shipping and Handling Co. S.A. v. Arab Republic of Egypt, Case No. ARB/99/6. https://www.italaw.com/cases/699. Accessed 03 August 2022

ICSID (2004) *MTD Equity Sdn. Bhd. and MTD Chile S.A. v. Republic of Chile*, Case No. ARB/01/7 https://www.italaw.com/cases/717. Accessed 03 August 2022

ICSID (2005) CMS Gas Transmission Company v. The Republic of Argentina, Case No. ARB/01/8. https://www.italaw.com/cases/288. Accessed 03 August 2022

ICSID (2005) *Tidewater Inc., Tidewater Investment SRL, Tidewater Caribe, C.A., et al. v. The Bolivarian Republic of Venezuela*, Case No. ARB/10/5. https://www.italaw.com/cases/1096. Accessed 03 August 2022

ICSID (2006) ADC Affiliate Limited and ADC & ADMC Management Limited v. The Republic of Hungary, Case No. ARB/03/16. https://www.italaw.com/cases/41. Accessed 03 August 2022

ICSID (2006) Azurix Corp. v. The Argentine Republic, Case No. ARB/01/12. https://www.italaw.com/cases/118. Accessed 03 August 2022

ICSID (2007) *CMS Gas Transmission Company v. The Republic of Argentina*, Case No. ARB/01/8, Decision of the ad hoc Committee on the Application for Annulment of the Argentine Republic https://www.italaw.com/cases/288. Accessed 03 August 2022

ICSID (2007) *Compañía de Aguas del Aconquija S.A. and Vivendi Universal S.A. v. Argentine Republic, Case No. ARB/97/3 (formerly Compañía de Aguas del Aconquija, S.A. and Compagnie Générale des Eaux v. Argentine Republic)* https://www.italaw.com/cases/309. Accessed 03 August 2022

ICSID (2007) *Siemens A.G. v. The Argentine Republic*, Case No. ARB/02/8. https://www.italaw.com/cases/1026. Accessed 03 August 2022

ICSID (2008) *Desert Line Projects LLC v. The Republic of Yemen*, Case No. ARB/05/17 https://www.italaw.com/cases/documents/357. Accessed 03 August 2022

ICSID (2009) *Bernardus Henricus Funnekotter and others v. Republic of Zimbabwe,* Case No. ARB/05/6 https://www.italaw.com/cases/467. Accessed 03 August 2022

ICSID (2009) *Waguih Elie George Siag and Clorinda Vecchi v. The Arab Republic of Egypt*, Case No. ARB/05/15. https://www.italaw.com/cases/1022. Accessed 03 August 2022

ICSID (2010) *Ioannis Kardassopoulos v. The Republic of Georgia*, Case No. ARB/05/18. https://www.italaw.com/cases/599. Accessed 03 August 2022

ICSID (2010) Suez, Sociedad General de Aguas de Barcelona, S.A. and Vivendi Universal, S.A. v. Argentine Republic, Case No. ARB/03/19, Decision on Liability https://www.italaw.com/cases/309. Accessed 03 August 2022

ICSID (2011) *Joseph Charles Lemire v. Ukraine, Case No. ARB/06/18* https://www.italaw.com/cases/614. Accessed 03 August 2022

ICSID (2011) *Señor Tza Yap Shum v. The Republic of Peru*, Case No. ARB/07/6. https://www.italaw.com/cases/1126. Accessed 03 August 2022

ICSID (2012) *EDF International S.A., SAUR International S.A. and León Participaciones Argentinas S.A. v. Argentine Republic,* Case No. ARB/03/23 https://www.italaw.com/cases/372. Accessed 03 August 2022

ICSID (2012) *Marion Unglaube v. Republic of Costa Rica*, Case No. ARB/08/1. https://www.italaw.com/cases/1134. Accessed 03 August 2022

ICSID (2012) *Occidental Petroleum Corporation and Occidental Exploration and Production Company v. The Republic of Ecuador*, Case No. ARB/06/11. https://www.italaw.com/cases/767. Accessed 03 August 2022

ICSID (2014) Gold Reserve Inc. v. Bolivarian Republic of Venezuela, Case No. ARB(AF)/09/1. https://www.italaw.com/cases/2727#sthash.VgHhhAmn.dpuf. Accessed 03 August 2022

ICSID (2015) *Bernhard von Pezold and Others v. Republic of Zimbabwe,* Case No. ARB/10/15 https://www.italaw.com/cases/1472. Accessed 03 August 2022

ICSID (2017) Karkey Karadeniz Elektrik Uretim A.S. v. Islamic Republic of Pakistan, Case No. ARB/13/1. https://www.italaw.com/cases/2024. Accessed 03 August 2022

ICSID (2017) *Marco Gavazzi and Stefano Gavazzi v. Romania*, Case No. ARB/12/25. https://www.italaw.com/cases/2026. Accessed 03 August 2022

ICSID (2017) Valores Mundiales, S.L. and Consorcio Andino S.L. v. Bolivarian Republic of Venezuela, Case No. ARB/13/1 https://www.italaw.com/cases/2127. Accessed 03 August 2022

ICSID (2018) *UP (formerly Le Chèque Déjeuner) and C.D Holding Internationale v. Hungary*, Case No. ARB/13/35. https://www.italaw.com/cases/6902. Accessed 03 August 2022

ICSID (2018) *UP (formerly Le Chèque Déjeuner) and C.D Holding Internationale v. Hungary*, Case No. ARB/13/35 https://www.italaw.com/cases/6902. Accessed 03 August 2022

Sapphire International Petroleums Ltd. v. National Iranian Oil (1963) Ad hoc Tribunal https://www.trans-lex.org/261600/_/saphire-award-ilr-1963-at-136-et-seq/. Accessed 03 August 2022

SCC (2010) RosInvestCo UK Ltd. v. The Russian Federation, Case No. V079/2005. https://www.italaw.com/cases/923634-635. Accessed 03 August 2022

SCC (2013) *Yuri Bogdanov and Yulia Bogdanova v. Republic of Moldova*, Case No. V091/2012 https://www.italaw.com/cases/2634. Accessed 03 August 2022

UNCITRAL (2003) CME Czech Republic B.V. (The Netherlands) v. The Czech Republic, https://www.italaw.com/cases/281. Accessed 03 August 2022

UNCITRAL (2014) *Yukos Universal Limited (Isle of Man) v. The Russian Federation*, PCA Case No. AA 227 https://www.italaw.com/cases/1175. Accessed 03 August 2022

International Legal Instruments

Agreement between the Government of Canada and the Government of the Hong Kong Special Administrative Region of the People's Republic of China for the Promotion and Protection of Investments Canada Hong (2016). https://investmentpolicy.unctad.org/international-investment-agreements/treaty-files/5094/download. Accessed 03 August 2022

Agreement between the Slovak Republic and the Islamic Republic Of Iran for The Promotion And Reciprocal Protection of Investments (2016). https://investmentpolicy.unctad.org/international-investment-agreements/treaty-files/3601/download. Accessed 03 August 2022

Agreement for The Promotion And Protection Of Investment between The Government of the Republic Of Austria and The Government Of The Kyrgyz Republic (2016). https://investmentpolicy.unctad.org/international-investment-agreements/treaty-files/5500/download. Accessed 03 August 2022

Agreement on Trade in Goods under The Framework Agreement on Comprehensive Economic Cooperation between The Republic Of India and The Association Of Southeast Asian Nations (2009). https://commerce.gov.in/trade/ASEAN-India%20Trade%20in%20Goods%20Agreement.pdf. Accessed 03 August 2022

Comprehensive Economic and Trade Agreement (CETA) between Canada, of the one Part, and The European Union, of the other part (2014). http://trade.ec.europa.eu/doclib/docs/2014/september/tradoc_152806.pdf. Accessed 03 August 2022

Energy Charter Treaty (ECT) (1994). https://www.europarl.europa.eu/meetdocs/2014_2019/documents/itre/dv/energy_charter_/energy_charter_en.pdf. Accessed 03 August 2022

Global Adaptation and Resilience Investment Working Group https://garigroup.com/. Accessed 03 August 2022

International Law Commission (2001) *Draft Articles on Responsibility of States for Internationally Wrongful Acts.* Supplement No. 10 (A/56/10), chp.IV.E.1, Art. 31

International law Commission (2001) Draft articles on Responsibility of States for Internationally Wrongful Acts, with commentaries 2001. Yearbook of the International Law Commission 2, Part 2, 86

Investment Promotion and Protection Agreement between The Government of the Federal Republic of Nigeria and The Government Of The Republic Of Singapore (2016). https://investmentpolicy.unctad.org/international-investment-agreements/treaty-files/5410/download. Accessed 03 August 2022

Kyoto Protocol to the United Nations Framework Convention on Climate Change (1997) 2303 UNTS https://unfccc.int/resource/docs/convkp/kpeng.pdf. Accessed 03 August 2022

North American Free Trade Agreement (NAFTA) (1994). https://www.nafta-sec-alena.org/Home/Texts-of-the-Agreement/North-American-Free-Trade-Agreement. Accessed 03 August 2022

OCDE (1972) Recommendation of the Council on Guiding Principles concerning International Economic Aspects of Environmental Policies. C(72)128 C/M(72)15 https://legalinstruments.oecd.org/en/instruments/OECD-LEGAL-0102. Accessed 03 August 2022

Pacific Agreement on Closer Economic Relations (PACER) (2017). https://dfat.gov.au/trade/agreements/not-yet-in-force/pacer/Pages/pacific-agreement-on-closer-economic-relations-pacer-plus.aspx. Accessed 03 August 2022

Reciprocal Investment Promotion and Protection Agreement Between The Government of the Kingdom of Morocco And The Government of the Federal Republic of Nigeria (2016). https://investmentpolicy.unctad.org/international-investment-agreements/treaty-files/5409/downloada. Accessed 03 August 2022

Trans-Pacific Partnership (2018). https://ustr.gov/sites/default/files/TPP-Final-Text-Investment.pdf. Accessed 03 August 2022

United Nations (1992) *Rio Declaration on Environment and Development 1992* 31 ILM, p. 874 https://www.jus.uio.no/lm/environmental.development.rio.declaration.1992/portrait.a4.pdf. Accessed 03 August 2022

United Nations Framework Convention on Climate Change (1992) FCCC/INFORMAL/84 GE.05-62220 (E) 200705 https://unfccc.int/resource/docs/convkp/conveng.pdf. Accessed 03 August 2022

United Nations Framework Convention on Climate Change, Adaptation Private Sector Initiative https://unfccc.int/topics/resilience/resources/adaptation-private-sector. Accessed 03 August 2022

United Nations Global Compact, Caring for Climate https://www.unglobalcompact.org/engage-locally/manage/engagement/caring-for-climate. Accessed 03 August 2022

United Nations Paris Agreement (2015) FCCC/CP/2015/L.9/Rev.1 https://unfccc.int/resource/docs/2015/cop21/eng/l09r01.pdf. Accessed 03 August 2022

Chapter 11
Conclusions

Summum ius summa iniuria[1] (*utmost law, utmost injustice*). This simple aphorism that Cicero presents in his book *De Officiis*[2] (*On Obligations)* captures the essence of the message that I have intended to deliver throughout this study, by focusing on the need to interpret investment treaty law in harmony with the global climate interest. This is not to say that the global climate interest has a natural superior value in comparison to the global interest of promoting the flow of investment, but to stress that interpreting a BIT's text isolated and trapped within its own rationale, and disregarding the context of global warming in investment treaty disputes, may lead to *summa iniuria.*

The context of global warming, as well as the position of the host state, the investor and third parties on climate change in investment treaty disputes, calls for an interpretation that, on the one hand, enhances the rationale of the investment treaty law system, when the protection of the foreign investment is in line with the promotion of the global climate interest, and, on the other hand, that profits from the deployment of the system's counter rationale mechanisms, when the rationale of the investment treaty law system undermines the global climate interest. Fostering the alignments and tempering the collisions between the ambitions of economic growth and the flow of investment that lies behind the adoption of BITs, with the protection of the global climate interest, can be achieved within the investment treaty law system.

Given the finding that the global climate interest might be *materially relevant* in investment treaty disputes, because of the impact of the investment treaty law system on the host state and investor's behaviour concerning global warming (ii), and given that the *formal relevance* of the climate change international legal framework is evinced in the system's operation through *multi(syncretic)lateralization* (i), it is reasonable to assume that investment treaty tribunals are compelled (allowed) to

[1] Cicero (1913), para. 33.
[2] *Ibid.*

T. Restrepo Rodríguez, *Investment Treaty Law and Climate Change*,
https://doi.org/10.1007/978-3-031-18655-4_11

search for harmonisation in investment treaty-climate change disputes. Harmonisation could be achieved through an orthodox approach that consists of the interpretation of the *FET* (iii) and the *Expropriation* (iv) standards' content in a climate change context (iv), as well as through the direct application of the *precautionary* (v) and the CBDRRC principles (vi). The study of *relevance, content-interpretation* and *law-ascertainment* in investment treaty-climate change disputes provides reasons to be optimistic in terms of tribunals changing their attitude towards the harmonisation of the investment treaty law and the climate change law (vii).

11.1 Investment Treaty Law Is a *Multi(syncretic)lateralized* System that Comprises the Climate Change International Legal Framework Without Losing Its Identity (*Formal Relevance*)

Investment treaty law could be perceived in functional terms as a system stemming from the process of multi(syncretic)lateralization, which embeds the protection against global climate change within its realm, even if this protection does not characterise the rationale of the system. Multi(syncretic)lateralization involves a double paradox that explains the formal relevance of the global climate interest in investment treaty law. The first paradox is Schill's finding that bilateral treaties are modelling international investment treaty law as a multilateral system that has community interests at its centre (multilateralization). The second paradox is that the multiverse of international law is modelling the investment treaty law system by constituting part of it (syncretism).

Multilateralization makes it possible to consider investment treaty law as a system and to conceive the interaction of this system with other systems that belong to the multiverse of international law, such as the climate change international legal framework. The *self-referentiality, closedness* and *autopoiesis* of the investment treaty law, are crucial to preserve the system's autonomy from the multiverse of international law, which constitutes its environment; and, simultaneously, this environment of international law makes it possible to characterise the identity of the investment treaty law by reference to it and makes it possible to envisage communication between them. The confirmation that investment treaty law is a multilateral system rather than a compound of isolated BITs and incohesive decisions, requires BITs not to be interpreted as mere bargains between the parties in the vacuum, but as instruments belonging to an autopoietic system with principles and rules that represent *community interests.*

Restricting the scope of the *community interest* that investment law promotes to the flow of foreign investment—or any complementary interest such as the prohibition of the investor's discrimination or the provision of a level-playing field—would confine the functioning of the system to radical neoliberal approaches that neglect any space to balancing the global interest of flow of investment with other global interests such as the global climate interest. Such an approach contradicts the notion of *multilateralization* and the real system's operation. It contradicts the concept of multilateralization, because the goal of optimally allocating resources between capital-exporting states and capital-importing states would be better achieved through isolated bargains, whose content and interpretation are not limited by systemic considerations, rather than through a multilateral investment treaty law system. It contradicts the real system's operation, because tribunals and BITs often refer to international law instruments and global interests that are different from the global interest of the flow of investments.

Amid the criticism against Schill's restrictive approach to the notion of *community interest*, it is important to clear from ambiguity the concept of *multilateralization,* by making explicit that investment treaty law is *syncretic. Multi(syncretic)lateralization* involves investment treaty law moving towards a multilateral system permeated by global public interests, including the climate change interest. Accordingly, if investment treaty law is defined as a framework for the functioning of the global economy—as Schill formulates—*syncretism* unearths that the functioning of this global economy is connected to and limited by the promotion and protection of global public interests. Naturally, the climate is within those global interests that the framework embraces.

Embracing *Multi(syncretic)lateralization* has some advantages over other approaches that attempt to integrate global interests into investment treaty law. By contrast to the *comparative public law* methodology, which only serves as a tool for the BIT's *content-interpretation, syncretism* legitimises the direct application of principles and rules from the climate change international legal framework to investment treaty disputes, because the international law provisions are regarded as a constitutive (basic) part of the system. Unlike substantive validity theories that envisage the application of (international) public law provisions to investment treaty based on the normative (moral) choice about their inherent superior value, *syncretism* justifies the application of the climate change international legal framework grounded on the system's operation. Moreover, while *syncretism* preserves the identity of the system by locating the axis of validity within principles and rules that characterise the investment treaty law system, substantive validity theories may erode the autonomy and the identity of the system by moving the axis of validity to (international) public law.

Multi(syncretic)lateralization also benefits from sidestepping the redundancy of eclectic theories that, in addition to a *positivist* approach, rely on public substantive validity theories or topical rhetoric for legitimising the application of international public law in investment treaty disputes. In this respect, the *Global Public Interest Theory*—designed by Kulick—could prescind from considering that the *relevance* of global public interests stems from the (allegedly) inherent public law character of the system, and still support the application of the climate change international legal framework, grounded on the copious evidence that the author provides concerning the tribunal's deployment of international law instruments and of global public interest arguments. Similarly, confining investment treaty arbitration under the *topic* of international public law is not a problem, as long as it is clear that international law rules derive their authority from the social practice of investment treaty tribunals—as De Brabandere clearly demonstrates in his study. Yet, if the rhetoric of investment treaty arbitration as international public law is taken to an extreme instead, the *positivist* approach might be clouded, and the *topic* could be unnecessarily identified as the reason for the application of the climate change international legal framework.

Nevertheless, Kulick and De Brabandere's findings coincide with my observation that the tribunals' practice is characterised by the deployment of international law instruments and the promotion of global public interests. This observation allows me to affirm that the climate change international legal framework is formally relevant in investment treaty-climate change disputes. This approach is confirmed by other authors—such as Vadi or Baetens—who recommend embracing the global climate interest stemming from the principles of interpretation of Article 31 VCLT.

The great virtue of *multi(syncretic)lateralization* is integrating into investment treaty law global public interests, without distorting the identity or menacing the autonomy of the system, by keeping the axis of validity in the dynamic of the principles that characterise the system's operation. From this perspective, investment treaty law could be defined as an *autopoietic system for the functioning of the global markets permeated by global public interests*—including the climate interest.

Multi(syncretic)lateralization is the backbone of this research, not only because it captures the formal relevance of the climate change international legal framework in investment treaty law, but also because it legitimises the direct application of the rules and principles that pertain to this framework (*osmosis*), in addition to the role that the climate change context may play in the activity of content-interpretation (*gravitation*).

Furthermore, *multi(syncretic)lateralization* is the backbone of this work because it leads to reshaping the tension that underlies the *(global) public interest challenge* in investment treaty law, from a static bilateral tension—private interest v. host state's (global)public interest—to a dynamic three-cornered tension, where the global public interest may overlap and/or conflict with either the investor's interest or the host state's interest. As a consequence of this reshaping, the challenge of harmonising the investment treaty law system with the climate change international legal framework in this study, attempts to balance the global interest with the interest of the parties to the dispute in two scenarios. These are, on the one hand, when the global climate interest comes into tension with the global interest of transnational investment, when the host state takes measures that promote the objective of levelling global warming by affecting the investor's interest, and, on the other hand, when the global climate interest and the global interest of transnational investments—in one corner—come into tension with other global interests—in the other corner—when the host state takes measures affecting investors in climate friendly industries to promote other interests.

11.2 Investment Treaty Disputes Comprise Climate Change Disputes that Stem from the Operation of the Investment Treaty Law (*Material Relevance*)

Investment treaty disputes can involve climate change disputes. Although the original idea of investment treaty arbitration did not envisage the emergence of conflicts and synergies between the global climate interest and the interest of protecting investments within the system, economic and socio-political studies show that the behaviour of host states and investors concerning climate change action, as well as the global and local governance of global warming, appears to be affected by the—not always neutral—operation of investment treaty law.

There is consensus in the literature on law and economics that the investment treaty law system fosters foreign investments by protecting investors against regulatory risks and, in accordance, it persuades host states not to modify the conditions of the regulation that were valid at the time of the investment. As a consequence, the system is vital for boosting investments in the rising green economy, but it chills the host state's will to regulate against the interests of investors in other industries that might be affected by climate change policies.

The green economy is gaining momentum. Host states might find it convenient to promote green investments, not only for complying with the international commitments on climate change, but also for boosting their economics and encouraging social development. However, large long-term investments in the green industries might be especially vulnerable *vis-à-vis* political and regulatory risks. After the investment is captured by the host state, the latter might be tempted to amend the regulation against investors, since the beneficial conditions of the investment may be seen as obsolete or the retribution to the investor could be regarded as too high, in light of new circumstances or new market conditions.

The European experience concerning renewable energy support schemes (RESSs) showed that renewable energy investments are especially vulnerable to legal and political risks. After these investments rocketed as a result of the benefits and subsidies provided to investors in the early 2000s, a set of hostile policies in the late 2000s disarticulated these schemes in several European countries. The measures were justified by host states for many reasons that ranged from the impact of the 2007–2008 global financial crisis to the decreasing costs of renewable energy technologies. More than 40 claims of investors in renewable energy that were filed under different BITs for this reason—many of them under the ECT—evince the relevance of the system of international treaty law for promoting and protecting investments in the green industries, and call for the action of arbitrators to strengthen the synergies between investment treaty system and climate change law.

Equally important to the action of arbitrators in strengthening the promotion of green investments, is their awareness that the *regulatory chill* might obstruct the switch to a green economy. The investment treaty law system—in itself—incentivises host states not to take climate change measures; it is within its rationale. Furthermore, the outrageously high value of the claims for compensation and the awards on environmental and climate change measures in investment treaty disputes, suggests that host states might not be willing to assume this risk in exchange for favouring the climate change action, neither to endanger the health of their economies nor the adoption of other policies that increase social welfare. Moreover, the *regulatory chill* to take climate change measures is also influenced by the high costs of litigation in investment arbitration, the unrealistic inflation of the claims submitted by investors and the distrust of host states in a system that is characterised by incohesive decisions that sometimes adopt radical pro-investor approaches.

Persuading the host state to not implement climate change measures—or withdraw regulation on climate change—acquires a higher intensity when the host state is aware that the investor is serious about presenting a claim or when a formal notice of arbitration has been served against the host state. The cases of *Transcanada* and *Vattenfall* (I) evince that the *threat chill* may be effective to induce host states to take back measures concerning the delay or the rejection of permits that were grounded on climate change concerns. Moreover, the case of New Zealand's suspension of the plain packaging regulation for tobacco until the disputes against Australia and Uruguay for restricting tobacco-related advertising were concluded, illustrates that the *regulatory chill* may be the result of a deliberate strategy of transitional corporations to instigate governments not to issue, suspend or withdraw regulations. The effectiveness of the *transboundary chill* in the tobacco industry could also be considered a strategy for big players in the fossil and the food industries—interested in land-grabbing and deforestation—to block climate change regulations.

Although I am aware that these cases do not constitute full proof on the accuracy of the mainstream models concerning the rationale of the investment treaty law and *regulatory chill*, I find it better to accept the mainstream approach rather than falling in the nihilism behind the skeptical approach. In fact, the skeptical approach (in general) is grounded on the idea that it is impossible to demonstrate that investment treaty arbitration increases foreign investments in the host state and/or to demonstrate that the investment treaty arbitration prevents countries from adopting public policy measures. The consequence of this skeptical approach would be denying the benefits of undertaking economic studies and disregarding the utility of rational models. On the contrary, I consider that rational models are crucial for interpreting the investment treaty law in context and avoiding the arbitrariness in the legal reasoning.

Investment arbitration tribunals should be aware of the economic analysis' finding that on some occasions, the investor's protection runs in parallel with the global interest of lowering greenhouse gas emissions, and that on other occasions, the investor's protection runs against this interest. This means that tribunals should be aware that their decisions could—positively or negatively—impact the achievement of the objectives of the climate change legal framework and act towards the promotion of these objectives within the limits traced by the same investment treaty law system.

Socio-political studies run in the same direction as economic studies concerning the material relevance of the climate change international legal framework in investment treaty law. This is not surprising since these studies also deal with social science/behavioural interaction and the centre of the inquiry that this book addresses is still the same conflictual or synergetic (three-cornered) relationship between the host state, the investor and the global public interest in levelling global warming.

Global Administrative Law studies (GAL) focus on revealing that investment treaty law is an instrument of global climate change governance. Investment arbitration tribunals have the power to dictate the global (and domestic) climate change policy to a great extent by influencing the behaviour of foreign investors and host states, as well as other tribunals, adjudging authorities and/or decisionmakers.

Adapting to the reasoning and the choices that investment treaty tribunals make concerning the correct behaviour of the host state according to BIT standards, increases the host state's probabilities to profit from the adoption of BITs, in the sense that the amount of investments that the host state might capture for this reason, would be higher than the potential losses for the costs of compensation to investors and of arbitration. The adaptation of the climate change regulation to these tribunals' 'good administration' standards evinces the factual governance of investment arbitration tribunals over host states' authorities. If investment tribunals are climate change governance authorities, it is natural that they should be compelled (allowed) to rely on the climate change international legal framework as a tool for decision-making. The application of this framework appears to be even more necessary, if it is considered that investment treaty tribunals might be one of the few in which climate change disputes can be solved.

The power of tribunals to shape the global climate change policy requires arbitrators to be attentive to tactics or strategies of big transnational corporations and the group of big oil-producer countries to turn the system into a machine of corporations' policymaking or oil-producers' policymaking, with the result of undermining global climate interest.

Post-modern critical studies affirm that the system is, in fact, a tool for corporations and developed countries to achieve or preserve power over the developing countries. Generally, these studies posit that the investment treaty law leans in favour of the investor's interests over (global) public interests, or that it is incompatible with the climate change interest. The origin of the disbalance is the asymmetry in the power of negotiation of the parties to BITs and the reproduction of imperialist patterns through the system, that preserves the power of foreign investors over decolonised countries and to commoditise the environment.

The race to capture investments in a globalised world might compel least developed and developing host states to lower environmental regulations and to adopt BITs. In this scenario, the burden of lowering costs of production, that initially lies with the transnational companies, is transferred to the citizens of the countries that offer the best economic conditions, through the detriment of their own environmental and climate interests. Furthermore, statistics on international investment arbitration show that vast majority of the claimants belong to developed countries and that most of the defendants are least developed or developing countries.

The imperial pattern of treating environmental interests from developing countries as simple assets under BITs allows private interests to rank higher than climate change interests. Given that investment treaty-climate change disputes are likely to increase in the coming years, the commoditisation of the environment in combination with the investor-centred origins of the system might suggest that the protection of the global climate interest is incompatible with investment treaty law.

The discouraging scenario that critical studies impose, not only underpins my argument that the climate change international legal framework is materially relevant to investment treaty-climate change disputes, but also highlights the need to reflect on the origins and on the rationale of the system when addressing investment treaty-climate change disputes to tackle inherent disbalances. Otherwise, the system would play against the achievement of the goal of levelling global warming.

However, the panorama is not as dramatic as critical studies put it. After 30 years of the BIT boom, a new generation of lawyers, NGOs and corporations are now aware of the dangers of commoditising global public interests in investment treaty disputes. Moreover, the application of rules of international law and the deployment (creation) of doctrines that represent the relevance of global public interests in international investment treaty law are within the practice of international investment tribunals.

The de-commoditisation of the global climate interest is crucial to rebalance any congenital (postcolonial) defect of the investment treaty law. The proposal—that is made in the chapter on formal relevance—to detach the investment treaty law from the private-public dilemma, by embracing a three-corner tension that pictures the synergies and collisions between the global climate interest and other global public interests that underlie the interests of the parties to the dispute, is useful to that end.

11.3 The *Sacrificial Lamb Theory* in Investment Treaty-Climate Change Disputes Harmonises the Operation of the Expropriation Standard with the Climate Change International Legal Framework

Harmonisation between investment treaty law and the climate change law can be achieved by interpreting the expropriation standard in a climate change context. Although it is tempting to adopt the police-power doctrine to this end, this doctrine disincentivises investments in the green industry and exceeds the pliability of the expropriation standard, by liaising the violation of the standard to the illegitimate character of the measure. The sole-effects doctrine, in contrast, encourages investments in the green industry and adapts to the ordinary text of the expropriation standard, since it defines the violation of the standard through the result of depriving the investment from its use or value, rather than to its illegitimate character. Yet, the application of the sole-effects doctrine chills the host state's intention to regulate against the threat of climate change. My proposal of applying the sacrificial-lamb doctrine tempers this failure of the sole-effect doctrine, as a result of recognising that host state's climate change measures would only constitute expropriation when the limitations to the investor are abnormal, and still promotes investments in the green industry by linking the expropriation only to the result of the measure, independently of its legitimate character.

Although the *police-power doctrine* incentivises the host state to take climate change measures by proposing that public interest measures are never against the FET standard, its general application to investment treaty-climate change disputes would disincentivise investors in green industries, because it puts them at risk of having to accept any regulatory change of the host state. Furthermore, they cannot be used for harmonisation in cases where the host state takes measures in favour of climate, because the (ordinary) expropriation standard excludes the possibility of including the legitimacy of the measure or its public interest as a condition for its violation.

Defining indirect expropriation through its *character,* is not only contradictory to the text that allows expropriations for the public interest in exchange for a fair compensation, but also against the purpose of maximising the protection to the foreign investor against the higher risk that it faces, which is the expropriation of the investment. Moreover, determining indirect expropriation through the illegitimate character of the measure would be incoherent with the system, as it would cancel any difference between the indirect expropriation and the FET standard, apart from the intensity of the investor's loss.

In contrast, the *sole-effects doctrine* has the virtues of defining indirect expropriation through the effect of depriving the investment from its use or value, rather than through the (public interest) character or the legitimacy of the measure, and of incentivising investments in the green industry. Nevertheless, this doctrine finds limitations in BITs that are tailor-made to mirror the *police-power doctrine* and in the context of climate change.

If tailor-made clauses are inspired by the *Methanex* award, hosts states' climate change measures would never violate the expropriation standard, since that award posits that the standard cannot curtail the host state's right to regulate in the public interest. Under a provision as such, the global climate interest would only be promoted when it matches the host state's interest, with the result of discouraging investments in the green industry. Similarly, if tailor-made clauses are inspired by the *Tecmed* award, it would be difficult for host states' climate change measures to breach the expropriation standard, given that this award postulates that the success of the expropriation is conditioned to *legitimate-backed expectations* of the investor. In a warming world, it is very difficult for an investor to argue that it holds the expectation that the host government would not modify its climate change regulations. Consequently, the *Tecmed*-inspired standard also discourages investments in the green industry, even if to a lesser extent than the *Metanex*-inspired standard. Nevertheless, both—the *Metanex*-inspired standard and the *Tecmed*-inspired standard—encourage the host state's determination to take measures in favour of the global climate interest, since its risk of facing international arbitration for the breach of the expropriation standard considerably decreases in this scenario.

The *sole-effects doctrine* cannot disregard the influence of global warming. This would be against Article 31 VCLT, which calls to understand the text of the BIT in context. The marked pro-investor inclination of this doctrine needs to be rebalanced to lessen the chilling effects on the host state to take climate change measures, and to reflect that the investor as an expert absorbs some of the commercial risks that its activity involves.

The suggestion of adopting the *sacrificial-lamb doctrine* to investment treaty-climate change disputes—that is made in Chap. 6—bridges the gaps of the *sole-effects doctrine* and aligns the interpretation of the expropriation standard with the climate change context. This doctrine—which is an adaptation of the *préjudice anormal et special* or *Sonderopfer*—posits abnormal limitations that destroy the investment to constitute indirect expropriatory measures. Abnormality for this purpose is understood as the disproportionality of the negative effects of the measure that the investor internalises as compared to the other members of the community. This concept, apart from matching the pattern that general regulatory measures are less likely to be considered expropriatory than targeted measures, entails a balanced interpretation of the expropriation standard that maximises the promotion of the global climate interest.

Under the *sacrificial lamb doctrine*, while general regulatory measures concerning climate change that destroy the value of the investment would rarely be regarded as expropriation, since they address the whole community, targeted measures that address a particular investor that destroy the investment would often be considered expropriation, since they might lead to a substantial disproportion between the burden that the investor bears for the climate interest and the impact of the measure on other stakeholders of the community.

The adoption of *sacrificial-lamb theory* ensures balance between the parties as entrenched in the expropriation standard and, at the same time, recognises the gravitational power of the climate change international legal framework. On the one hand, it incentivises the host state to take climate action, because its power to regulate would only be limited by the extraordinary situation where the substantial impact of the measure is abnormal to the investor in comparison to the community. On the other hand, the investor in green industries would be persuaded to invest, because it is aware that any substantial damage will be compensated, unless it is established that the public interest costs of the regulatory measure are shared with the community. Furthermore, this approach acknowledges that climate action is everyone's responsibility, and that the investor should bear the commercial risk that is natural in a context of climate change.

11.4 The Determination of the *Proportionality/Reasonableness* of Host State's Measures in the Context of Global Warming Harmonises the Operation of the FET Standard with the Climate Change International Legal Framework

Investment treaty law and climate change law can be harmonised by interpreting the FET standard in a climate change context. The notion of proportionality/reasonableness in the recent decisions on the cancelation of Renewable Energy Support Schemes (RESSs), that defines the breach of the FET standard by reference to the examination of the elements of arbitrariness, essentiality and predictability, has been shown to provide a balanced interpretation of the standard that incentivises investments in the green energy, without eroding the principle of deference that characterises the operation of system. Extending the application of this notion of proportionality/reasonableness to the disputes that stem from regulatory measures promoting the goal of levelling global warming, leads to an interpretation of the FET standard that is in line with the climate change context that lessens the regulatory chill to take climate action, by ensuring that a regulatory measure would not breach the standard unless it is capricious or alters the essential elements of the legal framework.

The concepts of (*dis*)*proportionality* and (*un*)*reasonableness* were definitive in the determination of whether the FET standard was breached in the awards concerning the RESSs. Although *reasonableness* and *proportionality* are different concepts, for the purposes of defining the breach in these awards, they were characterised with the same elements: *non-capriciousness*, *predictability* and *essentiality*. Accordingly, a *disproportional/unreasonable* measure was defined as a *capricious* change to the legal framework or an *unpredictable* change to the *essential* features of the legal framework.

Concerning the determination of *(non)capriciousness* in RESS disputes, it was found that a *capricious* measure is one that is either not taken for the public interest (*unnecessary*) or is useless in terms of achieving the purpose for which it was created (*unsuitable*), by reference to the threshold of arbitrariness.

The definition of *capriciousness* by reference to the threshold of arbitrariness not only ensures that the determination of the FET's breach conforms to the principle of deference to the local authorities, but also promotes the global climate interest. On the one hand, the threshold of arbitrariness encourages investments in the green industry, because investors know that they will be protected against host states' abuse of power. On the other hand, this threshold incentivises the host state to take climate action, since it confers a considerable discretion and flexibility to the host state's authorities to pursue the development of the objectives of the climate change legal framework without the fear of facing arbitration.

On the other two elements—*(in)essentiality* and *(un)predictability*—it is observed that they cannot be understood without reference to the creation of legitimate expectations. The determination of the elements from the legal framework that are *essential* and that the investor assumes to be maintained throughout the duration of the project, is the result of the effectiveness of the host state's representations to create the investor's expectation that those essential elements will not suddenly change. In the RESS awards, the representations capable of creating *legitimate expectations* were not required to fulfill any formalities, such as stability contracts, stability clauses, letters of intention or any other similar unambiguous formal declaration addressed to a particular investor. Yet, the creation of the *legitimate expectations* in these awards were not only buttressed by the investor's subjective understanding of the representations, but also by the objective circumstances and the context that make the representations credible.

The interpretation of *(non)essentiality* and *(un)predictability* stemming from the informal creation of *legitimate expectations* in context, encourages climate change action by two means. First, investors in RESSs can legitimately expect that the *essential* features of the legal framework that motivated the investments in the green industries will be maintained during the project, independently of the formalities of the representation that gave rise to the expectation. Under these conditions, investors feel protected against the regulatory risk and are incentivised to invest in renewable energies. Second, the expectations of investors in non-green industries might not be protected in the context of global warming, since they cannot—in principle—legitimately expect that in a warming world—and given the international commitments in this regard—host states will restrain from changing the regulation to protect the global climate interest. Devoid of the recognition of the investor's *legitimate expectations* in the stability of non-green investments, host states would be motivated to take measures to tackle climate change, as the risk of facing arbitration for taking these measures decreases considerably.

The analysis of the FET standard in investment treaty-climate change-related disputes mirroring the awards concerning RESSs, refines previous works of the scholarship by providing a systematic, balanced and methodic interpretation to the concept of *proportionality/reasonableness* that captures the gravitational power of the international climate change legal framework and, in particular, overcomes the systematic shortcomings of the public law-modelled (interest-weighing) proportionality test.

Against the pliability of the FET standard, the public law-modelled (interest-weighing) proportionality test entails a decreased level of deference. Apart from assessing the *proportionality* of the measure by reference to *alternative measures* better suited to serve the policy purpose, this approach deploys a test of *proportionality strictu sensu*, which confers to the tribunal the discretion to define whether the legitimate objective of the measure justifies the limitation to the investor. This exceeds the purpose of the FET, since the object of this standard is not transferring from the host state to the tribunal the competence to decide which climate change (or other) policy should have been adopted, to define which the most effective policy design was, or to establish the level of interference to the investor's interest that is justified for the public interest. In short, the role of the tribunal concerning the FET is not to decide whether the host state measure was the best possible policy-measure to tackle climate change.

In contrast, the model for assessing *(dis)proportionality/(un)reasonableness* that I have adopted based on the model of the RESSs' decisions, promotes the global climate interest, considering the limits of the FET's interpretation that stem from the purpose of the standard and the system's operation. Through the de-formalisation of the creation of legitimate expectations, my position conforms to the principle that the host state cannot induce investors to invest to then dishonour its commitments and, simultaneously, captures the climate change context, as it compels the tribunal to rely on the objective circumstances to assess the creation of *legitimate expectations*, rather than restricting their effectiveness to the subjectivity of the investor. Additionally, my approach is cautious in preserving a high level of deference as an expression of the hosts state's sovereignty and its right to regulate. These principles are deeply ingrained in the system's operation and conform to the intention of the state-members that are parties to BITs.

Furthermore, unfolding the notion of *unreasonableness/disproportionally* in *capriciousness*, *unpredictability* and *essentiality*, as well as defining the limits of these concepts through the threshold of arbitrariness and the determination of *legitimate expectations*, provides a clear methodology when interpreting the FET standard in investment treaty climate change-related disputes.

Deploying this methodology ensures the balance between the parties as entrenched in the FET standard and, at the same time, recognises the gravitational power of the climate change international legal framework. On the one hand, defining the limits of the *(non)capriciousness* of the measure through the benchmark of *arbitrariness,* captures the regulatory flexibility that the host state requires to be incentivised to pursue the development of the objectives of the climate change legal framework, while it protects investors in the green industries against policies that lack public interest motivations or are unfit to achieve the policy purpose. On the other hand, defining the limits of *(non)essentiality* and *(un)predictability* through the creation of legitimate expectations as a product of informal representations, encourages climate change action by determining that investors in the green industries can legitimately expect that the *essential* features of the legal framework will be maintained during the project, and by barring the creation of the investor's *legitimate expectation* that the investments in non-green industries will be stable against climate measures regulation.

11.5 The Direct Application of the Climate-Change Precautionary Principle Through the International Defence of Necessity and the Interpretation of General Exceptions in a Global Warming Context Harmonises the Investment Treaty Law with the Climate Change International Legal Framework

The investment treaty law and climate change law can be harmonised through the direct application of the climate change precautionary principle. The international defence of necessity (Art. 25 DASR) serves as a duct for integrating the precautionary principle to investment treaty-climate change disputes. Although this defence is recognised in the practice of the tribunals as part of the system's operation, its effectiveness to exclude the wrongfulness of climate-change precautionary measures requires a flexible approach to some of the elements of Article 25 DASR, namely, that the measure answers to an imminent peril, that it was the only way to safeguard the interest, and that the state did not contribute to the situation of necessity. Similar to the international defence of necessity, general exceptions could prevent the host state's liability for taking climate-change precautionary measures. However, unlike the operation of the former, the efficacy of general exceptions to legitimise climate-change precautionary measures depends on the particular model of BIT. General exceptions may range from self-judging clauses to clauses that require the verification of the necessity of the measure to protect the interest; as well as they may range from open exceptions protecting global public interests in general, to exhaustive exceptions that consist of a list of particular interests which could include measures to promote the environment, sustainable development or health.

The climate-change precautionary principle may apply directly through the international defence of necessity, with the result of precluding the effectiveness of the BIT obligations. Although the exceptional character of Article 25 DASR and the stringency of its elements continue to be recognised as a guiding principle of the investment treaty law system, the operation of *necessity* in investment treaty disputes reveals that the climate could be conceived within the interests protected under the scope of the article. Moreover, climate change precautionary measures might fulfill the requirements of Article 25 DASR, since investment treaty awards evince that these requirements withstand an interpretation that surpasses the literal-restrictive meaning of the text.

The evolution of decisions concerning the investment treaty disputes arising from the Argentinian crisis of the early 2000s indicates that the international defence of necessity (Art. 25 DASR) integrates global public interests that are different from the state's existence, such as national security, political and economic stability, ecology, environment, health and human rights. This trend of diversifying the catalogue of global public interests was reiterated by the annulment committees and dissenting opinions on the Argentinian crisis awards, as well as by investment treaty decisions on other matters, which added that the defence of necessity could be deployed to untie tensions between the compliance of host state's obligations under the BIT and the compliance of other obligations rooted in international law, such as those promoting anti-discriminatory action, preventing risks to health, protecting the autonomy and equality of indigenous communities, ensuring the provision of water and energy services to the citizens or preventing human rights violations and damages to the ecology and the environment.

Against this background, there is no inconvenience to conceive global climate within the interests protected under the international defence of necessity or to deploy Article 25 DASR to solve the tension between the obligation to take climate change precautionary measures and the obligation to conform with BIT standards.

Additionally, the recognition of the *essentiality* of the global climate interest as a requirement to trigger the international defence of necessity (Article 25 (1)(a)) does not admit much discussion. It is clear from the comments of the DASR's drafting commission that the concept of *essential interest* of Article 25 (1)(a) embraces global public interests, such as the global climate interest. Contrarily, the other requirements—*imminence of the peril, the only way to safeguard the interest and the non-contribution from the state to the situation of necessity*—demand a bigger interpretative effort to envisage climate-change precautionary measures within the realm of Article 25 DASR.

Tribunals have lessened the stringency of the international defence of necessity to a point that makes it possible to argue that the climate-change precautionary measures comply with the requirements of Article 25 DASR:

First, climate-change precautionary measures might conform to the requirement of *imminence of the peril,* given that the *evidence that is reasonably available* in an investment-climate change-related dispute could probably indicate that the industrial (or non-industrial activity) activity of the claimant represents an actual risk to the global climate interest. Here, absolute scientific certainty is not required to prove immediacy, but only *evidence reasonably available.*

Second, a climate-change precautionary measure may be regarded as the *only measure available,* as long as there is evidence that there are no reasonably available alternatives that would either not breach the BIT obligations or breach the BIT obligations in a less grave manner, and which are equally or more effective than the measure that is adopted to protect the climate interest. In this event, the notion of *only measure available* is not taken literally, but within the key of reasonably available alternatives.

Third, a climate-change precautionary measure could satisfy the requirement of *non-contribution to the situation of necessity* either when it is verified that the host state has not materially influenced climate change or when the effects of global warming were not foreseeable for the host state. Thus, it is not just any contribution of the host state that prevents the operation of the defence of necessity, only material contributions whose effects are foreseeable would do.

Apart from the operation of the international defence of necessity, the host state's precautionary climate-change measures may prevent liability by fitting into general exceptions. Contrary to the operation of *necessity*—which precludes the wrongfulness—general exceptions impede the breach of the BIT. The formula to determine the reception of climate change precautionary measures varies depending on the catalogue of subject complements that link the measure with the interest protected—which range from self-judging clauses to necessity clauses—and the catalogue of the public interests measures that the exception aims to cover through open exceptions or exhaustive exceptions, which often include references to health, environment or sustainable development.

General exceptions could be divided into two categories according to the subject complement. While one category embraces general exceptions that prevent liability when the measure is 'necessary for' the protection of the interest, the other category comprehends general exceptions that contain expressions that are not associated with *necessity*—such as 'directed to', 'appropriate to', 'designed and applied for', etc.

Unless the text of the BIT suggests something different, the general exceptions that contain the expression 'necessary for' or equivalent, should be distinguished from the international defence of necessity. Although there is no consensus in investment treaty law on the (non)relationship between the international defence of necessity and this group of general exceptions, the most suitable approach by which to harmonise the climate-change precautionary principle and the investment treaty law is the *primary-secondary application.* This method radically detaches general exceptions from the international defence of necessity, with the result of emancipating general exceptions from Article 25 DASR's strict requirements and, simultaneously, of opening the possibility of applying the international defence of necessity in cases where the text of the general exceptions rules out the climate change interests from its scope.

The chances for climate-change precautionary measures to prevent liability through general exceptions that are not directly related to the expression *necessity* are higher than for those that contain the expression 'necessary for' or equivalents. While the expression 'necessary for' suggests the need to prove the inexorability of the climate change measure to trigger the exception, other expressions grant the host state some level of discretion in choosing the adoption of climate change precautionary measures, such as 'directed to', 'in pursuance of', 'considers appropriate for', 'intended to', 'relating/ed to', 'in pursuing', 'imposed to', 'that provide for' or 'applied and designed to'. While some of these subject complements condition the legitimacy of the precautionary climate change measure—as general exception—to the verification of the effectiveness of the measure to protect the climate interest; others redirect the examination of the legitimacy of the measure to the host state's subjective considerations (self-judging clauses)—such as 'considers appropriate' or 'considers necessary'.

In addition to the subjective complement that links the measure with the interest protected, the effectivity of climate-change precautionary measures to prevent the breach of the BIT is also determined by their suitability to the interests that the exception aims to cover. There are two basic models that define the adequacy of climate-change precautionary measures to general exceptions: open exceptions and exhaustive exceptions. While the former embraces climate change precautionary measures by including any public interest measure within their scope through a general statement in that sense, or through a list of non-exhaustive measures characterised by the phrase 'such as'; the latter conditions the suitability of climate-change precautionary measures to a list of specific interests that may imply the protection of the climate interest as a rebound effect, such as environment, health or sustainable development.

Open exceptions are likely to legitimise the application of climate-change precautionary measures, unless these measures are arbitrary, discriminatory or disguise a restriction, or are excluded as an exercise of ponderation, depending on the explicit or implicit limits of the tailor-made provisions. Contrarily, exhaustive exceptions do not prevent the breach of the BIT, when they are restricted to public order, essential security interest and peace. The causal link between the effects of climate change and these interests is very weak in these events. Nevertheless, exhaustive exceptions comprehend climate-change precautionary measures, when they refer to 'human, animal, plant life or health', the environment or sustainable development, since the link between these interests and the climate interests is very strong. The recent trend of BITs to include this last set of interests within the general exceptions, in addition to the increasing relevance of the climate, the environment and sustainable development in BITs—as well as the relevance of international treaties promoting these interests in BITs—evinces that the recognition of precautionary climate change measures within the scope of general exceptions is gaining ground in the field of investment treaty law.

Although the counter rationale and the exceptionality of both—*necessity* and general exceptions—justify studying them together, their deployment in investment treaty-climate change disputes represents two different operative mechanisms of harmonisation that address different dimensions of conflict of interests and require different methods to solve the tensions. While the direct application of the climate-change precautionary principle represents an operation of law ascertainment, that addresses the tension between the global climate interest against the global interest of transitional investment—and other interests that are aligned with it—through a ponderation exercise that is systematised in Article 25 DASR the embracing of climate change precautionary measures in general exceptions represents an operation of text-interpretation, that addresses the tension between host state's interest and the investor's interest, through the examination of the adequacy of the measure to the to the BIT's exceptions in a climate change context. Briefly, while the direct application of the climate-change precautionary principle through *necessity* represents the osmosis of the climate change international legal framework into the investment treaty law system, the recognition of climate change precautionary measures as general exceptions represents the gravitational power of global warming context—including the climate change international legal framework over the investment treaty law system.

11.6 The Direct Application of the *Common but Differentiated Responsibilities and Respective Capabilities Principle (CBDRRC)* to the Compensation Through International Liability Norms Harmonises the Investment Treaty Law with the Climate Change International Legal Framework

The investment treaty law and climate change law can be harmonised through the direct application of the Common but Differentiated Responsibilities and Respective Capabilities Principle (CBDRRC). Although the operation of the investment treaty law ordinarily relies on the full compensation principle, it is possible to exceptionally adopt a differential approach to compensation that considers the circumstances of the parties regarding climate change. Indeed, the application of the CBDRRC to investment treaty-climate change disputes is conceivable, since the practice of tribunals shows that the assessment of compensation is occasionally tied to considerations that are different from the magnitude of the investor's loss, including retributory and distributive justice considerations. Furthermore, the double-characterisation of the functioning of both—investment treaty law and the CBDRRC—as instruments of liability and cooperation, clears the operationalisation of the CBDRRC in investment treaty-climate change disputes from the fears of incurring in topical contradiction. This operationalisation could be enhanced through the translation of the CBDRRC principle into *norms that are familiar to international civil liability: the principle of equity, the polluter should pay and Article 27 (b) DASR. However, this translation cannot be effected without considering that the ever-changing position of host states before climate change requires a dynamic approach to the application of CBDRRC in investment treaty disputes, that captures the conditions of the—developed or developing—host state at the time of the dispute, rather than relying on static models of the CBDRRC.*

The full compensation principle is not absolute in investment treaty law. The tribunals practice exceptionally admits restrictions to *full compensation* stemming from retributory and distributive justice considerations. Although the complete independence of distributive justice considerations from the compensatory justice considerations is not always clear in the rulings, there are some awards that openly embrace a differential approach to compensation in proportion to the particular conditions of the parties and in light of a given context. This practice makes it possible to embrace within the investment treaty law system, the assessment of the compensation in proportion to the conditions of vulnerability, responsibility and capability of the host state before climate change.

The fact that the compensation of damages in investment treaty disputes for lawful expropriation (market value) is lesser than the compensation for unlawful expropriation (full compensation), that the effectiveness of contributory negligence to lower the compensation is dependent on the illegitimacy of the investor's conduct, and that moral damages are subject to compensation, reveals that the assessment of compensation is occasionally impregnated with traces of retributive justice. Similarly, awards establishing that the compensation should reflect the investor's burden to bear part of the country/regulatory risk, or deploy the principle of equity as a means to exclude the application of the full compensation, evidence that the assessment of compensation is occasionally impregnated with distributive justice considerations.

The conditions of the host state in a given context have proven to be relevant to assess compensation in investment treaty disputes. Particularly, the situation of 'Third World' nations concerning natural resources, or the striking reality of civil war, have been reasons for tribunals to bar the application of the full compensation principle and to take instead a compensation that appears more equitable in those circumstances. The application of the CBDRRC to the compensation in investment treaty disputes stems from the same logic of these cases. The different degree of development of host states should serve as basis for defining a differential approach to compensation in a climate change context, especially when the application of the full compensation principle would result in discouraging the objectives of the climate change international legal framework.

The CBDRRC also has topical connections with the investment treaty law, that make it possible to envisage its operationalisation in investment treaty disputes. On the one hand—such as investment treaty law—the climate change international legal framework recognises that, although investors are relevant actors, they are not burdened with direct obligations and their role is accessory to the rationale of the system. On the other hand, the CBDRRC and the investment treaty law system share the same bidimensional liability-cooperative nature. In fact, although investment treaty law is a system of liability, BITs are adopted for the purpose of cooperating in the flow of investment and vice versa; although the climate change international legal framework is a cooperative mechanism, the CBDRRC is partly based on liability considerations, as the differentiated responsibility is intended to be proportional to each country's contribution to climate change. Moreover, the investment treaty law system is complementary to the CBDRRC, as it is one of the few avenues whereby the principle can transfer to investors part of the costs of climate change.

The translation of the CBDRRC into the language of international liability enhances the application of the principle to investment treaty law-climate change-related disputes. The principle of equity, the *polluter should pay* and Article 27 (b) DASR may shape the operationalisation of the CBDRRC in the determination of compensation.

The relationship between the principle of equity and the CBDRRC in investment treaty-climate change disputes is bi-directional. While the principle of equity defines the effectiveness of the CBDRRC—as the exceptional application of the former is accepted in the practice of the investment treaty tribunals—the CBDRRC shapes the content of the equity principle in investment treaty-climate change disputes with special regard to the concept of climate justice. Therefore, the content of the principle of equity takes the form of a differentiation approach to compensation between developed and developing countries, depending on their vulnerability, responsibility and capability in regard to climate change. Accordingly, the application of the principle of equity—under the influence of the CBDRRC—might result in the exceptional reduction of compensation grounded on the host state's condition as a developing country, as well as the imposition of compensation based on the host state's condition as a developed country.

The *polluter should pay* and Article 27 (b) DASR are, in the end, expressions of the principle of equity in the context of climate change. Yet, their application is still useful since these rules are more concrete than the principle of equity, while the *polluter should pay* makes it possible to transfer part of the costs of climate change from the developing host state to the investor in non-green industries through the reduction or cancellation of compensation; makes it possible to allocate—partly or fully—the burden of compensating investors in the green industries to developed host states, even if the wrongfulness of the breach was excluded.

However, the differentiation in compensation based on the CBDRRC needs to be adapted to a dynamic approach, that matches the development of the climate change international legal framework in the Paris Agreement. In this vein, the determination of compensation stemming from the bifurcation between developed and developing countries cannot neglect the real circumstances of the host state at the time of the award. Consequently, the *polluter should pay* may only reduce damages when the application of the full compensation principle would jeopardise the social or economic stability of the developing host state and it is conditioned to the verification that the emissions of the developing host state have not considerably increased. Likewise, the application of Article 27 (b) as a source for imposing compensation may only be possible if the imposition of a compensation does not threaten the developed host state's social or economic stability and it is conditioned to the verification that the emissions of the developed host state have not considerably decreased.

The osmosis of the climate change international legal framework into the investment treaty law system through the penetration of the CBDRRC, certainly inclines the balance in favour of the global climate interest. Yet, it should be noted that this approach to compensation has the virtue of—moderately—reaching a middle point in the tension between global interests in investment treaty-climate change disputes.

On the one hand, it provides the opportunity to reduce compensation for promoting the global climate interest rather than cancelling compensation, thus the global interest in the flow of investment is still promoted to a certain extent. On the other hand, it provides the opportunity to impose compensation in proportion to the circumstances of the parties for promoting the global climate interest rather than imposing full compensation, so that the other colliding global interests are not completely undermined.

Moreover, the osmosis of the CBDRRC into investment treaty law reaches this—moderate—middle point between global public interests without incurring in systemic contradictions. While the approach suggesting the proportionality of the compensation as a reflection of the proportionality of the host state's measure, enlarges the functionality of the latter beyond the definition of the breach or the preclusion of wrongfulness, against the system's operation; the modulation of the compensation through the direct application of the CBDRRC ensures the independence of the proportionality of the compensation from the proportionality of the host state's measure, avoiding systemic contradictions.

11.7 Concluding Remarks: Reasons to Be Optimistic About the Harmonisation of the Investment Treaty Law and Climate Change Law

International Arbitration Tribunals are soon expected to openly address climate change disputes in their rulings. The overwhelming narrative (reality) of global warming as a common concern of humanity makes it not only inconvenient, but impossible to understand BITs' texts and the functioning of the investment treaty law without climate change considerations. Either as an expression of its rationale or through counter rational elements, the investment treaty law system has been proven throughout this research to be a fertile ground for the promotion of the global climate interest. The osmosis and the gravitation of the climate change international legal framework over investment treaty law challenges arbitrators to balance colliding and overlapping global interests, with the natural intricacies of pondering principles and solving conflicts of norms within the limits of the system.

The harmonisation of the investment treaty law with the climate change law does not require a whole reengineering of the system, a (unrealistic) global multilateral treaty, or the modification of more than 3,000 BITs. This publication demonstrates that it is possible to promote the global climate interest stemming from the *autopoietic* and *syncretic* operation of the investment treaty law system. When viewed from certain angle, one could say that harmonisation could be achieved simply by the arbitrators' changing attitude towards the embracing of the global climate interest. Some concluding remarks might clarify why this attitude of ignoring climate change considerations is likely to change:

First, the recognition of global public interests that are different from the flow of foreign investment and the growth of global economy is gaining ground as a fundamental-operative feature of the investment treaty law. This book describes several practices of tribunals, thus: the adoption of *the police-power doctrine* (Chap. 6), the relevance of global public interests and international law in the analysis of *reasonableness/proportionality* and in the determination of *legitimate expectations* (Chap. 7), the embracing of international essential interests beyond the security interest within the scope of the international defence of necessity and the flexibilisation of Article 25 DASR's requirements (Chap. 5), BITs' trend of defining general exceptions by reference to global public interests (Chap. 9), as well as the acknowledgement of exceptions to the full compensation principle based on distributive and retributive justice considerations (Chap. 10).

Second, the global climate interest could be promoted within the operative limits of the investment treaty law system. In this respect, the harmonisation in this work succeeds in finding legal instruments that conform or belong to the investment treaty law system, instead of relying on substantive law theories, comparative-public law interpretation or topical rhetoric to legitimise the application of the climate change international legal framework. In this respect, the application of the *sacrificial-lamb doctrine* to the expropriation standard—which I propose as a formula for harmonisation—is a tempered version of the sole-effect doctrine that defines expropriation through the effects rather than through the legitimacy of the measure (Chap. 6). Likewise, the assessment of *proportionality/reasonableness* as a key to the harmonisation of the FET standard with the climate change international legal framework, is cautious in preserving the deference to the host state as much as possible, as well as the honouring of the investor's legitimate expectations; and, consequently, avoids deploying any public law-modelled (interest-weighing) proportionality test (Chap. 7). In the same vein, concerning the climate change precautionary principle in investment treaty disputes, this book highlights that the international defence of necessity operates as a filter to its application. Similarly, I posit that general exceptions embrace climate change precautionary measures as long as the exception's text comprises global interests that have a relationship of proximity with the global climate interest, such as environment, health or sustainable development (Chap. 9). Lastly, the exceptional application of the CBDRRC principle to compensation that I defend, was justified by the tribunal's deployment of differential approaches to compensation and the principle of equity (Chap. 10).

Third, the climate change international legal framework could permeate the investment treaty law system without jeopardising the identity of the latter. On the one hand, the harmonisation of the FET, the expropriation and general exceptions through the orthodox approach does not risk the identity of the system, because it is limited by the pliability of the text and the purpose of these provisions in the particular BIT. Thus, the deployment of the *sacrificial-lamb doctrine* as a tool for harmonisation, as suggested here, does not exceed the pliability of the expropriation standard by adopting the abnormality of the impact on the investor as basis for identifying the breach (Chap. 6). Nor is the pliability of the FET standard exceeded, when harmonisation is linked here to the definition of the *disproportionality/unreasonableness* of the measure as a *capricious* change to the legal framework or an *unpredictable* change to the *essential* features of the legal framework (Chap. 7). In the same vein, the limitation of open general exceptions, either to the *chapeau* of arbitrariness, discrimination and disguised restrictions or to ponderation, ensures that general exceptions do not exceed the counter rationale purpose of the provision (Chap. 9). On the other hand, the risk of eroding the identity of the system through the direct application of the climate change precautionary principle and the CBDRRC—the heterodox approach—is minimised by the exceptionality of its application. While the exceptionality of the application of the climate change precautionary principle is guaranteed by the stringency of the requirements of Article 25 (Chap. 9), the exceptionality of the application of the CBDRRC is ensured by conditioning the differential compensation to the verification that a full compensation would involve a threat to the developing host state's economic or social stability (Chap. 10). Moreover, the CDBDRRC does not jeopardise the identity of the system because it does not imply cancelling the compensation to the investor, but reducing the compensation in proportion to the circumstances of the parties before climate change (Chap. 10).

Fourth, there is a new generation of academics, arbitrators and practitioners that bring a breath of fresh air to the epistemic community of international investment lawyers in the direction of rebalancing the promotion of global public interests—including the climate and the environment—against the genetic pro-investor inclination of the system. This new generation of lawyers dealing with investment treaty law is represented by the emergence of the theory of *multilateralization*, theories identifying investment treaty law as public (international) law, the deployment of comparative public law methodologies to interpret the BIT, and positivist approaches—such as this one—that envisage the application of global public interests stemming from the system's operation (Chap. 3); as well as those of interdisciplinary studies that include law and economics and socio-political analysis revealing the existence of conflicts and synergies that need to be openly addressed by tribunals, such as GAL or International Critical Studies (Chap. 4).

Fifth, investment treaty awards ignoring climate change and other common concerns of humankind jeopardise the survival of the investment treaty law system itself. After more than a quarter of a century from the booming of neoliberalism and the proliferation of BITs, there is an increasing pressure for the preservation of the environment and the climate over the economic interest of transnational corporations, along with an increasing distrust of the system by the international community, which is grounded on the misapplication of BITs as tools for intimidating against the host state from taking action in the public interest (Chap. 4). The adaptation of the investment treaty law to the changing reality of climate change is part of the package of changes that investment treaty law needs to perfect to encompass the broader frame of sustainable development. Otherwise, host states will stop adopting BITs and even refuse to comply with the investment treaty awards.

Finally, ignoring the climate change in investment treaty disputes would lead to an unacceptable disconnection between the interpretation and the environment. To explain the aphorism *Summum ius summa iniuria* (*utmost law, utmost injustice*) that was recalled at the beginning of these conclusions, Cicero criticises the occasion when, despite an agreed thirty-day truce, one of the enemies would have devastated the fields of the other enemy every night, based on the idea that the truce was agreed for *days* but not for *nights*.[3] In a global warming context, interpreting BITs' texts as shielding the investor from any regulatory change that favours the global climate interest, with the argument that the system's purpose is to promote foreign investment, is as absurd as understanding in Cicero's example that a thirty-day truce only applies to stopping war during daylight. Separating the interpretation of the issue from its environment is but impossible.

[3] *Ibid.*

Reference

Cicero MT (1913) De Officiis. Walter Miller tr, William Heinemann/the Macmillan Co (34), para. 33